Life of Fred®
Beginning Algebra
Expanded Edition

Life of Fred®
Beginning Algebra
Expanded Edition

Stanley F. Schmidt, Ph.D.

Polka Dot Publishing

ISBN: 978-1-937032-11-1
Printed and bound in the United States of America

Polka Dot Publishing   Reno, Nevada

To order copies of books in the Life of Fred series,

visit our Web site PolkaDotPublishing.com

Questions or comments? Email the author at lifeoffred@yahoo.com

Seventh printing

Life of Fred: Beginning Algebra Expanded Edition was illustrated by the author with additional clip art furnished under license from Nova Development Corporation, which holds the copyright to that art.

for Goodness' sake

or as J.S. Bach—who was
never noted for his plain
English—often expressed it:

Ad Majorem Dei Gloriam
(to the greater glory of God)

If you happen to spot an error that the author, the publisher, and the printer missed, please let us know with an email to lifeoffred@yahoo.com

As a reward, we'll email back to you a list of all the corrections that readers have reported.

What Algebra Is All About

When I first started studying algebra in the ninth grade, no one in my family could explain to me what it was all about. My dad had gone through the eighth grade in South Dakota, and my mom never mentioned that she had ever studied algebra before she took a job at Planter's Peanuts in San Francisco.

My school counselor enrolled me in beginning algebra, and I showed up to class on the first day not knowing what to expect. On that day, I couldn't have told you a thing about algebra except that it was some kind of math.

In the first month or so, I found *I liked algebra better than* . . .

> ✓ physical education, because there were never any fist-fights in the algebra class.

> ✓ English, because the teacher couldn't mark me down because he or she didn't like the way I expressed myself or didn't like my handwriting or didn't like my face. In algebra, all I had to do was get the right answer and the teacher had to give me an A.

> ✓ German, because there were a million vocabulary words to learn. I was okay with der Finger which means *finger*. But besetzen, which means to occupy (a seat or a post) and besichtigen, which means to look around, and besiegen, which means to defeat, and the zillion other words we had to memorize were just too much. In algebra, I had to learn how to *do stuff* rather than just memorize a bunch of words. (I got C's in German.)

> ✓ biology, because it was too much like German: memorize a bunch of words like mitosis and meiosis. I did enjoy the movies though. It was fun to see the little cells splitting apart—whether it was mitosis or meiosis, I can't remember.

So what's algebra about? Albert Einstein said, "Algebra is a merry science. We go hunting for a little animal whose name we don't know, so we call it x. When we bag our game, we pounce on it and give it its right name."

What I think Einstein was talking about was solving something like 3x – 7 = 11 and getting an answer of x = 6.

But algebra is much more than just solving equations. One way to think of it is to consider all the stuff you learned in six or eight years of studying arithmetic: adding, multiplying, fractions, decimals, etc. Take all of that and stir in one new concept—the idea of an "unknown," which we like to call "x." It's all of arithmetic *taken one step higher*.

Many, many jobs require the use of algebra. Its use is so wide-spread that virtually every university requires that you have learned algebra before you get there. Even English majors, like my daughter Margaret, had to learn algebra before going to a university.

I also liked algebra because there were no term papers to write. After I finished my algebra problems I was free to go outside and play. Margaret had to stay inside and type all night. A lot of English majors seem to have short fingers (die Finger?) because they type so much.

Life is unlivable if it is confined to algebra.

Life is incomplete without it.

Common Questions that Students Have

MAY I USE MY CALCULATOR?

Yes. It is the addition and multiplication tables that you need to know by heart. Once you have them down cold, and you know that the area of a triangle is one-half times base times height, there is little else that you should have to memorize.

When I taught arithmetic, the tests I gave were always taken without the use of a calculator, but when I taught algebra/geometry/trigonometry/calculus/math for business majors/statistics, the tests were always open-book, open-notes, and use-a-calculator-if-you-want-to.

There are a lot of times in life when you may need to know your addition and multiplication facts and won't have access to a calculator, but when you are doing algebra or calculus problems you will almost always have a calculator and reference books handy.

WHAT KIND OF CALCULATOR WOULD BE GOOD?

A basic calculator has these five keys: $+, -, \times, \div, \sqrt{\ }$. Years ago I saw one of those advertised in a magazine for over $100. Recently, at one of those stores that sell everything for about a dollar I paid $1.07 including the sales tax.

Most top-rated universities want their applicants to have four years of high school math. (Beginning algebra is the first of those four years.)

The next three years will be advanced algebra, geometry, and trig. For those courses you will need a "scientific calculator." It will have sin, cos, tan, !, log, and ln keys. The most fun key is the "!" key. If you press 8 and then hit the ! key, it will tell you what $8 \times 7 \times 6 \times 5 \times 4 \times 3 \times 2 \times 1$ is equal to. Recently, I saw one of those calculators on sale for less than $12. That's the last calculator you'll need to learn all the stuff through calculus.[*]

You might as well get your scientific calculator now.

---

[*] Some schools require their calculus students to buy a fancy graphing calculator that costs between $80 and $120. I don't own one and I've never needed one. I spent the money I saved on pizza.

WHAT BACKGROUND DO I NEED TO START ALGEBRA?

If you are feeling unsure, you might try your hand at this quiz. The answers are given on the next page.

The questions are all taken from books that precede *Life of Fred: Beginning Algebra Expanded Edition*.

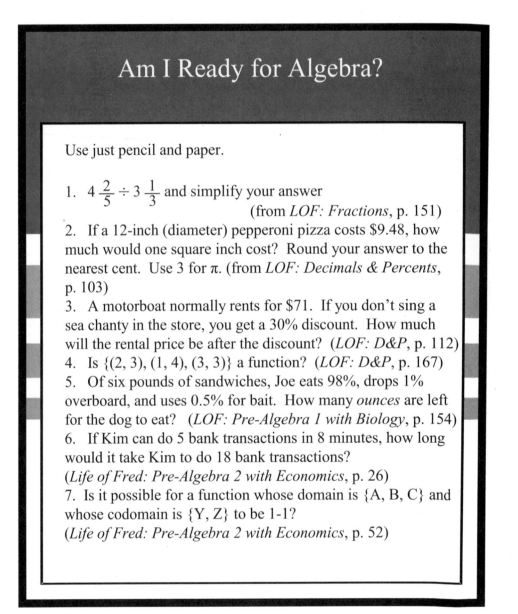

Am I Ready for Algebra?

Use just pencil and paper.

1. $4\frac{2}{5} \div 3\frac{1}{3}$ and simplify your answer
(from *LOF: Fractions*, p. 151)
2. If a 12-inch (diameter) pepperoni pizza costs $9.48, how much would one square inch cost? Round your answer to the nearest cent. Use 3 for π. (from *LOF: Decimals & Percents*, p. 103)
3. A motorboat normally rents for $71. If you don't sing a sea chanty in the store, you get a 30% discount. How much will the rental price be after the discount? (*LOF: D&P*, p. 112)
4. Is {(2, 3), (1, 4), (3, 3)} a function? (*LOF: D&P*, p. 167)
5. Of six pounds of sandwiches, Joe eats 98%, drops 1% overboard, and uses 0.5% for bait. How many *ounces* are left for the dog to eat? (*LOF: Pre-Algebra 1 with Biology*, p. 154)
6. If Kim can do 5 bank transactions in 8 minutes, how long would it take Kim to do 18 bank transactions?
(*Life of Fred: Pre-Algebra 2 with Economics*, p. 26)
7. Is it possible for a function whose domain is {A, B, C} and whose codomain is {Y, Z} to be 1-1?
(*Life of Fred: Pre-Algebra 2 with Economics*, p. 52)

The answers: (1) $1\frac{8}{25}$ (2) \$0.0877 which rounds to 9¢ (3) \$49.70
(4) Yes (5) 0.48 ounces (6) 28.8 minutes or $28\frac{4}{5}$ (7) No because
functions that are 1-1 must have at least as many elements in the codomain
as in the domain.

If you didn't get at least 70% (5 out of 7) right, the intelligent thing
to do might be to start with one of the earlier books in the series.

WHERE ARE **THE BRIDGES**?

At the end of each chapter are three Cities. They are not tests.
There is a lot of math to learn in this first year of high school math. The
Cities offer a much-needed chance to practice your algebra. Do not skip
any of them. By the time you do the third City in each chapter, you will be
doing the problems much more easily.

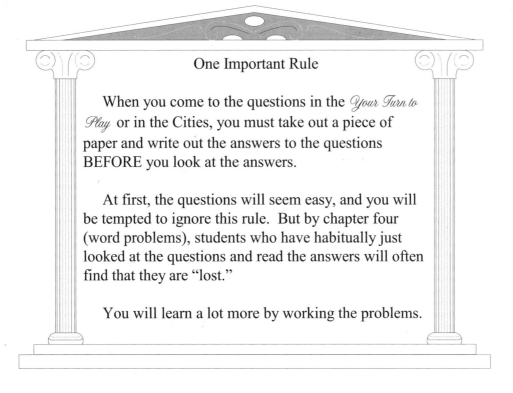

One Important Rule

When you come to the questions in the *Your Turn to
Play* or in the Cities, you must take out a piece of
paper and write out the answers to the questions
BEFORE you look at the answers.

At first, the questions will seem easy, and you will
be tempted to ignore this rule. But by chapter four
(word problems), students who have habitually just
looked at the questions and read the answers will often
find that they are "lost."

You will learn a lot more by working the problems.

Just before the Index is the **A.R.T.** section, which very briefly
summarizes much of beginning algebra. If you have to review for a final
exam or you want to quickly look up some topic eleven years after you've
read this book, the **A.R.T.** section is the place to go.

A Note to Parents

Your children are now on "automatic pilot." Each day they do one (or more) lessons. The reading in *Life of Fred: Beginning Algebra Expanded Edition* is fun. And because it is fun, they will learn mathematics much more easily.

Five-year-old Fred *first* encounters the need for mathematics in his everyday life, and *then* we do the math. This is true for all of the books in the series. The math is *relevant*. This is different than most math books.

I believe that mathematics should not be taught in a vacuum. It should not be compartmentalized. We are teaching children first, not just math. Other subjects are integrated into the text. I have not taken the oath: "Algebra, the whole algebra, and nothing but the algebra."

In this book we include some **English**. Do you know the complete "i before e" rule with its four classes of exceptions? It's in this book. The army chaplain is at a private library and he pulls a leather-bound book of poetry off the shelf and begins to read a poem. He thinks to himself, "A good example of enjambment." This word is then defined in a footnote.

Health. Fred and Jack LaRoad decided to head out for an afternoon jog. The other eleven decided to watch TV for five hours. On another occasion, when Fred and Jack were on a six-hour army leave in a town they had never been in before, they headed to a carrot juice bar.

Reading. But before that, they went to the public library. "He loved books and had heard of this library from the chaplain on the army base. It has more than 22,000 books, magazines, and audio tapes. Fred's eyes and fingers were itching to examine them all."

Vocabulary. In telling the story of Fred's life, I use a full adult vocabulary, for example, the words *eponymous, hebdomadal,* and *faux pas*. However, the vocabulary is kept simple when I'm explaining the math.

Students are expected to do ALL of the problems. It is really better for them if you don't help them with any of the problems. It is so important that they learn how to learn by reading. If it takes them two days to figure out a particular problem, that is perfectly fine.

There is an old story of someone who saw a butterfly trying to break out of its chrysalis. He felt sorry for the effort that the butterfly was making and tried to "help" it by breaking open the chrysalis. The butterfly could never fly. It needed to struggle and exercise to develop its wings. (In *Life of Fred: Butterflies* we learned that butterflies do not use cocoons.)

Contents

Chapter 1 Numbers and Sets.............................. 17

    Lesson 1: Finite/Infinite, Exponents, and Counting

    Lesson 2: Natural Numbers, Whole Numbers, Parentheses, Braces, and Brackets

    Lesson 3: Negative Numbers and Integers

    Lesson 4: Ratios and Adding Signed Numbers

    Lesson 5: The First City Aly, Arkansas

    Lesson 6: The Second City Elk, Washington

    Lesson 7: The Third City Ulm, Wyoming

Chapter 2 The Integers. 45

    Lesson 8: How to Show Multiplication

    Lesson 9: Multiplying Signed Numbers

    Lesson 10: Proportion and Inequalities in the Integers

    Lesson 11: Circumference of a Circle

    Lesson 12: The First City Troy, New York

    Lesson 13: The Second City Zion, Illinois

    Lesson 14: The Third City Weed, California

Chapter 3 Equations. 82

    Lesson 15: Continued Ratios

    Lesson 16: Adding $3x + 3x + 4x + 6x + 2x$

    Lesson 17: Rectangles, Trapezoids, Sectors, Symmetric Law of Equality, and Order of Operations

    Lesson 18: Consecutive Numbers and Solving Equations

    Lesson 19: Rational Numbers and Set Builder Notation

    Lesson 20: Distance-Rate-Time and Distributive Property

    Lesson 21: Reflexive Law of Equality

    Lesson 22: The First City Ogden, Utah

    Lesson 23: The Second City Peetz, Colorado

    Lesson 24: The Third City Xenia, Ohio

Chapter 4 Motion and Mixture. 136

    Lesson 25: Proof of the Distributive Law

    Lesson 26: A Second Kind of Distance-Rate-Time Problem

    Lesson 27: Coin Problems

    Lesson 28: Coin Problems with Unequal Number of Coins

    Lesson 29: Age Problems

    Lesson 30: The First City Larned, Kansas

    Lesson 31: The Second City Dugger, Indiana

    Lesson 32: The Third City Seward, Alaska

Chapter 5 Two Unknowns. 182

    Lesson 33: Transposing

    Lesson 34: Solving Systems of Equations by Elimination

    Lesson 35: Work Problems in Two Unknowns

    Lesson 36: Graphs

    Lesson 37: Plotting Points

    Lesson 38: Averages

    Lesson 39: Linear Equations

    Lesson 40: Graphing Equations

    Lesson 41: The First City Stigler, Oklahoma

    Lesson 42: The Second City Wyoming, Pennsylvania

    Lesson 43: The Third City Roswell, New Mexico

Chapter 6 Exponents. 239

    Lesson 44: Solving Systems of Equations by Graphing

    Lesson 45: Solving Systems of Equations by Substitution

    Lesson 46: Inconsistent and Dependent Equations

    Lesson 47: Factorial

    Lesson 48: Area of a Square, Volumes of Cubes and
                    Spheres, Like Terms, Commutative Laws

    Lesson 49: Negative Exponents

    Lesson 50: The First City Seabrook, Texas

    Lesson 51: The Second City Florence, South Carolina

    Lesson 52: The Third City Glenmora, Louisiana

Chapter 7 Factoring..282

    Lesson 53: Multiplying Polynomials

    Lesson 54: Monomials, Binomials, Trinomials

    Lesson 55: Solving Quadratic Equations by Factoring

    Lesson 56: Factoring: Common Factors

    Lesson 57: Factoring: Easy Trinomials

    Lesson 58: Factoring: Difference of Squares

    Lesson 59: Factoring: Grouping

    Lesson 60: Factoring: Harder Trinomials

    Lesson 61: The First City Philomath, Oregon

    Lesson 62: The Second City Owensboro, Kentucky

    Lesson 63: The Third City Slatyfork, West Virginia

Chapter 8 Fractions.....................................322

    Lesson 64: Job Problems

    Lesson 65: Solving Fractional Equations

    Lesson 66: Simplifying Rational Expressions

    Lesson 67: Adding Rational Expressions

    Lesson 68: Subtracting Rational Expressions

    Lesson 69: Multiplying and Dividing Rational Expressions

    Lesson 70: The First City Winnemucca, Nevada

    Lesson 71: The Second City Livingston, Montana

    Lesson 72: The Third City Darlington, Wisconsin

Chapter 9 Square Roots. 377

    Lesson 73: Pure Quadratics, Square Roots

    Lesson 74: Pythagorean Theorem

    Lesson 75: The Real Numbers, The Irrational Numbers

    Lesson 76: Two Laws: $3\sqrt{x} + 5\sqrt{x} = 8\sqrt{x}$
                       $\sqrt{7}\sqrt{8} = \sqrt{56}$

    Lesson 77: Fractional Exponents

    Lesson 78: Radical Equations, Rationalizing the
                Denominator

Lesson 79: The First City Scottsbluff, Nebraska

Lesson 80: The Second City Chamberlain, South Dakota

Lesson 81: The Third City Bloomington, Illinois

Chapter 10 Quadratic Equations. 421

Lesson 82: Quadratic Equations in Everyday Life

Lesson 83: Solving Quadratics by Completing the Square

Lesson 84: The Quadratic Formula

Lesson 85: Long Division of Polynomials

Lesson 86: The First City Marshalltown, Iowa

Lesson 87: The Second City Copperopolis, California

Lesson 88: The Third City Silver Spring, Maryland

Chapter 11 Functions and Slope. 459

Lesson 89: Functions

Lesson 90: Slope

Lesson 91: Finding Slopes from Equations

Lesson 92: Slope-Intercept Form of a Line

Lesson 93: Range of a Function, Graphing y = mx + b

Lesson 94: The First City Pleasantville, New York

Lesson 95: The Second City Upper Sandusky, Ohio

Lesson 96: The Third City Elizabethtown, Kentucky

Chapter 12 Inequalities and Absolute Value.. 499

Lesson 97: Fahrenheit-Celsius Conversion

Lesson 98: Graphing Inequalities

Lesson 99: Why You Can't Divide by Zero

Lesson 100: Absolute Value

Lesson 101: Solving Inequalities in One Unknown

Lesson 102: The First City Mechanicsville, Virginia

Lesson 103: The Second City Saint Augustine, Florida

Lesson 104: The Third City Fort Lauderdale, Florida

A.R.T. section (quick summary of all of beginning algebra). 531

Index. 541

He stood in the middle of the largest rose garden he'd ever seen. The warm sun and the smell of the roses made his head spin a little. Roses of every kind surrounded him. On his left was a patch of red roses: *Chrysler Imperial* (a dark crimson); *Grand Masterpiece* (bright red); *Mikado* (cherry red). On his right were yellow roses: *Gold Medal* (golden yellow); *Lemon Spice* (soft yellow). Yellow roses were his favorite.

Up ahead on the path were white roses, lavender roses, orange roses and even a blue rose.

Fred ran down the path. In the sheer joy of being alive, he ran as any healthy five-year-old might. He ran and ran and ran.

At the edge of a large green lawn, he lay down in the shade of some tall roses. He rolled his coat up in a ball to make a pillow.

Listening to the robins singing, he figured it was time for a little snooze. He tried to shut his eyes.

They wouldn't shut.

Hey! Anybody can shut their eyes. But Fred couldn't. What was going on? He saw the roses, the birds, the lawn, but couldn't close his eyes and make them disappear. And if he couldn't shut his eyes, he couldn't fall asleep.

You see, Fred was dreaming. He had read somewhere that the only thing you can't do in a dream is shut your eyes and fall asleep. So Fred *knew* that he was dreaming and that gave him a lot of power.

He got to his feet and waved his hand at the sky. It turned purple with orange polka dots. He giggled. He flapped his arms and began to fly. He settled on the lawn again and made a pepperoni pizza appear.

In short, he did all the things that five-year-olds might do when they find themselves King or Queen of the Universe.

Soon he was bored. He had done all the silly stuff and was looking around for something constructive to do. So he lined up all the roses in one long row.

They stretched out in a line in both directions going on forever. Since this was a dream, he could have an unlimited (**infinite**) number of roses to play with.

When Fred was three years old, he had spent some time studying physics and astronomy. He had learned that nothing in the physical universe was infinite. Everything was **finite** (limited). Every object could travel only at a finite speed. Even the number of atoms was finite. One book estimated that there are only 10^{79} atoms in the observable universe. 10^{79} means 10 times 10 times 10 . . . a total of 79 times, which is 10,000,000,000,000,000,000,000,000,000,000,000,000,000,000,000, 000,000,000,000,000,000,000,000,000,000,000. That is a lot of atoms! (The "79" is an exponent—something we'll deal with later.)

Now that he had all the roses magically lined up in a row, he decided to count them. Math was one of Fred's favorite activities.

Now, normally when you've got a bunch of stuff in a pile to count,

(← *These are Fred's dolls that he used to play with when he was a baby.*)

you line them up ....

and start on the left and count them   1   2   3

But Fred couldn't do that with the roses he wanted to count. There were too many of them. He couldn't start on the left as he did with his dolls. Dolls are easy. An infinite line of roses is hard.

Now it's your turn to play with some of the things covered thus far. Take out a piece of scratch paper and write out the answers for each of the following. **THiS iS iMPoRTaNT.** They've done the studies and have found that you learn and retain a lot more if you are actively involved in the learning process rather than just reading passively.

Your Turn to Play

1. Is there a finite or infinite number of grains of sand on all the beaches in the world?

2. 10^{79} means 10 times 10 times 10 . . . seventy-nine times. What does 3^4 equal?

3. Which is larger: 2^5 or 5^2?

4. In Fred's dream the set (collection) of roses was infinite. The set of even natural numbers, $\{2, 4, 6, 8, 10, 12, 14, \dots\}$, is infinite. The set of all possible melodies is infinite.

    You don't find infinite sets at the grocery store.

    You don't find infinite sets in your laundry basket.

    Where *is* a good place to find infinite sets?

5. What does 1^{8369} equal?

6. What would you multiply 10^2 by in order to get 10^5?

Intermission

Some people like to argue that infinite sets don't really exist. "After all," they say, "they're just a figment of your imagination. It's all in your head."

By that same argument I could prove that pain doesn't exist. When you cut your finger, the pain is experienced in your brain.

And the pleasure of a bite of warm pizza doesn't exist.

And the number three doesn't exist.

And truth doesn't exist.

Just because it is happening inside your skull doesn't mean that it doesn't exist.

7. When you want to count something, one of the easiest ways is to line them up in a row and count.

1   2   3

A hard question: Why doesn't it make a difference which order you line them up? Why do you always get the same answer?

.......**COMPLETE SOLUTIONS**.......

1. Nothing in the physical universe is infinite. There is a finite number of grains of sand.

2. $3 \times 3 \times 3 \times 3$ which is 81.

3. 2^5 is $2 \times 2 \times 2 \times 2 \times 2$ which is 32. 5^2 is 5×5 which is 25. So 2^5 is larger.

4. The set of roses in a dream, the set of even natural numbers, and the set of all possible melodies are all things that we can conceive. They are not things we can touch. To find infinite sets, one of the best places to look is your mind.

5. If you keep multiplying 1 times itself, you will always get an answer equal to 1.

6. $10^2 \times ? = 10^5$ is a restatement of the question.

$$100 \times ? = 100,000.$$

$$100 \times 1000 = 100,000$$

$$10^2 \times 10^3 = 10^5$$

7. Wow. That is something that most people never think about. They would say that it's *obvious* that the way you line up the items won't affect how many there are.

Could it be that it's obvious to them because that's what they've always experienced? But suppose the world were created a little differently. Suppose that the order in which you lined up the objects affected how many there were? Then everyone would go around saying that it's obvious that the way you line up objects affects how many there are. One of the enduring mysteries of mathematics is how well the stuff that goes on in our heads reflects what goes on out there in the "real world." *That didn't have to happen.*

One of the fun things I sometimes do in a calculus class (when we're studying infinite series) is to write on the board: $1 - 1 + 1 - 1 + 1 - 1 \ldots$ and ask the students what the sum is.

If I add together the pairs I get $(1 - 1) + (1 - 1) + (1 - 1) + \ldots$ which is $0 + 0 + 0 + \ldots$ which equals zero.

If I combine the second and third numbers together, the fourth and fifth numbers together, etc., I get $1 \underline{- 1 + 1} \underline{- 1 + 1} - \ldots = 1 + 0 + 0 + 0 + \ldots$ which equals one.

S o how do you count them? There wasn't even an obvious "middle" rose to start at. In some sense, every rose is in the middle since there is an infinite number of roses on each side of every rose. So Fred

just selected a rose and called it "1." From there it was easy to start counting . . . .   1 2 3 4 5 6 7 8 9 10 11 12 13 14 15. . . .

This **set** (collection, group, bunch) of numbers {1, 2, 3, 4, 5, . . .} is called the **natural numbers**. At least, Fred figured, with the natural numbers he could count half of all the roses.

What to do? How would he count all the roses to the left of "1"? Then Fred remembered the movies he'd seen where rockets were ready for blastoff. The guy in the tower would count the seconds to blastoff: "Five, four, three, two, one, zero!" So he could label the rose just to the left of the rose marked "1" as "0."

This new set, {0, 1, 2, 3, . . . } is called the **whole numbers**. It's easy to remember the name since it's just the natural numbers with a "hole" added. The numeral zero does look like a hole.

A set doesn't just have to have numbers in it. Fred could gather the things from his dream and make a set: {roses, lawn, birds, pizza}. The funny looking parentheses are called braces. Braces are used to enclose sets.   Left brace: {   Right brace: }

On a computer keyboard, there are actually three types of grouping symbols: Parentheses: ( )

Braces: { }   and

Brackets: [ ]

In algebra, braces are used to list the members of a set, while both parentheses and brackets are used around numbers. For example, you might write (3 + 4) + 9 or [35 – 6] + 3.

Braces and brackets both begin with the letter "b" and to remember which one is braces, think of braces on teeth. Those braces are all curly and twisty.

In English classes, parentheses and brackets are not treated alike. If you want to make a remark in the middle of a sentence (as this sentence illustrates), then you use parentheses (as I just did).

Brackets are used when you're quoting someone and you want to add your own remarks in the middle of their quote: "Four score and seven [87] years ago. . . ."

But brackets and parentheses weren't going to help Fred with counting all those roses. The whole numbers only got him this far:

0 1 2 3 4 5 6 7 8 9 10 11 . . .

What he needed were some new numbers. These new numbers would be numbers that would go to the left of zero. So years ago, someone invented negative numbers: minus one, minus two, minus three, minus four. . . .

Some notes on negative numbers:

♪#1: It would be a drag to have to write this new set as {. . . minus 3, minus 2, minus 1, 0, 1, 2, . . .}, or even worse, to write {. . . negative 3, negative 2, negative 1, 0, 1, 2, . . .}, so we'll invent an abbreviation for "minus." What might we use?

How about screws? The two most common kinds look like ⊕ (Phillips screws) and ⊖ (slotted screws). Okay. Our new number system will be written {. . . –3, –2, –1, 0, +1, +2, +3, +4, . . . }. We'll call this new set **the integers**.

♪#2: All these integers sit on the **number line**.

... -10 -9 -8 -7 -6 -5 -4 -3 -2 -1 0 1 2 3 4 5 6 7 8 9 10...

22

Of course, there are more numbers on the number line than just the integers. There is π and fractions like ½ and decimals like 2.7.

```
◄─┼────┼───┼───┼───┼───┼───┼───┼───┼───┼───┼───┼───┼───┼───┼───┼───┼───┼───┼───┼───┼───►
 ...-10  -9  -8  -7  -6  -5  -4  -3  -2  -1   0   1   2   3   4   5   6   7   8   9  10 ...
                                                 ½         2.7  π
```

Your Turn to Play

1. Match each item in the left column with one item in the right column.

    [ ]   parentheses
    ( )   braces
    { }   brackets

2. When you list the elements of a set, you enclose them in braces. Write the set that contains your three favorite numbers.

3. Write the set that contains the four smallest natural numbers.

4. Is it possible to name a natural number that is not a whole number?

5. This set has three elements in it: $\{7, 4, 0\}$.

    This set has one element in it: $\{5\}$.

    This set has one element in it: $\{0\}$. Zero is *something*. It's not nothing. Zero is a number. On the other hand, if you have a paper bag with zero things in it, there is *nothing* in the bag.

    Write the set that has zero elements in it. This is sometimes called the **null set** or the **empty set**.

6. In algebra, when you put parentheses (or brackets) around some expression, you are saying, "Do me first."

    For example, $(3 + 4) + 9$ means that you would add together the 3 and the 4 first.............. 7 + 9

    and then................. 16.

    $(3 + 4) + 9$ gives an answer of 16. Evaluate $3 + (4 + 9)$.

7. One of the fanciest laws of algebra is the **distributive law**. It states that for any three numbers, *a, b,* and *c,* it is always true that

$$a(b + c) = ab + ac$$

Evaluate $7(9 + 12)$ and evaluate $7 \times 9 + 7 \times 12$.

8. What does 6^2 equal?

9. Harder question: Write the set where the only element is the null set.

........COMPLETE SOLUTIONS.......

1.
```
[ ]          parentheses
( )          braces
{ }          brackets
```

2. Your answer will probably be different than mine. The set containing my three favorite numbers is $\{21, 0, \pi\}$.

   Note: π is a number. It is the number you get when you divide the circumference of a circle by its diameter. It is approximately equal to 3.14159.

3. $\{1, 2, 3, 4\}$

4. It isn't possible since every natural number is a whole number.

5. $\{ \}$

6. $3 + (4 + 9)$
   $= \ 3 \ + \ 13$
   $= \ \ \ \ 16$

   So $(3 + 4) + 9$ equals $3 + (4 + 9)$.

   It turns out that this is true for any three numbers. If a, b, and c are any three numbers, then it is always true that $(a + b) + c = a + (b + c)$. This is called the **associative law of addition**.

7. $7(9 + 12)$ $7 \times 9 \ + \ 7 \times 12$
       $7(21)$ $63 \ + \ 84$
        147 147

8. $6^2 = 6 \times 6 = 36$ (This is a problem from the previous lesson.)

9. $\{ \{ \} \}$ The null set is $\{ \ \}$. I have placed that set within braces to create a set where the only element is the null set.

Some more notes on negative numbers:

♪#3: Positive 7 can be written as either "7" or "+7"—it doesn't matter which you choose.

♪#4: Phillips screws ⊕ haven't been around forever. "Phillips Screws" was trademarked in the early 1930s.

♪#5: When it comes to counting moose, some of the integers don't work well at all. It's fine to say two moose:

Even zero moose makes sense. Here is a picture of zero moose:

zero moose

But how in the world could you ever have –3 moose? You can't. Negative numbers are useful some places and not others. I guess this could be considered a mathematical proof that moose aren't roses.

♪#6: When Mr. Fahrenheit was first fooling around making his thermometer, he had to figure out where to mark the zero on his scale. Without a zero, he couldn't have a 1°, 2°, 3°, . . .

He might have gone into his lab and tried to create the coldest possible temperature. That would make a perfect place for a zero on his thermometer. He might have taken his wife's fingernail polish remover (acetone) and put that into a blender with some ice. That would be plenty c... c... cold.

And all this would have been perfectly fine unless the Fahrenheits had decided to take a little winter vacation in everyone's favorite winter resort spot: North Dakota.

Besides being noted for famous North Dakotans such as Peggy Lee, Eric Sevareid, and Lawrence Welk, this winter playground offers the advantage of not having a whole lot of half-melted, slushy snow to ruin your tobogganing. On average, in seven years out of every ten in Bismarck, North Dakota, the thermometer falls 62° below the freezing point of water. (Be sure to wear a sweater!)

So on his scale, Mr. Fahrenheit would have to make 30 marks on his thermometer below the zero. From –30° up to the freezing point of water, which is +32°, is a 62° gain.

Fred, in his dream, had been running up and down the line of roses playing with the numbers as he passed them. For example, going from the rose marked 8 to the rose marked 13 he had to go five roses. That he knew from arithmetic: to go from 8 to 13 is $13 - 8 = 5$.

When Mr. Fahrenheit went from –30 up to +32, he had to go up 62. From –30 up to 0 is a movement upwards of 30 and from 0 up to 32 is an additional movement of 32.

Since going from + 8 to +13 means + 13 – (+ 8),

going from –30 to +32 means +32 – (–30) which we know is +62.

Here are some examples:

Going from 9 to 90 means $90 - 9 = 81$.

Going from 2 to +48 means $48 - 2 = 46$.

Going from –7 to +8 means $8 - (-7) = 15$.

Please remember the One Important Rule: Write out the answers to the questions BEFORE you look at the answers.

Your Turn to Play

1. Going from 44 to 55 means . . .

2. Going from –100 to +200 means . . .

3. Going from –20 to 70 means . . .

4. On May 29, 1953, the world's tallest mountain was climbed for the first time. Edmund Hillary and Tenzing Norgay made the ascent of Mt. Everest. The mountain was thought to be 29,002 feet above sea level at

that time. A year later the Surveyor General of the Republic of India set the height at 29,028 feet instead. By how much had he increased the height?

5. If my checking account was overdrawn by $12 and after making a deposit, it now has a balance of $888, how much was my deposit?

6. Name a number that is a whole number but is not a natural number.

7. Can you name a whole number that is not an integer?

8. If you started at the rose marked "0" and you went to the right four "steps" you would be at the rose marked "4." If you then went to the left one "step," what rose would you be at?

9. If you started at the rose marked "0" and you went to the right seven "steps," you would be at the rose marked "7." If you then went to the left nine "steps," what rose would you be at?

10. The previous problem could be written as the addition of two numbers: $7 + (-9)$

How would you write "Six steps to the right, followed by seven steps to the left" as an addition problem?

11. If you started at the rose marked "0" and you went two steps to the right and then eight steps to the left, what rose would you be at?

12. If the temperature at noon was 50° and it fell by 70° by midnight, what would the new temperature be?

13. Write the previous question as an addition problem.

14. If you started at the rose marked "0" and you went to the *left* seven steps you would be at the rose marked "–7." If you then went to the left again, nine more steps, what rose would you be at?

15. Write the previous problem as an addition problem.

16. If you gained $100 in the stock market on Monday and lost $40 on Tuesday, where would you be at the end of those two days?

17. If you gained $100 in the stock market on Monday and lost $120 on Tuesday, where would you be at the end of those two days?

18. $(3583 + 17) + 55 = 3583 + (17 + 55)$ is an example of what law of addition?

19. Which of these numbers are *not* whole numbers?   –7, 4, ½

20. Which of these numbers are not integers? 1.6, $\frac{8}{2}$, 0, 8972346

........**COMPLETE SOLUTIONS**........

1. $55 - 44 = 11$

2. $200 - (-100) = 300$

3. $+70 - (-20) = 90$

4. To go from 29,002 to 29,028 means $29,028 - 29,002 = 26$ feet

5. To go from -12 to $+888$ means $888 - (-12) = 900$. So I had made a deposit of \$900.

6. There's only one number that is a whole number and not a natural number. Since the whole numbers are $\{0, 1, 2, 3, 4, \ldots\}$ and the natural numbers are $\{1, 2, 3, \ldots\}$, that number is zero.

7. That would be silly. The integers are $\{\ldots -2, -1, 0, 1, 2, 3, \ldots\}$ so every whole number is an integer. Some people describe the whole numbers as the non-negative integers.

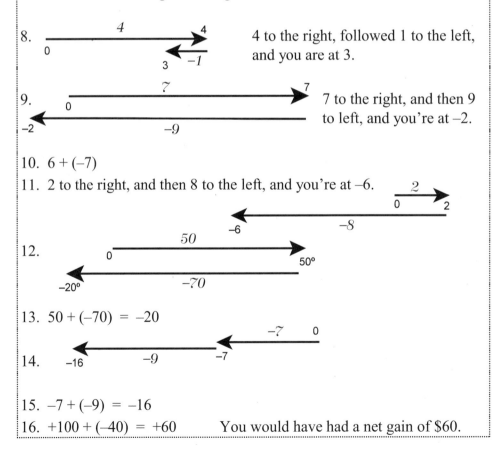

8.   4 to the right, followed 1 to the left, and you are at 3.

9.   7 to the right, and then 9 to left, and you're at -2.

10. $6 + (-7)$

11. 2 to the right, and then 8 to the left, and you're at -6.

12.

13. $50 + (-70) = -20$

14.

15. $-7 + (-9) = -16$

16. $+100 + (-40) = +60$ You would have had a net gain of \$60.

17. +100 + (–120) = –20   You would have had a net loss of $20.

18. (*a* + *b*) + *c* = *a* + (*b* + *c*) is the associative law of addition.

19. –7 and ½ are not whole numbers.

20. The only one that is not an integer is 1.6. (The number $\frac{8}{2}$ is the same as the number 4, which is an integer. If 4 were written as the Roman numeral IV, it would still be an integer.)

A Recapitulation
also known as a recap

　　Infinite sets have an unlimited number of members. Every set of things in the physical universe is **finite**.

　　4^6 means 4×4×4×4×4×4. The 6 is called an **exponent**.

　　$10^3 \times 10^4 = 10^7$　　　　10×10×10 × 10×10×10×10 = 10^7

　　Parentheses ( )

　　Braces { }

　　Brackets [ ]

　　The **natural numbers** {1, 2, 3, 4, . . .}

　　The **whole numbers** {0, 1, 2, 3, . . .}

　　The **integers** {. . . –3, –2, –1, 0, 1, 2, 3, . . .}

　　The **distributive law** a(b + c) = ab + ac　　5(3 + 9) = 5×3 + 5×9

　　The **empty set (null set)** is the set with no elements in it. { }

　　Pi (π) is a number. It is the circumference of any circle divided by its diameter.

　　The **associative law of addition**: (a + b) + c = a + (b + c)

　　To go from –4 to + 9 is a gain of 13.　9 – (–4)

Hey! Not bad for three lessons.

Is there a shortcut for problems like these?

When you work . . .

70 – (–20) and get an answer of 90

4 – (–6) and get an answer of 10

9 – (–20) and get an answer of 29

what you can do is change the two negative signs to plus:

70 – (–20) becomes 70 + (+20)

4 – (–6) becomes 4 + (+6)

9 – (–20) becomes 9 + (+20).

On the days when Fred taught algebra at KITTENS, he would write on the board:

To subtract a negative is the same as adding the positive.

There was a knock on the door. Alexander called out, "Professor Fred! It's time for our Thursday game night!" Our five-year-old awoke from his dreaming.

"I'll be right there, Alexander!" Fred responded. It had been a wonderful nap. He always liked it when he was dreaming and knew he was dreaming, but now it was back to the real world.

He had been sleeping where he normally had taken his naps for the last four years: under his desk in his office at KITTENS University. The stories of his birth and early years and how he got to KITTENS University (Kansas Institute for Teaching Technology, Engineering, and Natural Sciences) are all told in *Life of Fred: Calculus Expanded Edition*.

He got up carefully. He didn't want to hit his head against the underside of his desk. There are things you have to watch out for when you've reached the yard-tall mark in life.

"So you think you'll win today, do you?" Alexander asked Fred as Fred joined one of his favorite students to go to their Thursday game night. They walked down the hallway in keen anticipation of an evening of fun.

Fred looked at his watch. He had started his nap at five minutes to 5:00 and it was now 5:30. "I'm sorry, I must have overslept," Fred apologized. "My naps usually don't last 35 minutes." Fred thought of how long he had napped by visualizing 30 − (−5) as 30 + (+5) which is 35.

They played in the student recreation room which was at the far end of the hallway from Fred's office. Alexander was a little over six feet tall and the length of his pace was almost as long as Fred was tall. Fred was walking very fast. Or to be more precise, his speed wasn't that fast but his legs were moving very quickly.

For every step that Alexander took, Fred took about two. So the **ratio** of Alexander's steps to Fred's was one to two, which can be written 1:2. The colon can be thought of as a division sign.

So these all mean the same thing: 1:2

$$1 \div 2 \quad \text{and}$$

$$\frac{1}{2}$$

Betty, who's engaged to Alexander, takes steps that are about three-fourths as long as his. So she takes about four steps for every three that Alexander takes. The ratio of the number of her steps to the number of his steps is 4:3.

"There's our professor!" Joe exclaimed as Fred and Alexander entered the student recreation room. The table was all set up. Betty was getting the three phone books to put on Fred's chair. When they had first started playing poker last August, Fred had stood on his chair so he could see, but his little legs soon got tired and he asked for phone books to sit on.

The fifth member of the Thursday night poker group was Darlene. She was busy making the usual phone call to Stanthony's Pizza. Their routine was to play for an hour, until 6:30, and then the pizza would be delivered, and they'd break for dinner and conversation.

Now, at first blush, it may seem unusual that these four students were playing poker with their five-year-old teacher. Even Alexander's greeting to Fred, "So you think you'll win today, do you?" may have appeared like cruel teasing. But it wasn't. Playing poker with *this* five-year-old wasn't a matter of four students in their early 20s ganging up on a poor defenseless five-year-old.

It was a five-year-old "ganging up" on the four of them. He had read a half dozen books on poker before their first evening together, and that had given him quite an advantage.

There was a small jar of pennies in the student recreation room. It was Joe's job at the beginning of the evening to count out an equal number of pennies for each person. They used the pennies as poker chips. At the end of each evening, all the pennies would go back into the jar.

For fun, Fred had kept a diary and recorded how much he had won or lost for each Thursday night. His first eight entries in his book were:

+9   -2   -3   +6   +13   +2   -7   +8.

He used plus signs for the nights he won and negative signs for the nights he lost.

How far was Fred ahead? It was a matter of adding up these eight integers. Putting addition signs between the numbers:

$$(+9) + (-2) + (-3) + (+6) + (+13) + (+2) + (-7) + (+8).$$

First, Fred totaled up all the wins:

$$\begin{array}{r} +9 \\ +6 \\ +13 \\ +2 \\ \underline{+8} \\ +38 \end{array}$$

Then he totaled up all the losses:
$$\begin{array}{r} -2 \\ -3 \\ \underline{-7} \\ -12 \end{array}$$

Finally, he combined his wins and losses. He had a gain of 38 and a loss of 12, which gave him a net gain of 26¢. Not bad for eight nights' work.

The last step (when he combined his wins and losses) could be written: $(+38) + (-12) = +26$.

Joe's great-great-great-great-great-great-great grandfather was Joseph Priestley. Joe was named after his famous ancestor who had discovered the chemical element oxygen.

Joe's $(great)^7$-grandfather was probably a better poker player than Joe. (The little "7" in the previous sentence is an exponent.) Joe's poker record wasn't that hot. If he'd kept a diary, his entries would have read: –14 –8 +2 –9 +1 –23 –44 –3. Everyone liked to have Joe at the poker parties. Especially Darlene. Darlene likes Joe even though he's not the world's greatest poker player.

Adding up Joe's wins and losses, we first put in the addition signs: $(-14) + (-8) + (+2) + (-9) + (+1) + (-23) + (-44) + (-3)$.

Then we total up all the wins:
$$\begin{array}{r} +2 \\ \underline{+1} \\ +3 \end{array}$$

Then we total up all the losses:
$$\begin{array}{r} -14 \\ -8 \\ -9 \\ -23 \\ -44 \\ \underline{-3} \\ -101 \end{array}$$

Last, we combine Joe's wins and losses. He had a gain of 3 and a loss of 101, which gave him a net loss of 98¢

The last step could be written: $(+3) + (-101) = -98$¢.

Adding up a bunch of positive numbers is easy. You did that in elementary school.

Adding up a bunch of negative numbers is also easy. Just add them up and stick a negative sign in front. We did that for Joe's losses and you didn't bat an eye.

Adding a positive and a negative number together was the only thing that was really new. With Fred we had $(+38) + (-12) = +26$. With Joe it was $(+3) + (-101) = -98$.

If we total up Darlene's wins and losses, we find that she had a gain of 4 and a loss of 9. This could be written as: $(+4) + (-9)$, which gives a final result of -5¢. Losing five cents in eight nights of play is something that she could live with. Losing Joe is something that Darlene didn't want to even contemplate.

Instead, let's contemplate our progress so far with a *Your Turn to Play*.

Your Turn to Play

1. $(+22) + (-5) + (-9) + (+3)$
2. $(-8) + (-2) + (-4) + (+5)$
3. $(+8) + (-9) + (-66) + (+1) + (+13)$
4. $(+10) + (-2) + (-78) + (-8) + (+21)$
5. $(-33) + (-22)$
6. $(+50) + (-18)$
7. $(+50) + (-68)$
8. $(-100) + (-32)$
9. $5 + (-8)$
10. $-66 + (+88)$
11. Going from -88 to $+100$ is a gain of how much?
12. What number do you add to -88 to get an answer of $+100$?
13. Fill in the blank: $(-88) + \underline{\quad ? \quad} = +100$
14. Fill in the blank: $(-12) + \underline{\quad ? \quad} = 33$
15. If I'm somewhere in Death Valley, California, which is 120 feet below sea level, and I drive until I get to some point in San Francisco that's 43 feet above sea level, how much has my altitude changed?
16. If I drive in the opposite direction from that point 43 feet above sea level in San Francisco to the point in Death Valley that's 120 feet below sea level, how much has my altitude changed?
17. List the set (using braces) that contains your favorite female singer and your favorite food.
18. List the set that contains *, #, and %.
19. List the set that contains the names of all women over eighty feet tall.
20. List the set that contains a left bracket.
21. If you add a positive number and a negative number together, what will be the sign of the answer?
22. Can you name an integer that's not a whole number?

.......COMPLETE SOLUTIONS.......

1. Adding up the positive numbers: $+22 + (+3) = +25$

Adding up the negative numbers: $(-5) + (-9) = -14$

Combining: $+25 + (-14) = +11$

2. Adding up the positive numbers: $+5$ (There's only one positive number.)

Adding up the negative numbers: $(-8) + (-2) + (-4) = -14$

Combining: $+5 + (-14) = -9$

3. Adding up the positive numbers: $(+8) + (+1) + (+13) = +22$

Adding up the negative numbers: $(-9) + (-66) = -75$

Combining: $+22 + (-75) = -53$

4. Adding up the positive numbers: $(+10) + (+21) = +31$

Adding up the negative numbers: $(-2) + (-78) + (-8) = -88$

Combining: $+31 + (-88) = -57$

5. -55

6. $+32$

7. -18

8. -132

9. -3 Remember that $5 + (-8)$ is the same as $+5 + (-8)$. With positive numbers you can write either 5 or +5.

10. $+22$

11. Going from -88 to $+100$ means $+100 - (-88)$. This becomes $+100 + (+88) = +188$

12. To ask, "What number do you add to -88 to get an answer of $+100$?" is the same as asking how far it is from -88 to $+100$ which is the same question as the previous problem asked.

13. $(-88) + \underline{\ ?\ } = +100$ is the same question as the previous two questions. Later on when we get into heavy-duty algebra, instead of the "?" we'll use the letter "x" and write: $-88 + x = +100$, but that's getting ahead of the story.

14. "What do you add to -12 to get to 33" asks how far is it from -12 to $+33$, which means $+33 - (-12)$. This becomes $+33 + (+12) = +45$.

15. 120 feet below sea level can be written as -120 feet. To go from -120

to 43 means   43 – (–120), which is 43 + (+120) = 163 feet.

16. Going from +43 to –120 means –120 – (+43). Changing the two signs, this becomes   –120 + (–43), which is –163 feet. *Subtracting a positive is the same as adding a negative.*

17.   When I list my favorite female singer and favorite food as a set, I get {Jeanette MacDonald, pizza}. Now your answer may be different from mine. In fact, you may not have ever heard of Jeanette MacDonald, the red-haired soprano whose most famous movie was *Rose Marie* (1936), in which she sang "Indian Love Call" with Nelson Eddy.

18. {*, #, %}   The order in which you list those three items doesn't matter. You could have written {#, *, %} and it would still be correct. However, it's not considered acceptable to list an element of a set more than once. So {*, *, #, #, #, %} is not cool. The reason for this rule is that if you have a fairly large set with hundreds of elements in it, and you were trying to count how many elements the set has, it might drive you a little crazy if you found that some of the elements had been listed more than once. Imagine trying to count the number of *different* letters and symbols that have been used on this page. It would be much easier for you, if after I used the letter "e" for xampl, I didn't us it again on th pag.

19. There aren't a whole lot of women that tall. In fact, I don't think there are any. So the set would look like this: { }. It's the set with no elements in it. It's sometimes called the **empty set** or the **null set**.

20. This may look a little weird, but here goes: { [}. This is just silliness. What it illustrates is that a set can contain just about anything that you'd care to name. You can make a set that contains Miss America and a potato if you like. You can have a set that contains an infinite number of items such as the set of all the whole numbers: {0, 1, 2, 3, . . .}.

21. It depends. When you add 34 + (–2) you get +32.   ↵ *Example One* When you add 5 + (–18) you get –13.   ↵ *Example Two*

    Whether the answer is positive or negative depends on (this may get a bit technical) *which of the two numbers is farther away from zero.*

    In Example One, the 34 is farther away from zero than the –2, so the answer will be positive since the 34 is positive.

In Example Two, the –18 is farther away from zero than the +5, so the answer will be negative since the –18 is negative.

If this is too technical, forget it. You knew what the sign of the answer was going to be before I "explained" it.

22. There are lots of integers that are not whole numbers.

Since the integers are {. . . –4, –3, –2, –1, 0, 1, 2, 3, . . .} and

the whole numbers are {0, 1, 2, 3, . . .}, your answer could be any number like –723 or –22340939072973992. Any negative integer would work. It wouldn't be correct to answer this question with, "Any negative number." Why? Because –⅓, for example, is not an integer, but it is a negative number.

 It's traveling time now. This will happen at the end of each chapter.

We will have the opportunity to visit three cities in the next three lessons where you will get to show what you've learned.

These cities are not **Bridges** (which were in the previous *Life of Fred* books). They are necessary practice to help you really learn the material. The reason algebra teachers in school know their algebra so well is that they have been doing the same problems for *years*. You will have the chance to do them for *minutes*.

Please do all three cities. <u>Doing one each day</u> is not that much work.

Aly, Arkansas in Yell County

1. What is another word for *set* ?

2. Using braces, give an example of a set that has four elements in it.

3. The natural numbers are often used for counting the elements in a set. Can you name a set that can't be counted using a natural number?

4. Here are two sets that are equal: {*, 5, mouse} and {5, mouse, *}. Here are two that are not equal: {%, ¢, rat} and {¢, dog, %}. Complete the following definition: *Two sets are equal if . . .*

5. If two sets have the same number of elements in them, must the sets be equal?

6. If I measure something in cups and then measure it in quarts, the ratio of the two measurements is ? .   (Note: there are 4 cups in a quart.)

7. Here is a record of my wins and losses. What is my total?

        −4 −7 +800 −36 −9 +20 −3

8. What, if anything, is wrong with {#, house, star, tomato, star, pencil}?

9. Is −4 a whole number?

10. Is it correct to write {4 + 5} + 3?

11. Who's richer: the man whose net worth is −$100 or the one whose net worth is −$75?

12. (+4) + (−70) + (−2) + (+17) + (−30)

13. Are the following sets equal? The set of all whole numbers greater than +6 and the set of all integers greater than +6.

14. If I measure my life in days and then measure it in weeks, what is the ratio of the two measurements?

15. If the stock I own gains $6 one day and loses $11 the next, I have a net loss of $5. Another way I could say that in English is that I have a net gain of ? .   (Fill in the blank.)

16. If set A is the set of all whole numbers greater than +4 and set B is the set of all integers greater than +4½, is it true that A = B?

17. Is the set containing all the people now living on the earth a finite set or an infinite set?

answers

1. Group, bunch, collection. If it were birds, we might say "flock." If it were a set of ants, we might say "colony." For cattle, "herd."

2. {Sir Thomas More, starlight, #, Mickey Mouse}

(Your answer will probably differ from mine.)

3. There are two kinds of sets whose elements can't be counted using the natural numbers. The first is the empty set, { }. This needs the number zero, which isn't a natural number. The second kind of set is any infinite set. You can't use a natural number to name the number of elements in an infinite set.

4. Two sets are equal if they have the same elements in them.

5. No. For example {#} and {pen} each have one element in them but they are not equal.

6. 4:1   For example, 8 cups would correspond to 2 quarts.

7. For the gains: $+800 + (+20) = +820$

For the losses: $-4 + (-7) + (-36) + (-9) + (-3) = -59$

Combining the wins and losses:

$+820 + (-59) = +761$

8. It is not acceptable to list an element more than once. The element "star" was listed twice.

9. No. The whole numbers are {0, 1, 2, 3, 4, . . .}

10. No. We use either parentheses or brackets to group numbers together. It would be correct to write $(4 + 5) + 3$ or $[4 + 5] + 3$.

11. The guy whose net worth is –$75. If you gave $25 to the one whose net worth is –$100, he would then have a net worth of –$75.

12. Adding up the positive numbers: $(+4) + (+17) = +21$

Adding up the negatives: $(-70) + (-2) + (-30) = -102$

Combining: $+21 + (-102) = -81$

13. Yes. Each of the sets is {7, 8, 9, . . .}

14. 7:1   For example, 2100 days would correspond to 300 weeks.

15. –$5. A loss of $5 is the same as a gain of –$5.

16. Yes. Both sets are {5, 6, 7, . . .}.

17. It is finite. There are about seven billion on the planet right now. That's 7,000,000,000. Using exponents it would be 7×10^9 people.

Elk, Washington about 3000 people but no elk

1. Using braces, list the natural numbers.

2. Do +8 and 8 mean the same thing?

3. List, using braces, the set that contains all the numbers that are whole numbers that are not natural numbers.

4. Joe was thinking about the time till he saw Darlene next in days. She was measuring that same time in hours. What is the ratio of his measurement to her measurement?

5. Two sets are equal if they have the same elements in them. For example, {@, red, staple} = {red, @, staple}.

       If two sets are equal, must they have the same *number* of elements in each set?

6. Name two infinite sets that are not equal.

7. If it was –30° outside and you walked inside where it was 70°, how much warmer would it be?

8. (–28) + (–9) + (+33) + (+1)

9. Using braces, list the set of all integers less than –8½.

10. Would it be correct to enclose the elements of a set in brackets?

11. Using braces, list the integers.

12. Name a set that contains 348890 elements.

13. Is it possible to name an integer that is not a whole number?

14. 9 – (–9)

15. Going from –17 to +4 is how much of a change?

16. If I measure my weight in ounces and then measure it in pounds, what is the ratio of the two measurements? (Note: there are 16 ounces in a pound.)

17. (–25) + (–35) + (100) + (–60)

18. List, using braces, the set of all doorknobs that can speak English.

19. If for every $5 I make, you make $7, what is the ratio of my earnings to yours?

20. What is the ratio of the number of even natural numbers that are less than ten to the number of whole numbers less than 4?

answers

1. {1, 2, 3, 4, 5, ...} is the set of natural numbers.
2. Yes. It can be written either way.
3. { 0 }
4. 1:24
5. Yes.
6. The integers and the whole numbers (or the natural numbers) or the even whole numbers and the odd whole numbers. There are many possibilities. For two sets to be equal they must have exactly the same elements in each set.
7. Going from -30 to $+70$ means $70 - (-30) = 70 + (+30) = +100$. It would be a gain of $100°$.
8. Adding the gains: $+33 + (+1) = +34$.
   Adding the losses: $-28 + (-9) = -37$.
   Combing the gains and losses: $+34 + (-37) = -3$
9. Those are the integers to the left of $-8\frac{1}{2}$ on the number line, which would be $\{\ldots -11, -10, -9\}$.
10. No. We use braces { }.
11. $\{\ldots -4, -3, -2, -1, 0, 1, 2, 3, \ldots\}$
12. There are many possible answers. One of the easiest is
    $\{1, 2, 3, 4, \ldots, 348889, 348890\}$.
13. Yes. Any negative integer, $-1, -2, -3, \ldots$ will work.
14. $9 - (-9) = 9 + (+9) = 18$
15. -17 to $+4$ means $4 - (-17) = 4 + (+17) = 21$
16. 16:1
17. Adding the losses: $(-25) + (-35) + (-60) = -120$. Combining that with the gain of $+100$ gives $(-120) + (+100) = -20$
18. Since there aren't any doorknobs that can speak English, this would be the empty set, which is written: { }.
19. 5:7
20. The even natural numbers that are less than ten is the set {2, 4, 6, 8}. The whole numbers less than 4 is the set {0, 1, 2, 3}.

     Since there are four members in each set, the ratio would be 4:4. This could also be written as $4 \div 4$ or as $\frac{4}{4}$. Since $\frac{4}{4}$ is the same as $\frac{1}{1}$ we could have said that the ratio is 1:1.

Ulm, Wyoming named after a city in Germany

1. Name a number that is a whole number that is not a natural number.

2. Going from –10 to +20 is an increase of how much?

3. Using braces, give an example of a set that has three elements in it, one of which is an animal.

4. In Fred's dream, if he ran from the rose marked –42 to the rose marked 27, how far would he have gone?

5. If you owed someone $12 and that person both forgave your debt and handed you $100, how much richer would you be?

6. If I measure the time it takes to run a marathon in minutes and then measure that in hours, what is the ratio of the two measurements?

7. $(-37) - (-80)$

8. $(-4) + (-9) + (+20) + (-32) + (+12)$

9. If set A has 7 elements in it and set B has 8 elements in it, can A = B?

10. Using braces, give an example of a set that contains a color and a movie star.

11. Going from –21 to –7 is an increase of how much?

12. If my checking account is overdrawn by six dollars and I make a deposit of eighty dollars, how much will my account have in it?

13. Complete the following: *Subtracting a positive number is the same as adding a. . . .*

14. Express 22/7 as a ratio.

15. $(-61) - (-2)$

16. At 5:45 p.m. Joe had won a total of $126, but ten minutes later his total was –$17. What was his change in wealth during those ten minutes?

17. There are 428 sheep out in a field. What is the ratio of the number of left eyes to right eyes?

18. If one of the sheep in that field has two left eyes (and one right eye), what is the ratio of the number of left eyes to right eyes for all the sheep in the field?

One special sheep

19. If the ratio of marbles to nuts is 6:5, what is the ratio of nuts to marbles?

answers

1. 0   This is the only possible example.

2. $20 - (-10) = 20 + (+10) = 30$

3. {chicken, 85, Delaware}   Your answer may differ from mine.

4. $27 - (-42) = 27 + (+42) = 69$

5. Forgiving your debt of $12 would make you $12 richer. That, plus the gift of $100, would make you a total of $112 richer.

6. 60:1   For example, 180 minutes would correspond to 3 hours.

7. $(-37) - (-80) = (-37) + (+80) = +43$

8. -13

9. No, it is not possible. Since B has more elements than A, there must be an element in B that is not in A and since A = B only if they have the same elements in each set, it must be true that A is not equal to B.

10. {green, Jeanette MacDonald}   Your answer may differ from mine.

11. $-7 - (-21) = -7 + 21 = 14$

12. $-6 + (+80) = +74$. $74 will be in my account.

13. Subtracting a positive number is the same as adding a negative number.

14. 22:7

15. -59

16. Going from 126 to -17 is $-17 - (+126)$ His change was $-$143. This could also be expressed by saying that his net loss was $143 (or his net gain was $-$143).

17. 428:428 Since 428:428 means the same thing as $\frac{428}{428}$ which is the same as $\frac{1}{1}$ you could have written 1:1.

18. Since the number of left eyes is 429 and the number of right eyes is 428, the ratio would be 429:428.

19. The ratio of nuts to marbles is 5:6.

The five of them sat down at the poker table. When they had first started playing poker on Thursday nights back in August, they had used a square card table, but that didn't seem to work out very well with five players. The ratio of players to sides of the table was 5:4. Darlene didn't mind since she got to sit next to Joe.

Fred

Betty   Alexander

Joe   Darlene

But everyone else did mind since: 1) Darlene couldn't concentrate on the game with Joe so close to her and 2) Joe kept forgetting that you're supposed to keep your cards secret. Whenever he got a good hand, he'd proudly show it to Darlene.

To fix this situation, they got a new table. It was in the shape of a pentagon (*penta* means "five"). When Joe had ordered the table from a custom table manufacturer on the Internet, he had been asked if he wanted chip trays built into the top of the table. Since he figured that he might be winning lots of pennies, he told the table manufacturer to put in "lots and lots of trays." (He had told Darlene that someday he'd win at this game after he got the hang of it.)

So they made a table with "lots and lots of trays," which left very little room to deal out the cards.

The good news is that, because it was in the shape of a pentagon, the ratio of people to sides of the table was now 5:5. The bad news is that Darlene couldn't sit as close to Joe as she wanted to.

"Well, what kind of poker shall we play tonight?" Alexander asked as he shuffled the cards. In the past, they had played all the usual forms: draw poker, stud poker, deuces wild, and one-eyed jacks wild. But since these were all covered in the poker books that Fred had read, he continued to frequently be the winner. So the others at the table started to invent new poker games to try to eliminate Fred's advantage.

"Let's try MultiFlip," Betty suggested. No one else at the table had ever heard of that game.

"That's because I've just invented MultiFlip," she explained. "We play regular poker and after someone has won a really big pot with more than 20¢ in it, then we do a MultiFlip."

"What's that?" Fred asked.

"You'll see," Betty answered. Sometimes they invented the rules of the game as they played, and Fred found this a little disconcerting for his logical mind. He felt it was a little like what the Nazis used to do in passing a law today that made what you did *yesterday* illegal. You could be punished for doing something that was legal when you did it.

"First," said Betty, "everyone write a number less than ten on a piece of paper and fold it up and toss it into the center of the table. We'll pick one of those papers after someone's won a big hand."

After several hands of play, Darlene won a really big hand of 30¢.

"Okay," Betty continued her explanation, "Darlene, please shut your eyes and pick one of the papers out of the center."

Darlene got a paper and opened it. "It's a 3," she announced.

Betty finished her explanation of MultiFlip by saying, "Okay, you have won, not just 30¢, but three times 30¢!"

Darlene liked this game. She turned to Joe and said, "Honey, look how much I've won!"

He thought she meant that he should figure out how much that was, so he got out a piece of paper and wrote: $3 \times 30¢ = 90¢$.

Alexander looked at his work and frowned a little. Joe went back and checked his work and couldn't see any error. He got out his calculator and three times thirty cents was still ninety cents.

"What's wrong, Alexander?" Joe asked.

Alexander explained that there was a danger of confusion. In algebra they used "x" as an unknown and the times sign "×" that was used in elementary school isn't used for multiplication any more.

"Well, how do you write it then?" Joe wondered.

"You can use parentheses and write (3)(30). Sometimes they use a raised dot and write 3·30, but that can get confused with a decimal point."

Joe crossed out 3 × 30¢ = 90¢ and wrote (3)(30¢) = 90¢ and smiled.

Betty wrote on a slip of paper "(New)(York)" and passed it under the table to Alexander. He laughed at her joke about the famous newspaper.

Your Turn to Play

1. In algebra we will use both letters and numbers together. The most common letter that we'll use is *x*. The *x* will stand for some number that we don't yet know the value of. In elementary school you may have seen problems that look like:

$3 \times \square = 12$ and you had to figure out what number goes in the box.

In algebra, we'll replace the $\square$ by an *x*.

Explain why $3 \times x = 12$ could be really confusing.

2. In elementary school you learned that 3×4 meant $4 + 4 + 4$. What would $3(-4)$ mean?

3. $5(-2) = ?$

4. $3(-8) = ?$

5. $40(-5) = ?$

6. $(3.56 + 8) + 4\frac{1}{2} = 3.56 + (8 + 4\frac{1}{2})$ is an example of what law of addition?

7. What value of x will make this true? $\frac{5}{x} = \frac{17}{34}$

8. Which is larger: 2^4 or 4^2?

. . . . COMPLETE SOLUTIONS . . . .

1. If you are writing that equation by hand, it might come out looking like 3xx = 12. You could easily confuse the multiplication sign, ×, with the *x*.

Here are the rules
for showing multiplication:

If you want to multiply two *numbers*:
    put them in parentheses.   Example: (5)(4)

If you want to multiply a *number* by a *letter*:
    write them together.   Example: 7x

If you want to multiply two *letters*:
    write them together.   Example: xy

2. 3(–4) means –4 + (–4) + (–4) which would be –12.
3. 5(–2) = –10
4. 3(–8) = –24
5. 40(–5) = –200
6. The associative law of addition
7. x is equal to ten.
8. They are both equal to 16. Neither is larger.

D arlene also won the next big pot. This one was even larger than the first one: 40¢. There were four pieces of paper left in the middle of the table and Betty urged her to draw another one.

It read "½." It was in Fred's handwriting. Fred shrugged his little shoulders and said, "You said that it should be a number less than ten. One-half is less than ten since it's left of ten on the number line." He illustrated his remark on a napkin.

0   $\frac{1}{2}$ 10

Fred's illustration

Fred continued, "In algebra you can write that as ½ < 10, where the symbol < means *less than*. On many keyboards the less-than symbol is on the comma key."

Meanwhile, Joe was figuring out how much Darlene had won. He wrote, "½ × 40¢" and then crossed that out and wrote, "(½)(40¢) = 20¢."

Betty scribbled another note to Alexander, "20¢ < 40¢."

To everyone's surprise, Joe won the biggest hand of the night. It was 70¢. He reached in the center of the table and took a slip of paper. He already knew what it read since the paper was orange and he had written his number on orange paper.

"What's the number?" Darlene asked excitedly. She (and everyone else at the table) knew that Joe had picked his own paper. She was hoping that he had followed the rules and had remembered to pick a number less than ten.

Joe proudly announced, "The number is –2."

There were several groans at the table.

"But look," Joe cried, "–2 is on the left of 10, and so –2 is less than 10."

He drew on the table:

–2   0   10

Joe's illustration

And Joe was correct. It's true that $-2 < 10$. Everyone agreed with him. "So what's the problem?" Joe inquired.

Alexander finally broke the silence and said, "No problem at all. You *owe* 140¢."

Joe wrote on another part of the table -2×70¢, crossed that out, and then wrote $(-2)(70$¢$)$ and hesitated. He didn't know how to multiply a negative number times a positive number.

Darlene whispered to him that a negative times a positive gives a negative answer. She said, "Suppose you had a speeding ticket for $5 and you didn't pay the fine. What would happen?" Joe responded that they would triple the fine. That would make the fine $15. So $(-5)(3) = -15$.

Joe tried out some multiplication of signed numbers:

$(+3)(-7) = -21$

$(-2)(+12) = -24$

$(+5)(+7) = +35$ (That was easy.)

$(-8)(+9) = -72$

and then he wrote:

$(-4)(-5) =$ and he was stuck.

He knew it wasn't equal to -20 since $(-4)(+5)$ equaled -20, and he figured that $(-4)(-5)$ and $(-4)(+5)$ couldn't both have the same answer.

Betty told him that a negative times a negative has a positive answer: $(-4)(-5) = +20$. That's kind of strange (but true) and Joe didn't ask for an explanation since that often confused him. He simply liked to find out what the rule was and obey it. (For those of us who don't just want to be told what to do, but demand an explanation, one will be given before the end of this book. Once we get the distributive property (also called the distributive law), we'll be able to show that a negative times a negative gives a positive answer. If you're in a hurry to see the proof, read faster.)

Having fun with his new knowledge, Joe wrote out a lot of multiplications all over the table top:

$(-2)(+10) = -20$ $(-9)(-9) = +81$ $(+100)(-\frac{1}{2}) = -50$

$(-4)(-2.5) = +10$ $(+66)(+3) = +198$ $(-66)(-3) = +198$

$(-1)(+16) = -16$ $(-me)(-you) = +us$ $(+10)(-3.74) = -37.4$

When Fred taught algebra at KITTENS, he would write on the board:

For multiplication--------

        Signs alike ⇨ *Answer positive*

        Signs different ⇨ *Answer negative*

Your Turn to Play

1. $(-4)(+7)$

2. $(-20)(-3)$

3. $(-10)(+7.19)$

4. $(+10)(-7.19)$

5. $(-40)(-50)$

6. You have studied the multiplication of signed numbers. Here, now, are some examples of the division of signed numbers. What is the rule for the division of signed numbers?

$$+6/-2 = -3 \qquad\qquad -18/3 = -6$$
$$-20/+5 = -4 \qquad\qquad 36/9 = 4$$
$$-33/-11 = +3 \qquad\qquad +14/-7 = -2$$
$$+4/-8 = -\tfrac{1}{2} \qquad\qquad -10/-10 = +1$$

7. Fill in the blank to make the following true: $+30/\underline{\ ?\ } = -5$

8. The prefix *penta* means "five." The prefix *hexa* means "six." Draw a hexagon.

9. Which number is smaller: -8 or -293?

10. Joe was having trouble remembering that " $<$ " meant "less than." Betty gave him a memory aid. She said that the little pointy end of " $<$ " pointed to the smaller number and the big end of " $<$ " pointed to the bigger number.

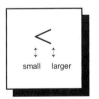

    With that memory aid, he could remember that " $<$ " meant "less than."

    Now for your question. What does " $>$ " mean?

11. What if we have three signed numbers multiplied together? For example, what does $(-8)(+2)(-3)$ equal?

12. How about $(-1)(-2)(-3)(-4)$?

13. $(-1)^6$ which equals $(-1)(-1)(-1)(-1)(-1)(-1)$ is equal to what number?

14. $(-1)^{1003} = ?$

15. If you're multiplying together a bunch of signed numbers and you're told that eight of them are negative and the rest are positive, what will be the sign of the final answer?

16. $-15/+2$

17. Zeno was a Greek who lived about 400 B.C. He loved to ask questions that no one could answer. For example, he "proved" that no one could ever walk across a room. In order to do that, he argued, you would first have to walk half way across the room. Then you'd have to go half of the remaining distance (which is 1/4 of the whole distance). Then you'd have to walk half of the remaining distance again (which is 1/8 of the whole distance). So, in order to cross the whole room, you'd have to make an infinite number of smaller trips: $1/2 + 1/4 + 1/8 + 1/16 + 1/32 + \ldots$. That, he concluded, would take an infinite amount of time. You'll never cross the room!

the room

His fellow Greeks were stumped. They didn't know how to answer him. Humankind held its collective breath waiting until Newton and Leibnitz came along and invented calculus before Zeno's question could be answered. They were alive in 1700 A.D. How long did we have to wait?

18. Fill in the blank with an integer that will make this true: $-2 < \underline{\ ?\ }$.

19. Is it possible to fill in the blank with a whole number that will make this true:

$$-44 < \underline{\ ?\ } < 0?$$

(Note: $-44 < ? < 0$ means both $-44 < ?$ and $? < 0$.)

.......**COMPLETE SOLUTIONS**.......

1. –28

2. +60   (or this could be written as 60)

3. –71.9

4. –71.9

5. +2000

6. It is the same as for multiplication. If the signs are alike, the answer is positive. If the signs are different, the answer is negative.

7. $30/-6 = -5$

8. Sometimes you see floor tiles with this shape: Bees make their combs with a hexagonal pattern.

9. –293, since it's to the left of –8 on the number line.

10. > means greater than. So, for example, 872 > 639.

11. (–8)(+2)(–3) is like a steak. You need to take it a bite at a time. We multiply the first two numbers together and get (–16)(–3), and then finish the problem to get +48.

12. (–1)(–2)(–3)(–4)
    = (+2)(–3)(–4)
       = (–6)(–4)
          = +24

13. +1

14. The first 1002 times you multiply (–1) by itself you arrive at +1, just as you did in the previous problem computing $(-1)^6$. Doing the last multiplication by –1, which is the 1003rd one, you arrive at a final answer of –1. In multiplying together a bunch of signed numbers, *pairs* of negative signs disappear (or more accurately, yield the same answer as if they were both positive).

15. The final answer will be positive. Those eight negative numbers could be thought of as four pairs of negative numbers. In each pair, the minus signs would eliminate each other.

16. $-15/+2 = -7½$ or –7.5   The signs are different and hence the answer is negative.

17. From –400 to +1700 is 1700 – (–400), which is 1700 + 400 = 2100 years. That's a long time to wait to have your question answered. (During

those 2100 years people continued to cross rooms.) Your wait won't be that long until you get to calculus. After this beginning algebra, there's advanced algebra, geometry, and trig, and then you'll be ready.

18. You could write –1, or zero, or +1, or +2, or +3, or any larger natural number.

19. No you can't. We can find integers that will work (such as –5), but no whole number will work.

F red won the next hand. The pot had 21¢ in it so he drew a slip of paper from the pile in the middle of the table. It read, "3" and Joe did a quick computation: (3)(21¢) = 63¢ and announced the result.

Where was the extra money to come from? There was only 21¢ in the pot. How was Fred to receive 63¢?

Betty, who invented this game she called MultiFlip, explained that not only is Fred's winning tripled in this case, but everyone else's losses are also tripled.

Darlene whipped out the new laptop computer that she had won at the beauty shop raffle. (She had purchased a ticket there when she was getting her hair done. She always wanted to look good for Joe.)

Opening a spreadsheet program, she entered:

Wins & Losses				
Fred	Betty	Alexander	Joe	Darlene
+21	–3	–5	–8	–5

This was the amount everyone either won or lost before the results were tripled.

Joe asked Darlene, "How come my number is bigger than yours?"

Darlene wanted to say four things to Joe:

① Say "why," not "how come."

② Your number is actually smaller than mine, because it's to the left of my number on the number line. $-8 < -5$.

③ If you'd learn to fold your hand (which, for those of you who don't play poker much, means to drop out of the game) when your cards aren't that good, you'd lose a lot less money. (Joe never folded his hands.)

④ Aren't you impressed with my new laptop . . . and my hair?

But, wisely, Darlene bit her tongue and said merely, "I guess it's the luck of the draw."

She keyed into the spreadsheet the command to triple the entries in the bottom row and enter them into a new row.

Wins & Losses				
Fred	Betty	Alexander	Joe	Darlene
+21	–3	–5	–8	–5
+63	–9	–15	–24	–15

While Joe was counting out the twenty-four pennies he owed Fred, we'll look at some notes.

♪#1: When we look at the original wins and losses, they add to zero as we would expect. $(+21) + (-3) + (-5) + (-8) + (-5) = 0$. What one person loses, another gains. In game theory (which is a later topic in math), this is called a zero-sum game.

In real life, in contrast to pushing pennies around on a table, we often find that we're not in a zero-sum game. If you share your algebra knowledge with a friend, you don't lose what your friend gains.

✓ The number of pennies on the table stays constant.

✓ The amount of knowledge in the world doesn't.

♪#2: When we triple $(+21) + (-3) + (-5) + (-8) + (-5)$

we get $3\big((+21) + (-3) + (-5) + (-8) + (-5) \big)$

which is $3(+21) + 3(-3) + 3(-5) + 3(-8) + 3(-5)$.

When we get to using letters in our algebra, we might write this as $3(a + b + c + d + e) = 3a + 3b + 3c + 3d + 3e$ and call this the distributive property. But it's too soon to do that now.

The doorbell rang. It was 6:30 and Stanthony's Pizza was making its usual Thursday night delivery. Alexander strode to the door (when you're as tall as Alexander, you don't just *walk*) and opened it.

It was Stanthony himself at the door. "Hello, yousa guys. Ima so happy to maka this delivery myself. My delivery boy, he sa sick and besides I wanted to meeta my best customers." (Stanthony had picked up his rotten Italian accent from all the Marx Brothers movies he'd seen.)

Stanthony used to have a pizza joint where all the gang from KITTENS would meet. But it closed after the big party celebrating Alexander and Betty's engagement was held there. The place—how shall we say it?—kind of melted. (See *Life of Fred: Calculus Expanded Edition* for the details.) Now he runs a pizza delivery service. The same fine pizza, only delivered hot to your home.

Alexander paid him and brought the pizza to the waiting crowd. Betty had already cleared the table. Darlene had brought out the glasses and poured the Sluice. (Not to be confused with Slice®, Sprite®, Storm®, Sierra Mist®, Squirt®, or Seven-Up® which all start with *S,* and which some people tend to confuse with each other. Sluice is different. Its bubbly clearness contains turtle spit, giving it a sweet lemon-lime flavor.) She dropped an ice cube into each glass.

After saying grace, they opened the box and the aroma of Stanthony's combination pizza filled the room. Joe took his pieces and proceeded to pick off the pepperoni and put them into one of the chip trays. He had never liked pepperoni.

Joe turned the box toward himself so that he could read the inside of the cover. "Look!" he exclaimed, "It's just like your computer."

Stanthony's Finest Combo Pizza				
"Justa Lika Howa You Like It"				
Serving Size	Protein	Carbs	Calcium	Pepperoni
one bigga slice	4g	12g	0.2g	6

"But I always eat six slices!" he continued. "How am I going to know how this works out for six slices? And what's the 'g' mean?"

Darlene told him the "g" stood for grams which is a metric measure of weight and that a raisin weighs about a gram. She wiped her hands and fired up her laptop. She did this every Thursday at about 6:43.

It was a ritual that she and Joe had. It kept him happy. She had learned to save it on her hard disk so she didn't have to enter the data each week.

After the screen came alive, she presented to him:

Just for Joe				
Serving Size	Protein	Carbs	Calcium	Pepperoni
one bigga slice	4g	12g	0.2g	6
Joe's six slices	24g	72g	1.2g	36 assuming you didn't pick them off!

"Joe, did you ever notice," Betty asked, "that the ratio* of protein in one slice to the protein in six slices is 4:24?"

"Ratio?" Joe frowned.

Darlene was going to say, "Yes, that's *radio* with a *t,*" but she again kept quiet.

"Yes," Betty continued, "4:24 which means the same as $4 \div 24$ or $\frac{4}{24}$."

At that point Joe "got it." Betty was a good and patient teacher, who for years had known she would go on to teaching math after she graduated from KITTENS University. She was one of the students at KITTENS when Fred started teaching there. As soon as Betty started taking his classes, she knew she wanted to teach math.

"I think I notice something," Joe said. "The ratio of the number of slices is 1:6 and the ratio of the grams of protein is 4:24. Those are equal. 1/6 = 4/24. And, wait a minute! For the carbohydrates, it's also 12:72, which is the same as the 1/6 and the 4/24. The ratios are equal. Isn't that a coincidence!"

"Not really, Joe," said Betty. "The amount of protein you get is

---

* *Ratio* is pronounced RAY-she-oh.

proportional to the number of slices you eat." She recalled the day Fred had written on the blackboard:

> A proportion is the equality of two ratios.
>
> $$\frac{a}{b} = \frac{c}{d}$$

Your Turn to Play

1. Insert either < or > in each of the following:
   A) 5 ___ 888
   B) –4 ___ 4
   C) 8.4 ___ 0
   D) 44¼ ___ 44½
   E) –7 ___ –3
   F) –1000 ___ –2000

2. Give an example of two ratios that are equal.

3. Here is a proportion. Fill in the correct value for "?".

$$\frac{\text{one bottle of Sluice}}{16\cent} = \frac{\text{five bottles of Sluice}}{? \cent}$$

4. Find the value for "?".   4:8 = 9:?

5. The number of pieces of pepperoni you get is proportional to the number of slices of pizza you take. (For example, if you take three times as many slices, you'll get three times as many pieces of pepperoni.)

   The number of gallons of gas your car uses on the freeway is proportional to the number of miles you drive. (For example, if you drive five times as far, your car will consume five times as much gas.)

   Suppose one night you sit down to memorize dates in history. It is *not* true that the number of history dates you memorize is proportional to the number of hours you study. Why isn't it?

6. –20 ÷ –4 = ?

. . . . COMPLETE SOLUTIONS . . . .

1. A) 5 < 888

   B) –4 < 4

   C) 8.4 > 0

   D) 44¼ < 44½

   E) –7 < –3

   −7 −6 −5 −4 −3 −2 −1 0 +1
   −7 is to the left of −3

   F) –1000 > –2000

2. Answers will vary. For example, you may have written 1:3 and 4:12. Or you may have written $\dfrac{6}{8} = \dfrac{3}{4}$

3. Replace the "?" with 80.

4. 4:8 = 9:18

5. Suppose in the first hour you were able to memorize 100 dates. Do you think that after six hours, you will have memorized 600 dates? For most human beings, fatigue starts to set in.

   Six one-hour sessions of study are much more effective than one six-hour session.

   Most of us can easily leap over a one-foot tall hurdle. Is it ten times as hard to leap over a ten-foot hurdle?

   No. It's a zillion times harder.

   The ease of the leap is *not* proportional to the height of the hurdle.

6. –20 ÷ –4 = +5

Recall that the rule for dividing signed numbers is the same as the rule for multiplying signed numbers:

If the two numbers have the same sign, then the answer is positive.
   (+6)/(+3) = +2   (–6)/(–3) = + 2

If they have different signs, then the answer is negative.
   (+6)/(–3) = –2   (–6)/(+3) = –2

J oe changed the subject. "You know," Joe said, "I wish when you order the combination pizza next time, that you'd ask Stanthony to put the pepperoni all on the outside edge of the pizza. That'd make it easier to pick them off."

Joe tore off the pizza box lid. He took little one-inch pepperoni slices he had saved from previous weeks and arranged them in a circle on the lid.

Although Joe didn't like the taste of pepperoni, he did like to play with his "little greasy poker chips" as he called them.

"I have too many chips to count," he announced. "There must be an infinite number of them. Don't you think so, Alexander?"

Alexander had been mildly attentive to what Joe had been doing. Laying them all out in a circle is one of Joe's less imaginative projects. Last week he had built a little fort out of them. He had called it Fort Francine, but when Darlene glared jealously at him, he changed it to Fort Darlene claiming it had just been a slip of the tongue.

"An infinite number? No," Alexander responded. "I'd guess there are about 75 of those one-inch slices."

"How'd you do that?" Joe asked.

"Well," Alexander began, "what size pizza do we normally order?"

Joe knew the answer to that. "It's Stanthony's medium."

"Do you remember how many inches that is?"

Darlene pointed to the lid of the box to help Joe get the right answer.

And since pizzas are measured by their diameters, we had:

25"

Alexander continued, "If I know the diameter of a circle, then I know its circumference. It's about

three times as long as the diameter. We had that last Sunday in our Bible class. Remember when they were talking about the round tank that was 10 across and 30 around?" (I Kings 7:23)

"But my circle isn't 10 across," Joe protested. "It's 25 inches across."

"You're right. So its circumference would be about three times that distance, which makes 75 inches. Since those are one-inch pepperoni slices, I figure there's about 75 of them." Alexander smiled.

"Oops, I lost count. I wonder if there will be exactly 75 slices," Joe said.

"Maybe not," Alexander said. "I said there would be about 75 slices. First, the slices may not be exactly one inch in diameter. Second, the pizza is only approximately 25 inches across. Third, the formula C = 3d is only approximately true."

In this formula that Alexander was using, the "C" stood for circumference and the "d" stood for diameter. If, for example, d were equal to 10, then C = 3(10), which is 30 (which is what they had in the I Kings passage). If d were equal to 25, then the circumference, C, would be equal to 3(25).

Joe frowned. "What do you mean when you say that C = 3d is only approximately true? Make it truer!"

"I'd be glad to. How about C = (22/7)d. That's more accurate."

Joe was happier. He tried the formula out and put in a value of 25 for d and found that C = (22/7)25. On part of the table that he hadn't already written on, he did the work with the fractions:

$$C = (22/7)25 = \frac{22}{7} \times \frac{25}{1} = \frac{550}{7} = 78\frac{4}{7}$$

"So," Joe said, "there are exactly $78\frac{4}{7}$ pepperoni slices on the circumference of a 25" pizza circle."

"Well, that's closer than when we used C = 3d," Alexander admitted.

"What do you mean 'closer'?" Joe asked. "Isn't that the true answer?" Then a little light dawned in Joe's memory. "Yes. I know. Don't tell me. The really correct formula is C = 3.14d." Then he wrote on the edge of the table (since the top was completely covered with his writing):

$$C = 3.14(25) = 78.5$$

Joe went on, "I remember the 3.14 formula. We had C = 3.14d taught to us a long time ago. My seventh grade teacher, Mrs. Angemessen, told us this is the formula we were to learn. So 78.5 is the exact answer, isn't it?"

Alexander shook his head slowly. "That was good for the seventh grade. It gave you what they used to call slide rule accuracy, which is about three significant digits, but the true answer might be 78.48 or 78.52 or something like that."

a slide rule: good to about three digits

"Make it right!" Joe shouted. "I don't want a seventh-grader's brain in my head for the rest of my life. If it isn't 3.14, what is it?"

"Okay, Joe, you asked for it. The really correct formula relating the circumference of a circle to its diameter is:

C = 3.141592653589793238462643d."

Joe asked, "You mean it has 25 digits as the coefficient of the d?" (**Coefficient** is the number that goes in front of one or more letters that are multiplied together. For example, 7 is the coefficient of 7xyz and 3908 is the coefficient of 3908w.)

"No, I'm afraid not, Joe," Alexander said with a sigh. "I just got tired of writing. After all, with ten decimal digits, if you took the diameter of the earth, which is about 8000 miles, you could compute the circumference of the earth to an accuracy of about one-eighth of an inch. I gave you 25 digits, Joe."

"I want more," Joe declared.

Alexander looked at his beloved Betty and said to her, "I've only got 25 digits memorized, Betty." He shrugged his shoulders.

Betty turned to Joe and asked, "But you've written all over the table. How can I write out more digits for you?"

Joe smirked and held out his felt tip pen and his right arm.

"Okay," Betty said with a resigned look on her face. She began at his right wrist and wrote:

C = 3.14159265358979323846264338327950288419716939937510582
09749445923078164062862089986280348253421170679821480865I3
28230664709384460955058223172535940812848III74502841027019
38521105559644622948954930381964428810975665933446I284756
4823378678316527120190914564856692346034861045432664821 33
936072602491412737245870066063155881748815209209628292540
9171536436789259036001I3305305488204665213841469519415II609
4330572703657595919530921861173819326II79310511854807446237 9
96274956735188575272489122793818301I94I29833673362440656b
43086021394946395224737190702179860943702770539217I762931
7675238467481846766940513200056812714526356082778577I3427
57789609173637178721468440901224953430146549585371050792 2
796892589235420199561I212902196086403441815981362977477I3099
605187072II3499999983729780499510597317328160963185950244 59
455346908302642522308253344685035261931I8817I01000313783 8
752886587533208381420617177669147303598253490428755468731I
5956286388235378759375195778I8577805321712268066130019278 76
61I195909216420198938095257201065485863278865936I53381827
9682303019520353018529689957736225994138912497217752834791
3151557485724245415069595082953311686172785588907509838I754
637649393192550604009277016711390098488240128583616035 63
707660I0471018194295559619894676783744944825537977472684 71
04047534646208046684259069491293313677028989152104752162
056966024058038150193511253382430035587640247496473263 91
4199272604269922796782354781636009341721641219924586315 03

028618297455570674983850549458858692699569092721079750930
29553211653449872027559602364806654991198818347977535663 6
98074265425278625518184175746728909777727938000816470600 16
14524919217321721477235014144197356854816136115735255213 3475
74184946843852332390739414333454776241686251898356948556 2
09921922218427255025425688767179049460165346680498886272 3
27917860857843838279679766814541009538837863609506800642 2
51252051173929848960841284886269456042419652850222106611 86
30674427862203919494504712371378696095636437191728746776 4
65757396241389086583264599581339047802759 01"

At this point Betty had covered most of Joe's body and was coming down to his left wrist. She said, "Well, I guess I'm running out of room. I'll have to put the 'd' on at this point." So she wrote . . . *275901* d. (Later that night when Joe gave a good-night hug to Darlene, he left a *58133904780275901* d on the back of her white blouse.)

Joe had a lot of questions at this point:

Joe's question #1: When does this number end?
    Answer: It doesn't.

Joe's question #2: When does it start to repeat itself? You know, like 1/7 = 0.142857142857142857142857142857. . . .
    Answer: This never repeats.

Joe's question #3: You were making up all these digits, weren't you Betty?
    Answer: No, these are the actual digits.

Joe's question #4: What about all those 9s? [See the previous page.]
    Answer: Those six 9s just happen.

Joe's question #5: Is there any shorter name for

3.14159265358979323846264338327950288419716939937510582097494459230781640628620899862803482534211706798214808651328230664709384460955058223172535940812848111745028410270193852110555964462294895493038196442881097566593344612847564822337867831652712019091456485669234603486104543266482133936072602491412737245870066063155881748815209209628292540917153643678925903600113305305488204665213841469519415116094330572703657595919530921861173819326117931051185480744623799627495673518857527248912279381830119491298336733624406566430860213949463952247371907021798609437027705392171762931767523846748184676694051320005681271452635608277857713427577896091736371787214684440901224953430146549585371050792279689258923542019956112129021960864034418159813629774771309960518707211349999998372978049951059731732816096318595024459455346908302642522308253344685035261931188171010003137838752886587533208381420617177669147303598253490428755468731159562863882353787593751957781857780532171226806613001927876611195909216420198938095257201065485863278865936153381827968230301952035301852968995773622599413891249721775283479131515574852424541506959508295331168617278558890750983817546374649393192550604009277016711390098488240128583616035637076601047101819429555961989467678374494482553797747268471040475346462080466842590694912933136770289891521047521620569660240580381501935112533824300355876402474964732639141992726042699227967823547816360093417216412199245863150302861829745557067498385054945885869269956909272107970509302955321165344987202755960236480665499119881834797775366369807426542527862551818417574672890977772793800081647060016145249192173217214772350141441973568548161361157352552133457418494

684385233239073941433345477624168625189835694855620992192221842725502542568876717904946016534668049886272327917860857843
838279679766814541009538837863609506800642251252051173929848960841284886269456042419652850222106611863067442786220391949
450471237137869609563643719172874677646575739624138908658326459958133904780275901 . . .?

Answer: Yes.

Joe's question #6: Well, what is it?
Answer: It is pi.

Joe's question #7: You misspelled it, Betty. Pi is spelled p-i-e.
Answer: No it isn't. This isn't like pizza pie. This is a letter from the Greek alphabet. It's written π.

Joe's question #8: But if it's a letter, how can we let it stand for a number?
Answer: We just let it. Take your choice, Joe. You can write π or you can write:

3.14159265358979323846264338327950288419716939937510582097494459230781640628620899862803482534211706798214808651328230664709384460955058223172535940812848111745028410270193852110555964462294895493038196442881097566593344612847564823378678316527120190914564856692346034861045432664821339360726024914127372458700660631558817488152092096282925409171536436789259036001133053054882046652138414695194151160943305727036575959195309218611738193261179310511854807446237996274956735188575272489122793818301194912983367336244065664308602139494639522473719070217986094370277053921717629317675238467481846766940513200056812714526356082778577134275778960917363178721468440901224953430146549585371050792279689258923542019956112129021960864034418159813629774771309960518707211349999998372978049951059731732816096318595024459455346908302642522308253344685035261931188171010003137838752886587533208381420617177669147303598253490428755468731159562863882353787593751957781857780532171226806613001927876611195909216420198938095257201065485863278865936153381827968230301952035301852968995773622599413891249721775283479131515574857242454150695950829533116861727855889075098381754637464939319255060400927701671139009848824012858361603563707660104710181942955596198946767837449448255379774726847104047534646208046684259069491293313677028989151210475216205696602405803815019351125338243003558764024749647326391419927260426992279678235478163600934172164121992458631503028618297455706749838505494588586926995690927210797509302955321165344987202755960236480665499119881834797753566369807426542527862551518418417574672890977772793800081647060016145249192173217212147723501414419735685481613611573525521334757418494684384852332390739414333454776241686251898356948556209921922218427255025425688767179049460165346680498862723279178608578438382796797668145410095388378636095068006422512520511739298489608412848862694560424196528502221066118630674427862203919494504712371378696095636437191728746776465757396241389086583264599581339047802759
01 . . .

Joe's question #9: Just one last question then. If Mrs. Angemessen's formula wasn't the exact one, what's the exact formula?
Answer: Easy. It's $C = \pi d$. So the number of one-inch pepperoni

slices needed to make a circle with a 25-inch diameter is 25π. If you want to approximate your answer, use as many digits of pi as you want.

Joe had asked all the questions he could think of. Betty had as many answers as Joe had questions. She was good at answering questions. Now it's your turn for a while.

Your Turn to Play

1. Just for fun: Before you look at the answer below, guess how many digits of π Betty wrote on Joe.

2. What's the gain in going from –25 to +160?

3. Name two infinite sets of numbers so that they have no members in common. (This may take a little thought.)

4. Which number is larger: –1234 or –123?

5. You order a bunch of identical pizzas from Stanthony's. Name something that is proportional to the number of pizzas you order.

6. Your test in German is tomorrow. You sit down to memorize all those German words that you've been putting off studying for several weeks. Is the number of German words you memorize proportional to the number of hours you study tonight?

7. If you measure the height of a mountain in feet and then measure it in yards, what is the ratio of the two measurements?

8. We know that $-10 < -3$ since –10 is to the left of –3 on the number line. If you were to double both numbers, would the resulting answers still be in the relation "less than"?

9. What if we take $-10 < -3$ and multiply it by +3 (or +4) (or +200) (or +½)? Will the resulting answers still be in the relation "less than"?

10. Is there a number that we can multiply both sides of $-10 < -3$ so that the sides become equal? (There is. Find that number.)

11. This is a tougher question. Can you think of a number (let's call it *a*) that, when you multiply both sides of $-10 < -3$ by that number, the "sense" of the inequality changes. Namely, find a number *a* so that $-10a > -3a$.

12. If the diameter of a circle is exactly 4 feet (that would make a nice-sized pizza), what is the exact circumference?

13. What is the coefficient of 34.7abc?

14. Now there's no number written in front of xwqz. Can you guess what the coefficient of xwqz is?

15. The amount of suntan I get is proportional to the number of rodeos I attend each month. If I attended four rodeos last month and I want to triple my suntan, how many rodeos should I attend this month?

16. Will 10π, or 100π, or 1000π, or 10000π, etc., ever be equal to a natural number?

17. If you start with a negative number and you multiply it by itself, will the answer be larger or smaller than the original number?

18. Now if *a* is any number, is it always true that $a < aa$? (We're continuing the previous question.) We know it's true for anytime that *a* is negative, since the left side of $a < aa$ will be negative and the right side will be positive. We'll try some non-negative numbers. If *a* is 5 we get $5 < (5)(5)$, which is true. If *a* is 18, we get $18 < (18)(18)$, which is true. If you can think of one value of *a* that makes $a < aa$ false, you're doing pretty good. If you can think of two different values of *a*, each of which make $a < aa$ false, you're on your way to fame and fortune. Fred, being a bright 5-year-old math teacher at KITTENS, can name many(!) values of *a* that will make $a < aa$ false. You are hereby challenged.

........COMPLETE SOLUTIONS.......

1. Betty hadn't really memorized all those digits of π. There was a big poster on the wall behind Joe that was entitled: 2000 Digits of Pi.

2. That's $160 - (-25)$, which is $160 + (+25) = 185$.

3. There are several possible answers. You might have said, "The even whole numbers and the odd whole numbers," or you might have said, "The negative integers and the positive integers," or you might have written, "$\{1/2, 1/3, 1/4, 1/5, \ldots\}$ and $\{6.1, 6.01, 6.001, 6.0001, 6.00001, \ldots\}$" or you might have written, "$\{0, 1, 2, 3, \ldots\}$ and $\{\pi, 2\pi, 3\pi, 4\pi, 5\pi, \ldots\}$."

4. -123 is larger than -1234 since it's to the right of -1234 on the number line.

5. There are several possible answers. The total price is proportional to the number of pizzas you order. If you order five times as many pizzas, you can expect to pay five times as much. The weight of all the pizzas is proportional to the number of pizzas you order. If you order three times as many pizzas, the whole package will weigh three times as much. The time it takes to get the pizzas is *not* proportional to the number of pizzas you order. If you order twice as many pizzas, it may take twice as long to put all the toppings on, but if they all bake at the same time (he has a very large oven), there may be very little extra time necessary to get them all baked.

6. Suppose you can memorize 35 words in your first hour of study. You memorize that *die Mutter* means mother and *der Nachbar* means neighbor and *die Raupe* means caterpillar and so on. In the second hour you may be able to memorize another 25. By the third hour, things are getting a little blurry and you stuff another 10 words in your head. By the fourth hour, you die of hypercerebral Germanitis (which, if there were such a phrase, would indicate an inflamation of the German-speaking part of your brain). This is why cramming doesn't seem to work very well.

7. There are three feet in a yard, so the ratio is 3:1. For example, if you measured a mountain and it was 30,000 feet, it would be 10,000 yards. The ratio would be 30000:10000 which is equal to the fraction 30000/10000, which reduces to 3/1 or 3:1.

8. Is $2(-10)$ less than $2(-3)$? Yes it is: -20 is less than -6.

9. Yes. Multiplying by any positive number preserves the sense of the inequality.

10. Any number times zero gives a zero answer. So multiplying both sides of $-10 < -3$ gives $0(-10)$ and $0(-3)$, both of which are now zero and hence equal.

11. If you got the answer to this one, you're good. Suppose we let a be the number -6. Then $-10a$ becomes $+60$ and $-3a$ becomes $+18$. So, in this case where $a = -6$, we have $-10a > -3a$. It turns out that if you let a be any negative number, it will change the sense of the inequality from "<" to ">".

12. Since the exact formula is $C = \pi d$, and $d = 4$, we know that the circumference is exactly $\pi 4$. We usually write that as 4π.

13. It's 34.7 since that's the number in front of the letters.

14. xwqz is equal to 1xwqz since multiplying anything by 1 doesn't affect its value. So the coefficient of xwqz is 1.

15. Since the suntan and rodeo attendance are proportional, to triple my suntan I would need to triple my rodeo attendance, which would mean tripling the four rodeos I attended last month. Hence, I should go to a dozen rodeos this month.

16. No, that will never happen. Suppose, for example, that $10^7\pi$ (which means 10 times 10 times 10 times 10 times 10 times 10 times 10 times π) were equal to a natural number. Multiplying π by 10000000 is the same as moving the decimal point in 3.14159265358979323384626433832 . . . seven places to the right. That doesn't give us a natural number since there will still be digits to the right of the decimal point after you've moved it. (Remember the natural numbers are $\{1, 2, 3, 4, 5, \ldots\}$ and they don't have anything except zeros to the right of their decimals. If we write the natural number "4" as "4.0000000000000" all it has to the right of the decimal is zeros.) Since π has an infinite string of non-repeating digits to the right of its decimal, no amount of multiplying by ten will ever turn it into a natural number.

17. Let's try it. Suppose we start with -12. If we multiply -12 by -12, we get $+144$. (If the signs are the same in multiplication, the answer is positive.) $+144$ is larger than -12 since it's to the right of -12 on the

number line. What if we start with –1? Multiplying –1 by –1 we get +1. Our answer, +1, is to the right of –1 on the number line, so it's larger than the original number.

Any negative number multiplied by itself will always give a postive answer.

18. The number that most people first think of as a value of a that makes $a < aa$ false is $a = 1$. This works since $1 < (1)(1)$ is false. The second number that people usually discover for a is $a = 0$. This works since $0 < (0)(0)$ is false. For a third value of a that works, try $a = 1/3$. (1/3) is not less than (1/3)(1/3) since 1/3 is not less than 1/9. In fact, when a is set equal to any number between zero and one, $a < aa$ will be false.

That was a pretty tough set of questions in *Your Turn to Play*. In many banal beginning algebra

> *banal* = boring and ordinary, lacking originality
> (Pronounced beh–NAL)

books, the problems the authors hand you look too much like elementary school math books.

They give you one or more examples and then assign you forty identical ones to do.

That gets boring in a hurry. You race through those forty problems so that you can slam the banal textbook closed and go do something more fun.

But in *Life of Fred,* the hope is that the problems are more like puzzles to solve rather than tasks to be repeated over and over again like the job of baking 400 dozen cookies.

example: 1/2 + 1/3

$$\frac{1}{2} + \frac{1}{3} = \frac{1 \cdot 3}{2 \cdot 3} + \frac{1 \cdot 2}{3 \cdot 2} =$$

$$\frac{3 + 2}{6} = \frac{5}{6}$$

Homework:
1. 1/3 + 1/4
2. 2/5 + 1/7
3. 3/8 + 4/9
4. 2/9 + 3/4
5. 2/7 + 2/9
6. 3/5 + 9/11   ☞ boredom sets in here
7. 5/6 + 1/7
8. 8/9 + 1/3
9. 4/7 + 3/4
10. 3/8 + 7/9
11. 2/5 + 3/7
12. 1/11 + 2/3
13. 4/5 + 3/10
14. 3/5 + 7/9   ☞ brain death occurs here
15. 1/3 + 2/5
16. 2/9 + 2/5
17. 7/8 + 3/4
18. 3/8 + 2/7
19. 4/5 + 1/8
20. 5/6 + 6/7
21. 6/7 + 7/8
22. 7/8 + 8/9
23. 3/8 + 1/2
24. 2/7 + 3/5
25. 3/7 + 4/9
26. 7/9 + 6/3
27. 3/8 + 2/9
28. 3/7 + 7/8
29. 1/8 + 1/9
30. 2/5 + 5/6
31. 2/7 + 3/4
etc.
etc.
etc.

What the typical banal elementary school math book seems to look like

Troy, New York   "Illium fuit, Troja est." (Illium was, Troy is.)

1. (−7)(+8)

2. −32/−8

3. Fill in the blank to make the following true: −20/_?_ = −5

4. You multiply together 20 different numbers. Five of them are positive and the rest are negative. What will be the sign of the answer?

5. (−6)(−5)(−4)(0)(−2)

6. Name a set that has no members in common with the set of integers.

7. Name a number whose decimal representation doesn't end.

8. Fill in the blank with an integer that will make the following true: (_?_)(−4) < (−3)(−6).

9. That thing that you've got at the end of your arm (known as your mitt or your grabber or your hand) is an appendage (something added on). In the fancy medical books, which of the following two names is your hand called:

    (A) a pentadactyl appendage or

    (B) a hexadactyl appendage?

10. If you know that some number times −4863 equals a positive answer, what can you say about that number?

11. Fill in the blank to make the following true: _?_ / −7 = −7

12. What is the coefficient of xy?

13. Name some reasons why Mrs. Angemessen (Joe's 7th grade teacher) gave Joe the formula C = 3.14d instead of C = πd? (Please think of some of the possibilities before you look at my answers. Psychologists have done the studies, and skills like creative thinking, which is what we may be doing here, can be strengthened with practice.)

14. Yesterday I spent $12 and bought 16 calendars. The number of calendars I can purchase is proportional to the amount of money I spend. Today I spent $36. How many calendars did I buy today?

15. Remember that a raised dot (·) can be used to indicate multiplication. So 4·3·2·1 means (4)(3)(2)(1), which is 24. We abbreviate 4·3·2·1 in mathematics by 4! (and call it "four factorial"). Using the raised dot notation, what is 10!

answers

1. –56
2. +4
3. +4
4. There are 15 negative numbers that are being multiplied together. (It doesn't matter how many positive numbers there are in the product.) The answer will be negative.
5. Zero times any number is equal to zero, so the final answer will be zero.
6. There are lots of possible answers. Sets that don't contain any numbers are the easiest, such as {all the stars in the sky} or {my Aunt Hilda}. But there are even sets containing numbers that will work, such as {π} or {6.1, 6.001, 6.0001, 6.00001, . . .}.
7. You might have thought of π or fractions like 1/3 (= 0.33333333. . .).
8. This is equivalent to

(_?_)(–4) < +18. You try various integers to see what will work. When you try –6, you find it doesn't work since (–6)(–4) is not less than 18. When you try zero it works. The set of integers that will work is {–4, –3, –2, –1, 0, 1, 2, 3, . . .}.
9. Hey, you probably have five fingers on your mitt. *Penta* means five. *Hexa* means six. The Martians may have six fingers and have hexadactyl appendages. Those seven-fingered dudes from Pluto would have heptadactyl appendages. *Hepta* = seven. The heptathlon (hep-TATH-lon) consists of seven track and field events for women.
10. That number had to be negative, since a negative times a negative yields a positive answer.
11. +49
12. It is 1 since xy = 1xy.
13. Here are some of my thoughts. Yours may be different.

(A) Mrs. Angemessen might have been a history teacher who was asked to teach the 7th grade math class since there was a shortage of qualified math teachers. She might not have known about π.

(B) The book said "C = 3.14d" and she just followed the book.

(C) Maybe she was trying to teach multiplication of decimals and the formula C = 3.14d would give them practice. She might say, "If d is 7.5, what is C?" and the kids would have to multiply out 3.14 times 7.5.

(D) She might have been afraid that if she wrote on the blackboard C = πd, some parents would complain that she was trying to teach the kids Greek instead of math. Then she'd have to explain to the parents that even though π is a Greek letter, it stands for a number.

(E) Mrs. Angemessen did teach Joe both C = 3.14d and C = πd, but Joe forgot about the second formula.

14. I spent three times as much (since $36 is three times as much as $12) so I could buy three times as many calendars. So I must have bought 48 calendars.

15. 10! = 10·9·8·7·6·5·4·3·2·1, which happens to equal 3,628,800. My calculator has a button that reads "x!" I typed in 10 and pressed the button and out came 3628800. I then typed in 50 and hit the "x!" button and the calculator gave me 3.0414093201713378043612600(10⁶⁴), which meant that I have to move the decimal 64 places to the right to get the answer, which is 30,414,093,201,713,378,043,612,600,000,000,000, 000,000,000,000,000,000,000,000,000. So 50! is a pretty big number.

Zion, Illinois the north-south roads have biblical names

1. $(-8)(-3)$

2. $-40/+10$

3. Is it true that $-4 > -2$?

4. You multiply together 18 different numbers. Seven of them are negative and the rest positive. What will be the sign of the answer?

5. Name a whole number a that makes the following true: $a < a + 3$.

6. Does the set of natural numbers and the set $\{\pi, 10\pi, 100\pi\}$ have any members in common?

7. Darlene was reading the novel *Gone With the Wind* in which the heroine, Scarlett, had a waist size of 18 inches. Assuming her waist was circular, approximately how thick was Scarlett at her waist?

8. Fill in the blank with a natural number that makes this true:

   $\underline{\ ?\ } + 16 < 22.$

9. Name a whole number a that makes this true:

   $20a < 20$

10. Fill in the blank with a natural number that makes this true:

    $\underline{\ ?\ } < 2$

11. $(-4)(-6)(2)$

12. $-1000/-500$

13. You multiply together 8 different numbers. Three of them are negative and four of them are positive. What can you say about the answer?

14. $(-10)(-\frac{1}{2})(-5)$

15. What is the coefficient of $-33xyz$?

16. If the earth had a diameter of exactly 8000 miles, how many miles would the equator (circumference) be?

17. The motto of Stananthony's Delivery Pizza is, "You give us the diameter and we'll come around with the pie." That has a double meaning. The first meaning is that if you phone and ask for a 16-inch pizza, they'll deliver it to you. What's the second meaning?

18. Fill in the blank with a whole number that makes the following true:

    $(-500)(\underline{\ ?\ }) < 17$

19. At my local Waddle Doughnuts store, $5 will buy six doughnuts. (They are fairly large doughnuts. It takes both hands to hold one.) Since

the number of doughnuts is proportional to the amount you pay, how much would 24 of these delicious and filling doughnuts cost?

20. Is it possible to fill in the blank with a whole number that will make this true: $-77 <$ _?_ < 0?

  (Note: $-77 <$ _?_ < 0 means both $-77 <$ _?_ and _?_ < 0.)

answers

1. +24

2. −4

3. No. −4 is not greater than −2.

4. It will be negative.

5. Any whole number will work. Try, for example, $a = 12$. This gives $12 < 12 + 3$ and certainly $12 < 15$. Another way to say this is: Every whole number is less than a number that is three larger than it.

6. No. We argued in question 16 of the most recent *Your Turn to Play* that $10^n \pi$ will never equal a natural number for any n.

7. About six inches. If we use the approximate formula $C = 3d$ and we know that $C = 18$, then we're looking for a value of d so that $18 = 3d$. Even without using algebra, you can answer the question, "Three times what number equals 18?" The exact answer (which we'll teach you how to get later) is $18/\pi$.

8. _?_ $+ 16 < 22$. You try various natural numbers. 3 works since $3 + 16 < 22$. 7 doesn't work since $7 + 16$ is not less than 22. Here is the complete set of natural numbers which work: $\{1, 2, 3, 4, 5\}$.

9. There's only one whole number that will work. When you try, for example, $a = 5$, it doesn't work since $5(20)$ is not less than 20. The whole number that does work is $a = 0$ since $0(20) < 20$.

10. 1 is the only possible answer

11. 48

12. +2

13. If three of them are negative and four are positive, then the last one must be zero, and zero times anything is zero. So the final answer will be zero.

14. −25

15. −33

16. 8000π. Now, of course, the earth isn't perfectly round, and any students who note this in their answer should get an "A" in some course like astronomy.

17. It's a play on the word *pie*. If you give someone the diameter of a circle, then πd is the circumference. You have "come around" (circumference) with the pi.

18. All the whole numbers, {0, 1, 2, 3, . . .}, work.

19. Since I bought four times as many doughnuts, it would cost four times as much. So I'd be paying $20 for the two dozen Waddles.

20. No, it can't be done. You could find integers that work, but none of the whole numbers work.

Weed, California site of the world's largest sawmill in the 1940's

1. (−44)/4
2. (−6)(−7)
3. Is it true that −50 < −7?
4. Does 22/7 equal π?
5. Someday Darlene wants to wear a diamond ring that Joe gives her. Her finger (fourth finger, left hand) is ½" in diameter. What would be the circumference of that ring?
6. Your teacher asked you to multiply out the following and give the exact answer using only pencil and paper. But as a bonus treat, your teacher said that you could fill in the blank with any number you liked. What number would you put in the blank?
(3.822899)(1000.3)(⅛)(32987.1)(_?_)
7. Fill in the blank with an integer that will make the following true:
(−12)(+3) < (−6)(_?_).
8. My lawyer is cheap. He only charges $3 for each minute of his services. (A lot of lawyers charge more than $200/hour nowadays.)

Since his bill is proportional to the number of minutes he spends, how much would 200 minutes of his time cost me?

9. Continuing the previous question: My lawyer and I realized that 6 of those 200 minutes that he had billed me for last month were minutes that we spent talking about fishing and not law stuff. Here is this month's bill that he sent me. Fill in the blanks.

Law bill	
Last month's bill	600.
Payment received	600.
Amount overdue	0.
This month's work	
430 minutes × ($3)	1290.
Correction for last month	
_?_ × ($3)	−18.
Total amount owing	_?_

10. Fill in the blank with an integer that will make this true: _?_ < −784.
11. (−2)(−2)(−2)(−2)

12. 30/(–15)

13. Fill in the blank to make the following true: _?_ / –4 = –4

14. You multiply together twelve numbers. Four of them are negative and the rest are positive. What is the sign of the answer?

15. What is the coefficient of –6abc?

16. If you take a positive number and a negative number and multiply them together, will the answer always be less than the positive number you started with?

17. How many different whole numbers can be used to fill in the blank to make the following true: (_?_)(–7) > –32.

18. The number of bubbles in a glass of Sluice depends on how many ounces of Sluice that you pour. If you pour only 4 ounces, there are a million bubbles in your glass. How many ounces should you pour to have three million bubbles in your glass?

19. Betty and Alexander are standing in a big circle with 88 other people all facing the center of the circle. Betty and Alexander are directly across from each other. The average shoulder width in this crowd is two feet. How far, approximately, are Betty and Alexander from each other?

answers

1. –11
2. +42
3. Yes.
4. No, it's just an approximation for π. $22/7 = 3.14285\ldots$ whereas $\pi = 3.14159\ldots$.
5. $(\frac{1}{2})\pi$
6. Stick a zero in the blank.
7. $(-12)(+3) < (-6)(\underline{\ ?\ })$ is the same as $-36 < (-6)(\underline{\ ?\ })$. Trying out various integers, you find that any integer in the set $\{\ldots -1, 0, 1, 2, 3, 4, 5\}$ will work.
8. It would cost me 200 times as much as one minute. So the total bill would be $600.
9. –6 and 1272
10. –785 or –786 or any integer less than these. $\{-785, -786, -787, \ldots\}$
11. 16
12. –2
13. 16
14. It will be positive.
15. –6
16. Yes since the answer will always be negative and any negative number is always less than any positive number.
17. There are five of them. Namely $\{0, 1, 2, 3, 4\}$.
18. Since you want three times as many bubbles, you should pour three times as much Sluice into your glass. Three times 4 oz = 12 oz.
19. There are 90 people in the circle. Average shoulder width is 2 feet. Circumference of the circle is 180 feet. Using $C = 3d$, the diameter (and hence the distance between Betty and Alexander) is approximately 60 feet. The super-exact answer would be $180/\pi$, but this would be overly exact given the nature of people standing in a circle.

The last of the pizza was eaten. It was getting late (7:45 p.m.) and tomorrow was Friday. There were classes for Fred to teach; classes for Darlene and Joe to attend. Betty and Alexander had become engaged after they had passed their comprehensive exams. They were in the Ph.D. program for math and now had their dissertations to write, which is the last major task to do before they would get their Ph.D.s.

Everyone pitched in to clean up the student recreation room. Betty put away the three phone books that Fred had been sitting on. Joe swept the floor. Alexander cleaned up the plates, cups, and pizza box. Darlene gathered together all the pepperoni slices that Joe had been playing with. She ate some of them and threw the rest in the garbage. Fred straightened the chairs and, as the last one out the door, he turned out the lights.

Two pairs left the building and Fred was alone. He didn't feel so bad since he was only five years old and thoughts of a real girlfriend had not entered his head yet. He walked down the hall heading back to his office, which is where he lived.

He got to the door that read **Prof. Fred Gauss** with a list of the classes he taught underneath: Arithmetic, Beginning Algebra, Advanced Algebra, Geometry, Trig, Calculus, Linear Algebra, and a Seminar in Biology, Economics, Physics, Set Theory, Topology, and Metamathematics. He opened the door and headed to his desk. With another hour of work to do before classes tomorrow, he hopped on his desk chair (with its three telephone books) and turned on the desk light.

He had always enjoyed teaching at KITTENS. The pay was good ($500/month) and no one objected to his using his office as his home, so his expenses were minimal. He donated $50 to Sunday School each month. He bought books that he wanted to read that were not available at the KITTENS library. He saved most of what he earned.

> "When I get a little money, I buy books; if any is left, I buy food and clothes."
> —Erasmus

One other item was his share of the pizza costs for the Thursday night poker party. Everyone paid according to the amount of pizza consumed. Betty and Darlene each usually had three pieces, Alexander had five, and Joe had six. Fred had one piece (half of which he ate and the other half he took back to his office so he could have it for breakfast the next morning).

So the ratio of pieces eaten for Betty, Darlene, Alexander, Joe, and Fred was 3:3:5:6:1. This is a shorthand way of writing out all the different pairs of ratios. It's called a **continued ratio**. For example, the Darlene-Alexander ratio is 3:5 and the Darlene-Joe ratio is 3:6, etc. Using 3:3:5:6:1, you can pick out any pair of people and get their corresponding ratio.

Stanthony's medium pizza cost $7.20 including the tip. This really isn't a bad price for a 25-inch combination pizza delivered right to your door. Last August when they first ordered the pizza, they had to figure out how much each person owed. They all looked at Fred.

Fred was about to do some algebra. With equations! Using an "x" as the unknown! Stuff you may never have seen before! Please don't panic. The first time that your elementary school teacher wrote $3 + 4 = 7$ on the blackboard, it was new to you, but after a while you got used to it and $3 + 4 = 7$ doesn't bother you much at all now. The procedure in learning math always seems to be the same:

Step One: This is new. I don't get it yet.

Step Two: I'm starting to get it.

Step Three: Oh. That's how it works.

Step Four: Boring. Anyone can do this.

Fred wrote on a napkin that the cost of the pizza was 720¢ and 3:3:5:6:1. (He used cents instead of dollars to avoid decimals.)

Now this isn't the simplest algebra problem in the world. Many banal beginning algebra books start off with something like, "Two times some number equals 6. What is the number?" and then they write $2x = 6$ and solve it. The problem with this approach is that you know the answer is 3 right from the start, and you don't see how algebra might be useful. With this problem that Fred faced, you do not know offhand how much Fred owed for his share of the pizza.

Here is what Fred wrote on the napkin in solving this problem. This is new to you. We'll explain all the steps *after* you see what he wrote.

3:3:5:6:1   720¢

Let x = the cost of my pizza.

Then 3x is cost of Betty's pizza.

Then 3x is the cost of Darlene's.

Then 5x = Alexander's cost.

Then 6x = Joe's cost.

The total bill was 720¢.

$$3x + 3x + 5x + 6x + x = 720$$
$$18x = 720$$
$$x = 40$$

Since x is the cost of my pizza, and 3x is Betty's cost, and Darlene's is 3x, and Alexander's is 5x, and Joe's is 6x:

Betty/Darlene/Alexander/Joe/me

120¢, 120¢, 200¢, 240¢, and 40¢

Now let's explain all the parts to this solution.

This is traditionally called a word problem. The hardest part is getting from the words to the equation. Notice that half the napkin is used before Fred arrived at the equation.

You can't just read the problem and write down the equation.

Here Are the Four Steps Getting Word Problems into an Equation

1. Read the problem until you have it almost memorized. Get the whole picture in your head.

2. On a sheet of paper write Let x = the thing you are trying to find out.

3. Below that line, write Then statements. Look at the napkin on the previous page. Based on the Let x = cost of my pizza, Fred could write four Then statements. The Then statements come from two sources: (1) the previous line(s) you have written on your paper and (2) the information given in the problem.

4. If you pay attention to what you have written down, the equation will almost write itself. What Fred had written was that his cost was x and his friends' costs were 3x and 3x and 5x and 6x. So the total cost was 3x + 3x + 5x + 6x + x. He also knew that the total cost was 720¢.

So 3x + 3x + 5x + 6x + x had to equal 720.

Your Turn to Play

1. Suppose the continued ratio of the number of pizza pieces that Betty, Darlene, Alexander, Joe, and Fred had eaten was 2:2:5:4:1. Suppose the price of the pizza is 1120¢. (That's the same as \$11.20, but I'm trying to keep the arithmetic as simple as possible.) Let x equal the cost of the pieces that Fred had. Find the equation. (In the next lesson, we will show how to solve it.)

2. Suppose Betty, Darlene, Alexander, Joe, and Fred take a trip in a hot air balloon and they are charged by how much they weigh. The continued ratio of their weights is 5:5:9:6:2. Suppose their total bill is \$54 for the ride. Find the equation. (In the next lesson, we will show how to solve it.)

3. Darlene has 90 hair ribbons. For every 5 red ribbons, she has 6 blue ones and 4 green ones. Find the equation.

```
.......COMPLETE SOLUTIONS.......
```

1. *Let x = the cost of the piece Fred had.*
   Then 2x = the cost of the of pieces Betty had.
   Then 2x = the cost of the of pieces Darlene had.
   Then 5x = the cost of the of pieces Alexander had.
   Then 4x = the cost of the of pieces that Joe had.

The total cost was 2x + 2x + 5x + 4x + x.

The total cost was 1120¢.

   The equation is 2x + 2x + 5x + 4x + x = 1120.

2. *Let 5x = Betty's cost.*
   Then 5x = Darlene's cost.
   Then 9x = Alexander's cost.
   Then 6x = Joe's cost.
   Then 2x = Fred's cost.

The total cost is 5x + 5x + 9x + 6x + 2x.

The total cost is $54.

   The equation is 5x + 5x + 9x + 6x + 2x = 54.

3. *Let 5x = the number of red ribbons.*
   Then 6x = the number of blue ribbons.
   Then 4x = the number of green ribbons.
   Then 5x + 6x + 4x is the number of ribbons.

The number of ribbons is 90.

   The equation is 5x + 6x + 4x = 90.

Most four-year universities ask for four years of high school math (beginning algebra, advanced algebra, geometry, and trig). What you are doing now is high school work. Each of these three problems required writing about eight short lines to get to the equation.

The real point of this lesson is not just to get the equation but to learn the procedure. Many readers can just look at the problem and write the equation, but when we get to Chapter 4, you will not be able to "do it in your head."

S olving $3x + 3x + 4x + 7x + x = 720$ is pretty straightforward. Here is how to do it.

In elementary school math we learned that:

3 apples plus 3 apples plus 4 apples plus 7 apples plus 1 apple is 18 apples.

3 oranges plus 3 oranges plus 4 oranges plus 7 oranges plus 1 orange is 18 oranges.

3 eggs plus 3 eggs plus 4 eggs plus 7 eggs plus 1 egg is 18 eggs.

3 ducks plus 3 ducks plus 4 ducks plus 7 ducks plus 1 duck is 18 ducks.

So, it's no surprise that: $3x + 3x + 4x + 7x + x = 18x$

$$3x + 3x + 4x + 7x + x = 720$$

gets replaced by $\qquad 18x = 720.$

This is called **combining like terms**.

Now, what to do with $18x = 720$? Let's go back to elementary school math just once more:

What's one-tenth of ten elephants? It's one elephant.

What's one-fifth of five marbles? It's one marble.

What's one-thirtieth of thirty puppies? It's one puppy.

So it's no surprise that one-eighteenth of 18x is one x (or just plain x).

We know that 18x and 720 are *equal*. If I take one-eighteenth of two numbers that are *equal*, won't the results be equal? In fact, if I do the same thing—no matter what—to two equal numbers, the results will always be equal.

So Fred took $18x = 720$ and took one-eighteenth of each side. He got $x = 40$. Here's the work he did in his head:

$$18 \overline{)\, 18x} \quad\quad \begin{array}{r} 40 \\ 18 \overline{)\, 720} \end{array}$$

(Small note to you, my reader: How fast did you read the previous page? People get "lost" in math when they try to zoom through new material, thinking to themselves, "Yeah, I think I get the general drift of this." If it took you 20 minutes per page to fully comprehend this new material, that is not unusual. Learn it well enough that you could explain it at a blackboard. Learn it well enough that if you were given just 3:3:5:6:1 and 720¢, you could get to $x = 40$ without looking at the book for help.)

In the previous *Your Turn to Play*, in the first problem we had the equation

$$2x + 2x + 5x + 4x + x = 1120$$

Combine like terms $$14x = 1120$$

Divide both sides by 14 $$x = 80$$

In the second problem we had the equation

$$5x + 5x + 9x + 6x + 2x = 54$$

Combine like terms $$27x = 54$$

Divide both sides by 27 $$x = 2$$

Actually, there were many months in which Fred's total expenses were well under $80. And to Fred's delight, he had been recently told by the administration at KITTENS that his salary would be $600/month when he turned six years old. Fred felt himself to be very rich.

If you ever get a chance to do a bunch of these ratio problems, they will often follow the same basic pattern that Fred wrote on his napkin. And here is your chance.

Your Turn to Play

1. Solve $20x + 7x + 3x = 510$

2. In Fred's rose garden dream, he gathered a bouquet of 132 roses. There was one red rose for every seven yellow roses for every four white roses. How many roses of each kind were in his bouquet? (Hint: Let x = the number of red roses.)

3. At Stanthony's pizza, for every six pepperoni pizza orders, there is one order of anchovy and three orders for mushroom pizzas. On one Tuesday evening, Stanthony noted that he had baked 250 pizzas that were pepperoni, anchovy, or mushroom. How many anchovy pizzas did he bake that night?

4. In Fred's college algebra class, for every three students who use fountain pens, there are four who do not. There are 91 students in his class. How many use fountain pens? (Hint: Let 3x = number of fountain pen users.)

We'll include some more of these ratio problems when we get to Cities at the end of this chapter. If you attempted to do these *Your Turn to Play* problems *on paper* before you looked at the answers, you can now say that *you are doing algebra*. If you wish (and if this is your book, not someone else's) you may cut out this award to keep.

cut here

If you merely just read the *Your Turn to Play* problems and looked at the answers without first trying to do them yourself, then (gulp!), I'm afraid that you don't get the award. Sorry.

.......COMPLETE SOLUTIONS.......

1. $$20x + 7x + 3x = 510$$

Combine like terms $$30x = 510$$

Divide both sides by 30 $$x = 17$$

2. *Let x = the number of red roses.*
   Then 7x = the number of yellow roses.
   Then 4x = the number of white roses.

We know the total number of roses is 132.

So $$x + 7x + 4x = 132.$$

Combining like terms $$12x = 132.$$

Dividing both sides by 12 $$x = 11.$$

So the number of red roses in his bouquet is x, which is 11.

Yellow roses: 7x, which is 7(11) or 77.

White roses: 4x, which is 4(11) or 44.

3. *Let x = the number of anchovy pizzas.*
   Then 3x = the number of mushroom pizzas.
   Then 6x = the number of pepperoni pizzas.

The total number of pizzas is 250.

So $$x + 3x + 6x = 250$$

Combining like terms $$10x = 250$$

Dividing both sides by 10 $$x = 25$$

He baked 25 anchovy pizzas. (The question did not ask about the other pizzas.)

4. *Let 3x = the number of fountain pen users.*
   Then 4x = the number of non-users.

The total number of students is 91.

So $$3x + 4x = 91$$

Combining like terms $$7x = 91$$

Dividing both sides by 7 $$x = 13.$$

The number of users of fountain pens (which was what was asked for) is 3x, which is 3(13), which is 39.

Fred was sitting on his three phone books at his desk writing his lecture for the trig class he had at noon tomorrow. He drew a right angle (which looks like the corner of a square) and cut it into two angles which he called A and B. It looked like angle B was about twice as large as angle A. That is to say, the ratio of the degrees in A to the degrees in B was 1:2. Since a right angle is 90 degrees, Fred wrote in his lecture notes:

Let x = number of degrees in angle A.
Then 2x = number of degrees in angle B.
The total number of degrees is 90°.

$$1x + 2x = 90$$
$$3x = 90$$
$$x = 30$$

so A = x = 30° and B = 2x = 60°

There was a knock on the door. Fred thought It's 8:30 at night. I wonder who's there. He called out, "Come in!"

It was Sam, the janitor. You could tell that by reading his name badge:

Samuel P. Wistrom
Chief Educational Facility Math Department Building, KITTENS University Inspector/Planner/Remediator
for offices 225—324.

The "Inspector/Planner/Remediator" part of his name badge provoked a lot of questions from the faculty. As Sam explained it, "They told me when I took this job, that I was to find full wastebaskets (*inspector*), think about emptying them (*planner*), and then empty them (*remediator*)." Sam was a cool guy. Everyone liked him.

"Sorry to disturb you, Professor Gauss," Sam said. "Your faculty mailbox got so full that the department secretary said I should bring it all to you."

Sam placed 120 letters wrapped with a rubber band on Fred's desk. Fred had already picked up his morning mail, and these 120 letters were the afternoon mail. Fred had been receiving a lot of mail since the news of his distinctive teaching style hit the national newspapers, magazines, television, radio, and the Internet. Darlene had even overheard something about the "Fred teaching style" in the beauty shop.

A typical newspaper front page read:

New Math Teaching Method Sweeps Nation!

Profs Burn Their Old Lecture Notes

KANSAS

Thousands of teachers of mathematics gathered yesterday outside the main quad at KITTENS University to honor Prof. Fred Gauss and his new teaching method.

Each teacher threw their old teaching notes into the bonfire and pledged to teach "the new Fred way."

"It's so revolutionary!" exclaimed one teacher. "My students have stopped asking me 'What good is this stuff?' "

Another teacher has signed up to teach summer school for the first time in her life. She explained, "I really love teaching the Fred way, and my students love learning."

Taking their cue from the stories in the best-selling *Life of Fred* series, teachers have had their classrooms packed with eager students. (continued on p. 12)

The Fred Way–Details Revealed

In an exclusive interview with Prof. Gauss (rhymes with "house") he revealed the secrets of his success in teaching math.

"All I do," he confided to our reporter, "is tell stories in which the uses of the mathematics naturally arise. My students have a reason to learn the material. No thirty boring, meaningless exercise problems. (continued on p. 14)

Our Fred: the Sweetheart of Our Nation

Heartthrob

Postal workers report that hundreds of marriage proposals have poured into KITTENS University. The sale of *Love* stamps has doubled recently and a new romance novel is (continued on p. 18)

Fred quickly dealt the letters into four piles: (1) the marriage proposals (which were perfumed and had "S.W.A.K." on the back), (2) job offers, (3) requests for interviews, and (4) letters from appreciative students. Looking at the piles, he noted that there were seven proposals for every one job offer for every three requests for interviews for every four letters of appreciation. How many of each kind of letter were there? Since this is old algebra which you're well acquainted with, we'll just stick it in a box so that you can ignore it if you wish.

Let x = job offers.
Then 7x = the number of proposals.
Then 3x = interview requests.
Then 4x = letters of appreciation.

$$x + 7x + 3x + 4x = 120$$
$$15x = 120$$
$$x = 8$$

x = 8 job offers
7x = 7(8) = 56 proposals
3x = 3(8) = 24 interview requests
4x = 4(8) = 32 letters of appreciation

At the age of five, Fred was hardly interested in the 56 "smelly" letters. Half of his friends were female, but he hardly paid attention to which half.

When the proposals started to come in, Fred had asked the department secretary to send a letter to each of them, thanking them for their proposal. In order that they would not take Fred's rejection personally, an explanation was included that ". . . his romantic life was currently full." That was, shall we say, technically true since his romantic-life needs were like that of any five-year-old, namely zero.

The supermarket tabloids had a field day with ". . . his romantic life was currently full" and published his picture surrounded by a half dozen beautiful women with the headline, "Which One Is It?" This greatly embarrassed Fred.

The job offers and interview requests were also passed on unopened to the department secretary who mailed out form responses. Fred, himself, read and responded to the letters of appreciation from students.

As Fred was turning over each envelope to check for "S.W.A.K." (sealed with a kiss), he smiled, thinking that every envelope is a rectangle. Each has four sides and all right angles in the corners.

I wonder he thought to himself is it legal to mail envelopes in other shapes? He drew some crazy-shaped envelopes.

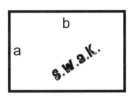

This is your ordinary rectangular envelope. If the width is a and the length is b, then the area (to write S.W.A.K. on) is A = ab. The distance around the edge of the envelope (which is called the **perimeter**) is p = a + b + a + b (or you can write it as p = 2a + 2b).

Or what about an envelope in the shape of a **trapezoid**? (Four-sided figure with exactly two sides parallel.)

trapezoid

Fred's favorite envelope shape he decided was that of a **sector** (which is the shape of a piece of pie). He thought Stanthony could use that envelope to mail out his advertisements.

sector

He was having all kinds of fun playing with the mathematics. His story was ready for his trig lecture tomorrow, the mail had been sorted, and it was playtime. It wouldn't be fair to let Fred have all the fun. Get out a piece of paper (please, please, please) and enjoy . . .

Your Turn to Play

1. The area, A, of a rectangle whose width and length are a and b, is A = ab. If a = 5 inches and b = 7 inches, what does A equal?

2. Suppose we have a rectangular envelope with an area of 56 square inches and the width of the envelope is a = 7 inches. Substitute these numbers into the formula A = ab and solve the resulting equation to find the length of the envelope.

3. If the area of a rectangle is 47.88 square inches and the width is equal to 6.3, using the formula A = ab, find the length of the envelope.

4. One rectangular envelope that Fred examined was 5 inches wide and 8 inches long. The next envelope was two square inches larger than this one. What was its area?

5. One rectangular envelope had dimensions of 3" x 5" (" means inches) and a second had dimensions of 4" x 7". What was the total combined area of the two envelopes?

6.

For any trapezoid, the area is equal to A = ½ h(a + b) where h is the distance between the parallel lines, and a and b are the lengths of the parallel sides. What is the area of a trapezoid where the lengths of the parallel sides are 4" and 6" and the distance between the parallel sides is 5"?

. . . . **COMPLETE SOLUTIONS** . . . .

1. A = (5)(7), which is 35 square inches. You could have written it as A = 5·7 = 35 using the raised dot notation.

2. A = ab becomes 56 = 7b. Dividing both sides by 7, we get 8 = b. (We can write the equation 8 = b as b = 8. This is called the **symmetric law of equality**, which we'll get to much later in algebra.)

> **Symmetric Law of Equality**
> If c = d, then d = c.

3. Now this problem is exactly like the previous one except that the numbers are more "real life" numbers. In the previous problem where the area was 56 and the width was 7, you might have thought to yourself, "Aw, I don't need algebra for this. It's obvious that the length has gotta be 8 to make the area equal to 56." Well, if the solution to this problem (where the area is 47.88 and the width is 6.3) is obvious to you, you have a great aptitude for arithmetic. The whole idea is that we start with a simple example to show you how the algebra is done and then give you a real-life example where the answer isn't so obvious. Here we go: Putting 47.88 and 6.3 into A = ab, we obtain 47.88 = 6.3b and dividing both sides by 6.3,

we get 7.6 = b. (I used a calculator to divide 47.88 by 6.3 and the answer came out even.)

4. The first envelope had an area of (5)(8) square inches. Since the second envelope was two square inches larger, it had an area of (5)(8) + 2. Now what is the answer to "Five times eight plus two"? It may be really obvious to you, but every beginning algebra book has to s-p-e-l-l i-t o-u-t so we all follow the same ground rules. The question is: Do we do

    Five times eight . . . . plus two *or*

    Five times . . . . eight plus two?

In the first case we get the (correct) answer of 42.

In the second case we get a (wrong) answer of 50.

    The rule is that we do all multiplications and divisions first (going from left-to-right) and then we do all the additions and subtractions. Here are three examples:

$4 \cdot 9 + 10$	$2 \cdot 18 \div 9 - 3$	$7 - 20 \div 5 \cdot 8$
$36 + 10$	$36 \div 9 - 3$	$7 - 4 \cdot 8$
46	$4 - 3$	$7 - 32$
	1	-25

    We need this rule of doing multiplication and division first, and then addition and subtraction so that we will all arrive at the same answer. The rule is really arbitrary but if we all follow the same procedure, we'll all get the same answer for $5 \cdot 8 + 2$. On another planet, they might have a different **order of operations**. In Kansas, the rule is that everyone drives on the right side of the road. Without the rule, there would be a lot more head-on collisions.

5. $3 \cdot 5 + 4 \cdot 7 = 15 + 28 = 43$ square inches.
We did all the multiplication first and then the addition.

6. A = ½h(a + b) becomes A = ½(5)(4 + 6). Things inside parentheses get done before anything else. Then multiplication and division (left-to-right) and finally addition and subtraction. Here are all the steps:

$$A = ½(5)(4 + 6)$$
$$A = ½(5)(10)$$
$$A = (2.5)(10)$$
$$A = 25 \text{ square inches for the trapezoid.}$$

It was getting near bedtime. Fred flossed and brushed his teeth, changed into his nightclothes, climbed into bed, said his prayers, and fell asleep.

The next morning, our hero greeted his sixth birthday with a dawn jog. The vending machines down the hall from his office furnished some chocolate milk. In his imagination, he pictured a slice of birthday cake in one of the machines, but reality dictated otherwise.

Fred had a rich and varied imagination. From his thoughts of weird envelope shapes last night to his dreams of rose gardens, these moments of reverie* provided a sweet background to his everyday existence.

And he was going to need it.

Heading to his first class, he stopped off at the department secretary's office to give her the unopened 56 marriage proposals, the 8 job offers and the 24 interview requests. Standing there were 503 pounds of military police: two men, very tall, with guns and helmets.

"Are you Fred Grass?" one of them barked.

"Gauss. Rhymes with *house*," Fred responded from his three-foot height.

"Come with us, boy. Uncle Sam wants to have a word with you."

The men each grabbed one of Fred's arms and carried him away.

The department secretary, Miss Belinda Bright, was aghast. Why were they taking Professor Fred away?

She ran down the hall and found Betty, who had known Fred as long as anyone else at KITTENS.

"Betty!" cried Belinda, "They've taken Fred away. Arrested him or something. Two big army-type men just grabbed him."

---

* Reverie (REV-eh-ree) = day dreaming, pleasant contemplation

Belinda showed Betty the card the men had left:

Sgt. S. Snow
U.S. ARMY

Induction Evaders
Investigator

"Induction . . . Army?" Betty shrieked. "Today is his sixth birthday. We've got a surprise birthday celebration ready for him in his trig class. The pizza, the birthday cake, the Sluice . . . it's all waiting for him. And he's been kidnapped or something. I'll get Alexander and we'll find out what's happened."

Betty rushed down the hallway, and Belinda Bright headed back to her office to call the chancellor of KITTENS to tell him what happened.

The chancellor's phone extension number was three digits (as was everyone's at KITTENS. Fred's was 314. He had selected that because it was easy to remember. Hint: π.) Fred had once remarked to Belinda that the chancellor's number was easy to remember. "It's three consecutive integers that add to 21," he told her.

Consecutive numbers are numbers like 8491, 8492, 8493 that come one after the other on the number line.

If *n* is the first number, then *n* + 1 is the next number, and *n* + 2 is the number after that.

Somehow, what Fred had said to Belinda had stuck in her head ("Three consecutive integers that add to 21") even though she tended to forget the actual phone extension number.

Belinda grabbed a piece of scratch paper and wrote:

LET N BE THE FIRST NUMBER.
THEN N + 1 IS THE NEXT NUMBER (SINCE THEY ARE CONSECUTIVE).
THEN N + 2 IS THE THIRD NUMBER.
THE SUM OF THE THREE NUMBERS EQUALS 21, SO N + (N + 1) + (N + 2) = 21.
SO N + N + 1 + N + 2 = 21. You can remove parentheses if there's just a plus sign in front. For example, + $(87x^2y)$ is the same as $87x^2y$.

$$3n + 3 = 21$$
$$3n = 18 \quad \text{She subtracted 3 from each side of the equation.}$$

If two things are equal and you subtract the same amount from each of them, won't the results have to be equal?

$$n = 6 \quad \text{This we've seen before; she divided both sides by 3.}$$

SINCE n = THE FIRST NUMBER AND SINCE n = 6, THE FIRST NUMBER IS 6.
n + 1 IS THE SECOND NUMBER. SO THE SECOND NUMBER IS 7.
n + 2 IS THE THIRD NUMBER. n + 2 = 8.

Belinda dialed 6-7-8 and said she needed to talk to the chancellor directly.

The chancellor, after hearing what happened, spouted, "Our little Fred. They won't get away with this!" He called out the campus cops to apprehend "those kidnappers."

So with Betty running down the hall to find her fiancé, the chancellor calling out the university police, Fred being hauled away somewhere, a trig class waiting to celebrate Fred's sixth birthday, and Belinda fretting over the new stack of envelopes that Fred had just given her, we take a break for . . .

Your Turn to Play

1. Find four consecutive numbers that add to 870.

2. By the symmetric law of equality, $4x = y^2$ is equivalent to what?

3. Evaluate $4(5 + 2) - 32$ showing all the steps.

4. In the order of operations, what is done first, second, and last?

5. Three military policemen have weights (in pounds) that are consecutive natural numbers. Their total weight is 753. How much does the heaviest of the three weigh? (Hint: to make things easier, let x = the weight of the lightest policeman.)

6. Solve $39n + 24 = 687$

7. Solve $20x + 133 = 393$

8. Solve $18y - 30 = 312$

9. **Consecutive even integers** are numbers like 724, 726, 728, 730. . . . If n is the first consecutive even integer, then n + 2 is the second. Find three consecutive even integers that add to 678.

10. If n is the first of four consecutive *odd* integers, what are the next three numbers?

.... **COMPLETE SOLUTIONS** ....

1. *Let n = the first number.*
   Then n + 1 = the second (consecutive) number.
   Then n + 2 = the third.
   Then n + 3 = the fourth.

Since these four numbers add up to 870, we write

$$n + (n + 1) + (n + 2) + (n + 3) = 870$$

Removing the parentheses
$$n + n + 1 + n + 2 + n + 3 = 870$$

Combining like terms
$$4n + 6 = 870$$

Subtract 6 from both sides
$$4n = 864$$

Divide both sides by 4
$$n = 216$$

The first number is 216.
The second number is n + 1. The second number is 217.
The third number is n + 2. The third number is 218.
The fourth number is n + 3. The fourth number is 219.

2. The symmetric law of equality states that if c = d then d = c. In this case since we know that $4x = y^2$, we may write $y^2 = 4x$.

3. $4(5 + 2) - 32$
    $= 4 \cdot 7 - 32$
    $= 28 - 32$
    $= -4$

4. First, we do whatever is inside parentheses.
Second, multiplication and division, working left-to-right.
Last, addition and subtraction.

5. *Let n = the weight of the lightest guy.*
   Then n + 1 = the weight of the middle guy.
   Then n + 2 = the weight of the heaviest guy.

Since their total weight is 753, we have $n + (n + 1) + (n + 2) = 753$
$$n + n + 1 + n + 2 = 753$$

Combine like terms
$$3n + 3 = 753$$

Subtracting 3 from both sides of the equation
$$3n = 750$$

Dividing both sides of the equation by 3
$$n = 250$$

The question asks for the weight of the heaviest guy who weighs n + 2, where n = 250. The heaviest guy weighs 252 pounds.

6. $39n + 24 = 687$

Subtracting 24 from both sides $39n = 663$

Dividing both sides by 39 $n = 17$

7. $20x + 133 = 393$

Subtracting 133 $20x = 260$

Dividing by 20 $x = 13$

8. $18y - 30 = 312$

What to do? We've never had an equation with a subtraction to solve. How do we get rid of the "–30"? As long as we do *the same thing* to both sides of an equation, the results will be equal. We'll add +30 to each side of the equation and obtain: $18y = 342$

Then divide by 18 $y = 19$

9. Let n = the first consecutive even integer.

Then n + 2 = the second consecutive even integer.

Then n + 4 = the third.

Since the sum equals 678, $n + (n + 2) + (n + 4) = 678$

$$3n + 6 = 678$$

Subtracting 6 from both sides $3n = 672$

Dividing both sides by 3 $n = 224$

The three consecutive even integers, n, n + 2, and n + 4 are 224, 226, 228.

10. If n is the first consecutive odd integer, the next one will be n + 2. Is that surprising? Here are some consecutive odd numbers: 501, 503, 505, 507, 509. To get from one of them to the next, you add 2.

Four consecutive odd integers are n, n + 2, n + 4, n + 6.

T he trig class waited 20 minutes. Then they cut a giant piece of the cake, put Fred's name on it, and took it to the rec room fridge. They looked at each other and asked, "What would Professor Gauss want us to do now?" The answer was clear. They headed back to the classroom and consumed the rest of the cake and the pizza and drank toasts to Fred with their glasses of Sluice.

Joe offered the toast: "To our beloved teacher who has more wonderful stories than there are natural numbers!"

"Wait, Joe, that's crazy," another student said. "The natural numbers are {1, 2, 3, 4, . . .}. There's an infinite number of them. How are you going to have a set, like the set of all Fred stories, with more elements in it than the natural numbers?"

An English major pleaded on Joe's behalf, "Permit Joe his little bit of hyperbole.* After all, it's only a toast."

Joe, trying to defend himself, argued, "Hey, there are sets with more elements than the natural numbers. How about the whole numbers, which are {0, 1, 2, 3, 4, . .}?"

Who was right? The answer: No one was, except the English major who asked for tolerance. Here are the corrections to the errors:

1. Joe wasn't correct that Fred had an infinite number of stories. As we pointed out in chapter one, there are only about 10^{79} atoms in the observable universe, so anything tangible like the ink needed to write all of Fred stories is necessarily finite.

2. When the other student said that there weren't sets with more

---

* For non-English majors, we note that hyperbole [pronounced high-PER-ba-lee] is an exaggeration not meant to be taken literally, such as "I bet you a million bucks that. . . ."

elements in them than the natural numbers, he was wrong. There are such sets.

3. When Joe said that the set of whole numbers has more elements than the set of natural numbers, he was wrong.

How can these things be true?

  The truth is:

A) The set of natural numbers, the set of whole numbers, the set of integers, and the set of rational numbers (which are defined below) all have the same number of elements in them.

B) There exist sets that have a greater number of elements than any of the sets listed in statement A.

How is this possible? There are many real surprises that await you in your study of mathematics. This is one of them. Less than a century ago when mathematicians discovered that statements A and B were both true, many of them had a fit. "Impossible!" some of them screamed. It wasn't that some math guy said that he *thought* A and B were true. Instead, some math guy *proved* that there are sets that have more elements than the natural numbers. These statements are all laid out and proved in the first chapter of *Life of Fred: Calculus Expanded Edition*, which isn't that far away for you. After the book you're holding in your hands, there's only advanced algebra, geometry, and trig before you open the pages of *Calculus*.

Rational numbers? That's the set of all possible fractions including 1/2, 34/97, –3/8, 44/3, 0/13, 1/100000, and 7/21. The rational numbers also include any numbers that can be written as a fraction. So 5 is a rational number because it can be written as 5/1 (or 20/4). The rational numbers do *not* include π, since that can't be written as a fraction.

Officially, the **rational numbers** is the set of all numbers that can be written as an integer divided by a non-zero integer.

It would be hard to use braces and just list all the rational numbers. One try might look like: {1/2, 1/3, 1/4, . . .}, but that wouldn't work since 3/8 would be left out. It would be hard using braces to just list the set of all four-legged animals. If I tried it might look like: {horse, mouse, cat, dog, rabbit, ibex, elephant}, but I'm sure I wouldn't be able to name them all. (Would ibexes have been on your list?)

Here's a new way of writing sets that will help us out. In English, the rational numbers is the set of all x such that x = a/b where a and b are integers and b is not equal to zero.

Using symbols, the rational numbers are defined as { x | x = a/b, where a and b are integers and $b \neq 0$}. The symbol $\neq$ means *not equal to*.

The three symbols "{ x |" are read as "The set of all x such that. . ." This is called **set builder notation**.

So that's what the trig class was up to.

The chancellor had called out the campus police. It was against university policy for anyone to kidnap faculty members without the permission of the administration. This regulation had been patterned after Article 114 of the Weimar Constitution of Germany: "*Personal freedom is inviolable. . . . No restraint or deprivation of personal liberty by the public power is admissible, unless authorized by law.*"

The campus cops caught up with Sergeant Snow and his buddy as they were sticking six-year-old Fred in the military transport vehicle. A very brief argument ensued, but was resolved as many arguments have been resolved in the history of the human race. Namely, Sergeant Snow had a gun and the campus cops didn't. The university police made Sergeant Snow a temporary member of the administration of KITTENS and thereby, no university regulation was violated.

So that's what the chancellor and the university police were up to.

And Belinda Bright? She headed back to her office and worked on the new stack of envelopes that Fred had handed her just before his abduction.

She had been working more efficiently ever since she took that "Work Speedier and Get Richer" seminar for $789 last summer. She never opened any of the letters that Fred gave her. She just copied the return addresses off the "smelly" envelopes and sent out the form letter: "Sorry, he's already taken," and tossed the unopened envelope into the garbage.

She copied the return addresses off the job offers and sent out the form letter, "Sorry, he's happy where he is."

She copied the return addresses off the interview requests and sent out the form letter, "For answers to all your questions, check out Fred's website FredGauss.com "

Somewhere in the above description of Belinda's work-speedier approach is the key to what happened to Fred.

So that is what Belinda was up to.

Your Turn to Play

1. Is 1.5 a rational number?

2. Is 1000 a rational number?

3. Using set builder notation, write the set of all animals in the San Francisco zoo.

4. Is π a rational number?

5. What is the name of the law that allows us to go from 3x – y = 888 to the equation 888 = 3x – y?

6. Name a rational number that is not an integer.

7. Belinda sorted Fred's mail. For every 6 marriage proposals Fred received, he received 4 job offers and 2 requests for interviews. If he received 720 letters, how many of each kind did he get?

8. Is every integer a rational number?

9. Solve 9x – 4 = 41

10. Solve 26 = 3y + 14

11. The set { 1, 1/2, 1/3, 1/4, 1/5, . . . } could be written using set builder notation as { 1/x | x is a natural number}.

Using set builder notation, how could you write {1, 1/4, 1/9, 1/16, 1/25, . . . }?

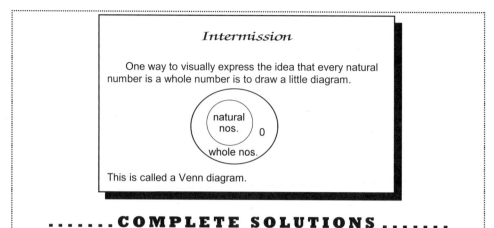

Intermission

One way to visually express the idea that every natural number is a whole number is to draw a little diagram.

This is called a Venn diagram.

.......COMPLETE SOLUTIONS.......

1. A rational number is any number that can be written as an integer divided by a (non-zero) integer.

    1.5 is equal to 1½ which is equal to 3/2. So 1.5 is rational.

2. 1000 can be written as 1000/1 so it is a rational number. (It could also be written as 2000/2 or 7000/7. The main idea is that if the number can somehow be expressed as a/b where a and b are integers and $b \neq 0$, then it's rational.)

3. $\{ x \mid x$ is an animal in the San Francisco zoo$\}$

4. π can't be expressed as a/b where a and b are integers and $b \neq 0$, and so π isn't a rational number. The number π is not exactly equal to 22/7 nor is it exactly equal to 3.14. These are only approximations.

5. The symmetric law of equality.

6. Answers will vary. Any of these will work: 1/5, 6.7, –4.2, 7/13 or 0.01. These would not work: 9, 0, –33, 8/2. (8/2 does not work since it is equal to 4 which is an integer.)

7. *Let 6x = the number of marriage proposals.*
    Then 4x = the number of job offers.
    Then 2x = the number of requests for interviews.

$$6x + 4x + 2x = 720$$
$$12x = 720$$
$$x = 60$$

Fred received *(6x)* 360 marriage proposals, *(4x)* 240 job offers, and *(2x)* 120 requests for interviews.

8. Every integer, c, can be written as c/1. Therefore, every integer is a rational number.

9. $9x - 4 = 41$

Add 4 to both sides $9x = 45$

Divide both sides by 9 $x = 5$

10. $26 = 3y + 14$ If it bothers you that the y is on the right side of the equation, then one possibility is to use the symmetric law of equality and switch the two sides and get $3y + 14 = 26$. Since it doesn't bother me, I'll leave the y on the right side.

   $12 = 3y$ Subtract 14 from both sides of the equation

   $4 = y$ Divide both sides by 3

11. $\{\, 1/x^2 \mid x \text{ is a natural number}\}$

A Larger Venn Diagram

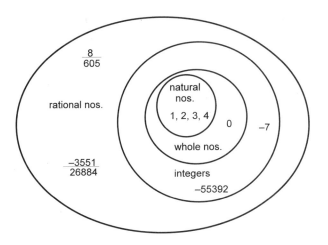

❧❦❧❦❧❦❧❦❧❦❧❦❧❦❧❦❧❦

Betty was the most distraught by Fred's disappearance. She had known him since he first arrived at KITTENS at the age of nine months. Betty used to carry him to class because, otherwise, it took him too long to get down the hall from his office to the classroom.

She was in tears by the time she found Alexander in the student rec room working on his Ph.D. thesis. She told him the story, and his muscles involuntarily flexed as adrenaline poured into his system. Actually, *adrenaline* is the old name for the hormone that's secreted in response to fright/flight/fight situations. The new name is *epinephrine* (EP-eh-NEFF-rin). With epinephrine, your heart beats faster, your blood pressure rises, and your blood clots more quickly. That's great if you're going to wrestle a lion, but not so good nowadays. One possible sequence of events is *stress* ⇒ *epinephrine* ⇒ *increased blood pressure and faster blood clotting* ⇒ *stroke/heart attack*.

But Alexander wasn't thinking about heart attacks right now. He ran to Belinda's office and then to the chancellor's and found out who had taken Fred and where they were headed.

Then Betty and Alexander met at the south end of campus. They had a long drive ahead of them.

"Did you find out where they're taking him?" Betty asked as they headed toward the only freeway through town.

"Yeah, Lampasas. It's in Texas and they've got a one-hour head start on us," Alexander replied. "Apparently the army has rounded up a whole bunch of induction evaders and a whole caravan of paddy wagons is heading south with their prisoners." She pushed her car to the speed limit (65 miles per hour) as they climbed onto the freeway in pursuit of Fred's abductors.

Now, as everyone knows, army caravans travel at 45 mph, but they had two advantages over Alexander and Betty: a one-hour head start and guns.

How long before our friends of Fred will catch up with the army caravan?

KITTENS

A. and B.

army caravan

Lampasas
Texas

We need a formula for motion. If you travel at, say, 30 mph for 5 hours, you will go 150 miles. The formula is $d = rt$, where d stands for distance, r stands for rate, and t stands for time.

We want to know how long it will take Alexander and Betty to catch up with the army vehicles, so we let t = the number of hours they will have to drive to catch up with the caravan. (*In many algebra problems, often the first step is to assign a letter to the quantity we are trying to find the value of.* ☞ important!*)

What else do we know?

After Alexander and Betty have driven t hours (and caught up with the army), the army will have driven $t + 1$ hours. How do we know that? We're told that the army had a one-hour head start. If, for example, A. and B. had to drive 18 hours, then the army would have driven 19 hours.

How far will A. and B. have driven? Using $d = rt$, they will have driven $65t$. They will have gone 65 mph for t hours.

How far will the army have gone? Again, using $d = rt$, the army will have gone $45(t + 1)$. They will have gone 45 mph for $t + 1$ hours.

HERE'S HOW IT IS SUPPOSED TO LOOK ON YOUR PAPER

Let t = number of hours Alexander and Betty will have driven.

Then t + 1 = number of hours the army will have driven.

Then 65t = miles Alexander and Betty will have driven.

Then 45(t + 1) = miles the army will have driven.

They will have both driven the same number of miles so $65t = 45(t + 1)$

To solve this equation, $65t = 45(t + 1)$, we're going to need the **distributive law**.

Please look at these examples of the distributive property and see if you can guess how it works:

$$7(2 + 11) = 14 + 77$$
$$9(8 - 4) = 72 - 36$$
$$\pi(4 + 7) = 4\pi + 7\pi$$

(We tend to write numerals before anything else, so we write 4π instead of $\pi 4$. They mean the same thing.)

$$\pi(5 + \pi) = 5\pi + \pi^2$$

(π^2 means $\pi\pi$. We'll talk about that in detail when we get to exponents.)

$$a(b + c) = ab + ac$$

That last line *is* the distributive property: $a(b + c) = ab + ac$. We'll show *why* the distributive property is true in Chapter 4, but right now we've got to find out the solution to $65t = 45(t + 1)$ so we can find out when A. and B. will catch up with Fred.

$$65t = 45(t + 1)$$
$$65t = 45t + 45 \qquad \text{using the distributive property}$$

Now, we're stuck again. $65t = 45t + 45$

We've never had the "unknown" (in this case, t) in two different spots in an equation. The trick is: Get the terms containing the unknown on one side of the equation and the numbers on the other. We have $65t$ on one side, and $45t$ on the other. Watch what happens when we subtract $45t$ from both sides of the equation:

$$20t = 45 \qquad \text{That's a lot nicer, isn't it?}$$

$$t = 2.25 \qquad \text{dividing both sides by 20.}$$

You could have written this last line as $t = 45/20$ or as $t = 2\frac{1}{4}$. They all mean the same thing.

So Alexander and Betty will have to drive two and a quarter hours to catch up with the army vehicles. That gives us plenty of time for . . .

Your Turn to Play

1. Solve $40x = 27x + 52$

2. Solve $6y + 21 = 9y$

3. Solve $2z - 11 = 5z$

4. Solve $7w + 16 = 9w - 8$

5. Use the distributive property to write this without parentheses: $8(x + 9)$

6. Solve $4(x + 12) = 6x$

7. Which of the following are rational numbers?

    682   4/5   0.0001   0 $\pi - \pi$

8. Solve $3(7 + x) = 5(-2 + x)$

9. Using set builder notation, express {2, 4, 6, 8, 10, . . .}.

10. Darlene and Joe are playing a game of tag. They start at the same spot, and Joe starts running east at 6 mph. After one hour Darlene starts out after him running at 8 mph. How soon will Darlene be happy?

11. Is the number of grains of sand on all the beaches in the world a finite or an infinite number?

12. Darlene and Joe are standing at opposite ends of a football field (300 feet apart). They were practicing for a movie that Darlene said she was going to make in which two lovers madly dash toward each other to embrace. (Actually, Darlene was using this as a pretext to implant in Joe's mind the thought of his being in love with her. Joe thought of this all as "a fun running game.") At the start of the movie scene, they both start running toward each other with arms outstretched. Joe runs at 22 feet/sec and Darlene, who is more motivated, runs at 28 feet/sec. How soon before they are in each other's arms?

28 feet/sec →   ← 22 feet/sec

300 feet

As a start let's say that Darlene runs for t seconds.

·······COMPLETE SOLUTIONS·······

1.
$$40x = 27x + 52.$$

We want the unknown on one side
of the equation so we'll first
subtract 27x from both sides $13x = 52$
Divide both sides by 13 $x = 4$

2. $6y + 21 = 9y$

Subtract 6y from each side $21 = 3y$

Divide both sides by 3 $7 = y$

3. $2z - 11 = 5z$
Subtract 2z from both sides $-11 = 3z$

Divide both sides by 3 $\dfrac{-11}{3} = z$

(You can write $-11/3$ as $-3\frac{2}{3}$ if you wish. Either way is fine.)

4. $7w + 16 = 9w - 8$

Subtract 7w to get all the terms containing w on one side $16 = 2w - 8$

Add +8 to each side to get all the numbers on the other side $24 = 2w$

Divide both sides by 2 $12 = w$

(Or, using the symmetric property of equality, $w = 12$.)

5. $8(x + 9) = 8x + 72$

6. $4(x + 12) = 6x$

Distributive property $4x + 48 = 6x$

Subtract 4x from each side $48 = 2x$

Divide each side by 2 $24 = x$

7. They all are rational numbers. The last number on the list, $\pi - \pi$, is equal to zero and zero is a rational number. Recall, the definition of a rational number is any number that *can be* written as an integer divided by a (non-zero) integer. And the number $\pi - \pi$ can be written, for example, as $0/17$.

8. $3(7 + x) = 5(-2 + x)$

Using the distributive property $21 + 3x = -10 + 5x$

Subtract 3x from both sides
to put all the terms containing x on one side $21 = -10 + 2x$

Add +10 to both sides $31 = 2x$

Divide by 2   15.5 = x

This could also have been written as

31/2 = x or as 15½ = x

or, using the symmetric property of equality as x = 15.5

9. {2, 4, 6, 8, 10, . . .} could be written as {x | x is an even natural number} or it might be written as {y | y is an even natural number} since the variable used doesn't make any difference.

If you wanted to be a little tricky, you could write {2, 4, 6, 8, 10, . . .} as {2x | x is a natural number} or {x | x is a positive even integer}. All of these will work.

10. We want to find out how long Darlene runs before she catches up with her guy.

Let t = how long (in hours) that Darlene runs.

Recall: Often the first step is to assign a letter to the quantity we are trying to find the value of. 🐭 important!

Then t + 1 = the number of hours Joe runs. Darlene runs for t hours and Joe runs for an hour longer.

Then 8t = the distance (in miles) that Darlene runs. Darlene runs for t hours at 8 mph.

Then 6(t + 1) = the distance that Joe runs. He runs for $t + 1$ hours at 6 mph.

Since, when Darlene overtakes Joe, they will have run the same number of miles, we have $8t = 6(t + 1)$.

$$8t = 6t + 6 \qquad \text{Distributive property}$$
$$2t = 6 \qquad \text{Subtract 6t from each side}$$
$$t = 3 \text{ hours} \qquad \text{Divide both sides by 2}$$

11.   There are a lot of grains of sand. There are more than any one person could count in a lifetime. The question is whether you could go on counting them forever. The answer is no. In an hour or so, you could count the number of grains of sand in a handful of sand. In a hundred hours, you could count the grains of sand in a large bucket. There's only a finite number of buckets of sand in a truckload of sand. The depth of the sand on an average beach isn't really that much. There's only a finite number of truckloads of sand on a beach (maybe 10,000 or so). And there's only a finite number of beaches in the world. If you lived an arbitrarily long time and you started counting 1, 2, 3, 4, . . . you would eventually come to the end.

Another way to look at this problem is to recall that there are approximately 10^{79} atoms in the observable universe and since each grain of sand contains zillions of atoms, there are clearly less than 10^{79} grains of sand. A finite number.

12. *Let t = the number of seconds that Darlene runs.*

Then t = the number of seconds that Joe runs.

Then 28t = the number of feet Darlene runs.   She is running 28 ft/sec for t seconds. d = rt

Then 22t = the number of feet Joe runs.

Together they run a total of 300 feet.

$$28t + 22t \;=\; 300$$
$$50t \;=\; 300$$
$$t \;=\; 6 \text{ seconds}$$

Darlene liked this game a lot better than when she had to run three hours to catch up with Joe.

In chapter two I promised you, "Once we get the distributive property, we'll be able to show that a negative times a negative gives a positive answer." Today I keep that promise.

Many banal beginning algebra books never really *prove* that. They sort of wave their hands and say something like, "Golly, take it from me, a negative times a negative is positive," and many algebra students go through the rest of their lives shrugging their shoulders. When their children ask them, "Why does $(-)(-) = +?$" their only response is that that is what their algebra teacher told them twenty years ago.

This proof isn't especially easy and you won't be quizzed on it in Cities. You can skip it if this is your first time reading this book, but realize that someday your kids may ask you at their bedtime, "Why does a negative times a negative equal a positive?"

Here goes.

We'll use **three facts** in this proof.
First: Any number or expression is equal to itself. "A is A" as Ayn Rand loved to say. (Ayn Rand is the author of *The Fountainhead* and *Atlas Shrugged*, two books that are among the top five books named by college graduates when they are asked to name a book that has most changed their outlook on life.)

In mathematics few people argue with $83 = 83$. Or $7.11\xi = 7.11\xi$. Or $\int \sec^2\theta \, d\theta - \log_2 33 - \pi^3 = \int \sec^2\theta \, d\theta - \log_2 33 - \pi^3$. If you want a fancy name for this, it's called the **reflexive property of equality**: $a = a$.

Second: Zero times anything is equal to zero.

Third: The distributive property, which is $a(b + c) = ab + ac$. Actually, we will use it in the symmetric form: $ab + ac = a(b + c)$.

We'll prove that $(-7)(-8)$ equals $+56$. Instead of the -7 and the -8, you could use any two negative numbers.

Step one: $7(8) + 7(-8) + (-7)(-8) = 7(8) + 7(-8) + (-7)(-8)^*$

by the reflexive property
of equality

Step two: $7(8 - 8) + (-7)(-8) = 7(8) + (-8)(7 - 7).$

by the distributive property

Now to figure out how this equation came from the previous equation may take you a whole minute or more. Math by its very nature is much more condensed than, say, reading English novels. Your English teacher may ask you to read 40 pages of *Ivanhoe* tonight, where your math teacher may ask you to read four pages of algebra.

Looking at the left side of step one, we have $7(8) + 7(-8)$. This is equal to $7(8 - 8)$ in step two.

This step was probably the hardest step in the proof.

Now doing arithmetic step two becomes $7(0) + (-7)(-8) = 7(8) + (-8)(0)$
Since $8 - 8 = 0$ and $7 - 7 = 0$.

And some more arithmetic: $0 + (-7)(-8) = 56 + 0$
Since zero times anything is zero.

And some more arithmetic: $(-7)(-8) = 56$
Since adding zero to anything doesn't change it.

Hey! We did it.

---

* Wow. That's a mouthful.

Why is it true? Because of the reflexive property of equality. What's on the left side is exactly the same as what's on the right side.

Where did I get $7(8) + 7(-8) + (-7)(-8) = 7(8) + 7(-8) + (-7)(-8)$ you may ask. I invented it.

Why am I starting with $7(8) + 7(-8) + (-7)(-8) = 7(8) + 7(-8) + (-7)(-8)$ you may ask. Because it works.

Your Turn to Play

1. Simplify as much as possible:
$$0(4 + \sqrt{3} - \pi \sin 30° + \log_e 15)$$

2. Supply a reason for each step in the following proof.
Here's a list of the possible reasons you might use:
   - ✓ symmetric property of equality
   - ✓ reflexive property of equality
   - ✓ Zero times anything is equal to zero.
   - ✓ distributive property
   - ✓ Any number subtracted from itself is equal to zero.
   - ✓ Zero plus any number is equal to that number.

To prove that (–4)(–9) = 4(9)

Statement	Reason
1. $4(9) + 4(–9) + (–4)(–9) = 4(9) + 4(–9) + (–4)(–9)$	1. ?
2. $4(9 – 9) \quad + (–4)(–9) = 4(9) + (–9)(4 – 4)$	2. ?
3. $4(0) \quad\quad + (–4)(–9) = 4(9) + (–9)(0)$	3. ?
4. $0 \quad\quad + (–4)(–9) = 4(9) + \quad 0$	4. ?
5. $\quad\quad\quad (–4)(–9) = 4(9)$	5. ?

3. If I know that $3 = 7w – 9$, by what reason can I say that $7w – 9 = 3$?

4. By what reason can I assert that $4\beta – 7\theta = 4\beta – 7\theta$?

5. What would you multiply 7^5 by in order to get 7^{10}?

·······COMPLETE SOLUTIONS·······

1. Zero times anything is equal to zero. Those other things in the expression are things you'll learn about in advanced algebra and trig.

$\sqrt{3}$ is called the square root of 3. Some square roots are easy to work with. For example, $\sqrt{4} = 2$ $\sqrt{9} = 3$ $\sqrt{81} = 9$ $\sqrt{100} = 10$.

But the square root of 3 isn't as easy to work with. $\sqrt{3}$ asks the question, "What number times itself is equal to 3?"

To get an approximate answer, you could type 3 into your calculator and then hit the $\sqrt{}$ key. You would get an answer of something like 1.7320508. This is saying that 1.7320508^2 (which means 1.7320508 times 1.7320508) equals 3. Watch. I'll multiply it out and see if it equals three.

$$
\begin{array}{r}
1.7320508 \\
\times \quad 1.7320508 \\
\hline
138564064 \\
000000 \\
8660240 \\
00000 \\
34641016 \\
51961524 \\
21243556 \\
17320508 \\
\hline
2.999922031498064 \\
\end{array}
$$
Almost, but not quite 3.

2.

Statement	Reason
1. $4(9) + 4(-9) + (-4)(-9) = 4(9) + 4(-9) + (-4)(-9)$	1. reflexive property of equality
2. $4(9-9) + (-4)(-9) = 4(9) + (-9)(4-4)$	2. distributive property
3. $4(0) + (-4)(-9) = 4(9) + (-9)(0)$	3. Any number subtracted from itself equals zero.
4. $0 + (-4)(-9) = 4(9) + 0$	4. Zero times any number equals zero.
5. $(-4)(-9) = 4(9)$	5. Zero plus any number equals that number.

3. Symmetric property of equality: If A = B, then B = A.
4. Reflexive property of equality: A = A.
5. $\qquad 7^5 7^?\;\; = 7^{10}$

$\qquad (7 \cdot 7 \cdot 7 \cdot 7 \cdot 7)(7^?) = 7^{10}$ $\qquad\qquad$ 7·7 means 7 × 7 means (7)(7)

$\qquad 7^5 7^5\;\; = 7^{10}$

The proof of $(-7)(-8) = +56$ may have been the first math proof you've ever seen. It takes a different style of reading than what you may be used to. The more of these you encounter, the easier they become. When you study geometry, you'll read proofs involving triangles and squares and eventually you'll be asked to invent proofs of your own. It isn't impossible. It just takes time and effort. Time and effort? Who wants to spend time and effort?

A better question is who can *avoid* spending time and effort? Those who ski spend a lot of time and effort:

⟍⟍⟍They buy their skis.

⟍⟍⟍They drive for hours to get to the snow.

⟍⟍⟍They pay for lessons and practice.

⟍⟍⟍They try to stay warm and avoid trees on the way down.

⟍⟍⟍They visit their ski friends in the hospital.

And why do they do that? For the moments of pleasure going down the hill.

Studying proofs in geometry takes time and effort, but it doesn't cost as much as skiing nor do many geometry students end up in the hospital because of running into triangles. And for many people, it is more pleasurable than skiing.

If you'd like to see the proof of $(-)(-) = +$ shortened up and in a more general form, it could be written as:

For any two positive numbers a and b,

1. $ab + a(-b) + (-a)(-b) = ab + a(-b) + (-a)(-b)$ $\qquad$ reflexive

2. $a(b - b)\;\; + (-a)(-b) = ab + \;\;(-b)(a - a)$ $\qquad$ distributive

3. $\qquad\qquad (-a)(-b) = ab$ $\qquad$ arithmetic

The "arithmetic" reason covered both A − A = 0 and 0A = 0 for any number A.

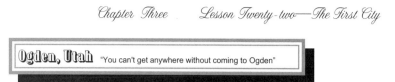

1. On my new diet I have two apples for every three oranges for every seven muffins I eat. Apples, oranges, and muffins are all I eat. Yesterday I had 156 of these permitted items. How many oranges did I have?

2. Solve $21y = 14y + 70$

3. The formula for the perimeter of a rectangle is $p = 2a + 2b$ where a and b are the width and the length. If the lengths of the four sides of a trapezoid are a, b, c and d, what would be the formula for the perimeter of a trapezoid?

4. Evaluate $4 \cdot 4 + 5$

5. Evaluate $6(1 + 5)$

6. Evaluate $60 \div (20 \div 2)$

7. Find three consecutive integers which add to 171.

8. Darlene and Joe went bowling. Standing next to him, Darlene bowled her favorite Red Spin ball down the alley. It was traveling at 12 ft/sec. Two seconds later, Joe using his slingshot, shot a marble at Darlene's ball. The marble flew at 60 ft/sec. If we let t equal the number of seconds the marble traveled, what is the equation for the time for the marble to hit the bowling ball? Also, solve the equation.

9. Suppose that Betty and Alexander had gone 75 mph instead of 65 mph in pursuing the army caravan. (This is, of course, purely hypothetical.) The army caravan had left an hour before they did and was going 45 mph. How long before Betty and Alexander caught up with them?

10. Solve $4x - 12 = 16x$

11. Joe bought a new car. The owner's manual said that there were five types of things he had to do to keep the warranty in effect. For every three oil changes, he should have the wheels rotated twice and the transmission pan dropped once and the windshield wipers replaced four times and the car painted once. He did 154 of these things in the first year he owned the car. How many times did he have his car painted?

12. Use the symmetric law of equality on $60 = 5 + 11x$.

13. For this sector, where the radius of the circle is r and the length of the curved arc at the edge of the sector is s, what is the formula for the perimeter of the sector?
(In calculus, the letter s is almost always used to measure arc length.)

14. Solve $10(x + 3) = 18x$

15. The chancellor headed out on his weekly bike ride. He pedaled at 18 mph. Three hours later, his wife realized that he forgot to take his lunch, and she started out after him on her motor scooter at 40 mph. How long will it take her to catch up with him?

16. Continuing the previous problem, how long will the chancellor have bicycled before his wife catches up with him?

17. A bee spots me and heads toward me at 3 ft/sec. Two seconds later I spot the bee and start heading toward it (with a flyswatter) at 5 ft/sec. We were originally 14 feet apart. Let t = the number of seconds that the bee is flying toward me. Then $t - 2$ is the number of seconds I'm heading toward the bee. What is the equation for the time until we meet? Solve that equation.

answers

1. 39

Let 3x = the number of oranges I ate. You always start with a "Let" statement.
Then 2x = the number of apples I ate.
Then 7x = the number of muffins I ate.

Since the total number of items is 156	$2x + 3x + 7x = 156$
Combine like terms	$12x = 156$
Divide both sides by 12	$x = 13$

We want to find the number of oranges (3x) $3x = 39$ oranges

2. $21y = 14y + 70$

Subtract 14y from both sides $7y = 70$

Divide both sides by 7 $y = 10$

3. $p = a + b + c + d$

4. $4 \cdot 4 + 5$

           $16 + 5$

            21

trapezoid
two sides parallel

5. There are two different ways this could be done.

$6(1 + 5)$	OR	$6(1 + 5)$
$6(6)$		$6 \cdot 1 + 6 \cdot 5$
36		$6 + 30$
		36

6. $$60 \div (20 \div 2)$$
Inside the parentheses is done first $$60 \div 10$$
$$6$$

7. *Let x = the first consecutive number.*
Then x+1 = the second consecutive number.
Then x+2 = the third consecutive number.

Since the numbers add to 171 $x + (x + 1) + (x + 2) = 171$

Since there is a + sign in front
of the parentheses $x + x + 1 + x + 2 = 171$

Combine like terms $3x + 3 = 171$

Subtract 3 from both sides $3x = 168$

Divide both sides by 3 $x = 56$

The three consecutive numbers are x, x+1, and x+2, which are 56, 57, 58.

Time Out!

Some students like to do problems like this "in their head." It saves a lot of labor. They don't have to write down all the steps—just fuss with the numbers until they find 56, 57, and 58.

If all the problems in life were as easy as finding three consecutive numbers that add to 171, then "doing it in your head" would make a lot of sense . . . and algebra would be worthless.

What you are learning here is the form. You are learning how to set up the equation using those *Let x =* statements.

When the word problems get much tougher (in Chapters 4 and 5), then "doing it in your head" is not going to work.

I have deliberately kept the numbers easy in this chapter so that you could concentrate on learning the way to get from the English to the equation. "Doing it in your head" defeats that goal.

8.   *Let t = seconds that the marble traveled.*
      Then t + 2 = seconds that Darlene's ball rolled.
      Then 60t = distance marble went.
      Then 12(t +2) = distance ball went.

Since the distance that the marble went is equal to the distance that the ball rolled

$$60t = 12(t + 2)$$
$$60t = 12t + 24$$
$$48t = 24$$
$$t = \frac{24}{48} \text{ seconds } \text{which reduces to } \frac{1}{2} \text{ second or } 0.5$$

The second way to approach this problem is to use "six pretty boxes" as you saw in Chapter 16 of *Life of Fred: Pre-Algebra 2*.

Start by putting the variable in the appropriate box.

	distance	rate	time
bowling ball			
marble			t

Then since the bowling ball rolled two seconds more . . .

	distance	rate	time
bowling ball			t + 2
marble			t

Fill in the speeds that were given in the problem.

	distance	rate	time
bowling ball		12	t + 2
marble		60	t

Fill in the last two boxes using d = rt.

	distance	rate	time
bowling ball	12(t + 2)	12	t + 2
marble	60t	60	t

Since we know the two distances are equal, we have the equation 12(t + 2) = 60t.

It is almost always the case that the equation comes from the last two boxes that are filled in.

9.
Let t = the number of hours until Betty and Alexander caught up with the army caravan.
Then t + 1 = the hours the army caravan drove.
Then 75t = the distance Betty and Alexander drove.
Then 45(t + 1) = the distance the army caravan drove.

Since they will have driven the same distance:	$75t = 45(t + 1)$
Distributive property	$75t = 45t + 45$
Subtract 45t from both sides	$30t = 45$
Divide both sides by 30	$t = \dfrac{45}{30}$ hours
	or 1½ hours or 90 minutes

Using the "six pretty boxes" approach . . .

We want to know how long Betty and Alexander drove, so we put the variable t in that box.

	distance	rate	time
Betty & Alexander			t
army			

The army drove one hour longer.

	distance	rate	time
Betty & Alexander		75 mph	t
army		45 mph	t + 1

Using d = rt, we fill in the last two boxes.

	distance	rate	time
Betty & Alexander	75t	75 mph	t
army	45(t + 1)	45 mph	t + 1

Since we know the distances are equal, $75t = 45(t + 1)$.

10. $4x - 12 = 16x$

Subtract 4x from both sides $-12 = 12x$

Divide both sides by 12 $-1 = x$

11.   *Let x = the number of times Joe had the car painted.*

Then 3x = the number of oil changes.

Then 2x = the number of wheel rotations.

Then x = the number of times the transmission pan was dropped.

Then 4x = the number of times the windshield wipers were replaced.

Since the total number of items is 154 $x + 3x + 2x + x + 4x = 154$

Combine like terms $11x = 154$

Divide both sides by 11 $x = 14$

Joe had his car painted 14 times.

12. $5 + 11x = 60$

13. $p = 2r + s$

14. $10(x + 3) = 18x$

Distributive property $10x + 30 = 18x$

Subtract 10x from both sides $30 = 8x$

Divide both sides by 8 $\dfrac{30}{8} = x$

Symmetric property of equality $x = \dfrac{30}{8}$

or $\dfrac{15}{4}$ or 3 ¾ or 3.75

15.   *Let t = the number of hours she will be on her motor scooter.*
      Then t + 3 is the number of hours he will be on his bike.
      Then 40t = the distance she travels.
      Then 18(t + 3) = the distance he travels.

Since they travel the same distance $40t = 18(t + 3)$

$t = 54/22$ or $27/11$ or $2\ 5/11$ hours.

16. Since he travels three hours longer than she, he goes $5\frac{5}{11}$ hours.

17.   *Let t = the number of seconds the bee is flying toward me.*
      Then t – 2 = the number of seconds I am heading toward the bee.
      Then 3t = the distance the bee flies.
      Then 5(t – 2) = the distance I travel.

Since the total distance is 14 feet, $3t + 5(t - 2) = 14$

$t = 3$ seconds

Using the "six pretty boxes" approach . . .

	distance	rate	time
bee		3	t
me		5	t − 2

And filling in the last two boxes using d = rt . . .

	distance	rate	time
bee	3t	3	t
me	5(t − 2)	5	t − 2

The distances are *not* equal. Instead, we know that the sum of the distances is 14 feet. $3t + 5(t - 2) = 14$

Peetz, Colorado home to 227 people and the Peetz Wind Farm

1. In the science section of the KITTENS library, for every chem book there are five math books, three physics books, and four biology books. Altogether there are 169 in these categories. How many biology books are there? (Hint: Start with Let 4x = number of biology books.)

2. Solve $6x + 9 = 21$

3. If an envelope is 9" long and 4" wide, what is the perimeter of that rectangle?

4. Evaluate $8 \div 4 \cdot 2$

5. Find five consecutive odd integers that add to 185.

6. In any triangle the sum of the three angles always adds up to 180°. Suppose in some particular triangle the angles are in the continued ratio of 3:4:5. What are the sizes of the three angles? (Hint: Let the smallest angle = 3x.)

7. Using set builder notation, write the set of all natural numbers that are not divisible evenly by seven.

8. Joe was getting ready for duck hunting season. He asked Darlene to throw her favorite plastic flying saucer into the air, which she did at 21 ft/sec. Two seconds later, Joe blasted at it with his favorite shotgun. The pellets from the gun traveled at 1000 ft/sec. How long did it take for the pellets to hit the saucer?

9. Solve $3x + 40 = 8x$

10. Joe once counted all the toppings on a Thursday night pizza. For every four red meat toppings, there were nine veggie toppings and three anchovy toppings. There were 240 toppings in all. How many red meat toppings were there?

11. Evaluate $6 \cdot 8 \div 2$

12. If three sides of a trapezoid are 11", 15", and 9", and the perimeter is 43", how long is the fourth side?

13. Using set builder notation, write the set of all pizza toppings.

14. Using the symmetric law of equality, what is $5w = 30$ equivalent to?

15. Two bull elephants are two miles apart and they charge toward each other in a mad testosterone rage. The bigger one charges at 45 mph and the littler elephant moves at 15 mph. How long before they collide?

16. The Royal Mounted police are chasing a Royal Mounted robber. The

robber has a two-hour head start and rides at 15 mph. The police ride at 16 mph. How long will the police have to ride to catch the bad guy?

17. Solve $6(x - 3) = 9(x + 5)$

answers

1.   *Let 4x = number of biology books.*
      Then x = number of chem books.
      Then 5x = number of math books.
      Then 3x = number of physics books.

Since there is a total of 169 books, $4x + x + 5x + 3x = 169$

$$13x = 169$$
$$x = 13$$

The number of biology books $(4x) = 52$

2. $6x + 9 = 21$

Subtract 9 from both sides $6x = 12$

Divide both sides by 6 $x = 2$

3. $p = 2\ell + 2w = 2(9) + 2(4) = 26"$ where ℓ is the length and w is the width

4. $8 \div 4 \cdot 2$

          $2 \cdot 2$

          4

First, we do whatever is inside parentheses.
Second, multiplication and division, working left-to-right.
Last, addition and subtraction.

5.   *Let x = the first number.*
      Let x + 2 = the second odd consecutive number.
      Let x + 4 = the third.
      Let x + 6 = the fourth.
      Let x + 8 = the fifth.

The five numbers add to 185.

$$x + (x + 2) + (x + 4) + (x + 6) + (x + 8) = 185$$
$$5x + 20 = 185$$
$$5x = 165$$
$$x = 33$$

The five consecutive odd numbers (*x, x + 2, x + 4, x + 6, and x + 8*) are 33, 35, 37, 39, 41.

6. *Let 3x = the smallest angle.*
    Then 4x = the medium-sized angle.
    Then 5x = the largest angle.

    We are told that these three angles add to 180°.

    $$3x + 4x + 5x = 180°$$
    $$12x = 180$$
    $$x = 15$$

The three angles (3x, 4x, and 5x) are 45°, 60°, and 75°.

7. {x| x is a natural number that is not divisible by 7}

8. *Let t = the number of seconds the pellets flew.*
    Then t + 2 = the number of seconds that the saucer flew.
    Then 1000t = distance pellets went.
    Then 21(t + 2) = distance saucer flew.

They went the same distance.

	$1000t = 21(t + 2)$
Distributive property	$1000t = 21t + 42$
Subtract 21t from both sides	$979t = 42$
Divide both sides by 979	$t = \dfrac{42}{979}$ seconds

9.

	$3x + 40 = 8x$
Subtract 3x from both sides	$40 = 5x$
Divide both sides by 5	$8 = x$

10. *Let 4x = number of red meat toppings.*
    Then 9x = number of veggie toppings.
    Then 3x = number of anchovy toppings.

Since the total number of toppings is 240,

	$4x + 9x + 3x = 240$
Combine like terms	$16x = 240$
Divide both sides by 16	$x = 15$

The number of red meat toppings (4x) is 60.

11.
    $$6 \cdot 8 \div 2$$
    $$48 \div 2$$
    $$24$$

> First, we do whatever is inside parentheses.
> Second, multiplication and division, working left-to-right.
> Last, addition and subtraction.

12.

By definition of perimeter	$11 + 15 + 9 + x = 43$
	$35 + x = 43$
	$x = 8$ inches

13. {x | x is a pizza topping} It would be much more difficult to just list all the toppings using braces. You might start the list with {pepperoni, onions, anchovies, bell peppers, ground beef}, but it's almost certain that you'd leave out some of the more unusual ones like sliced lamb or honey or peppercorns. Actually peaches and salami is an excellent combination. It would not be correct to write {pepperoni, onions, anchovies, . . .} using three dots, since it would not be clear how to continue this list. In contrast, it is clear how to continue {1, 2, 3, 4, 5, . . .}.

14. 30 = 5w

15. Let's use "six pretty boxes" to find the equation.

	distance	rate	time
bigger one		45	t
smaller one		15	t

Then using d = rt . . .

	distance	rate	time
bigger one	45t	45	t
smaller one	15t	15	t

The problem states that the total distance run by both elephants is equal to 2.

$$45t + 15t = 2$$
$$60t = 2$$
$$t = \frac{2}{60} \text{ hours (or two minutes)}$$

16. *Let t = the number of hours the police must ride to catch up with the bad guy.*
Then t+2 = the number of hours the robber will ride.
Then 16t = the distance the police will ride..
Then 15(t+2) = the distance the robber will ride..

Since they will ride the same distance $16t = 15(t + 2)$

Distributive property $16t = 15t + 30$

Subtract 15t from both sides $t = 30 \text{ hours}$

17.
$$6(x - 3) = 9(x + 5)$$
$$6x - 18 = 9x + 45$$

Subtract 6x and 45 $-63 = 3x$ Yes, we did two steps at the same time.
$$-21 = x$$

Xenia, Ohio   the only incorporated city in the US that begins with the letter X

1. During one hour in the army truck, Fred counted radio songs. There were two country western songs for every three rock songs for every two heavy metal songs. All together there were 329 songs in these three categories. How many rock songs did Fred have to listen to? (These were fairly short songs.)

2. Solve $8y - 12 = 3y + 3$

3. The perimeter of a rectangle is the distance around the outside edge. If a and b are the width and length, then the perimeter is $2a + 2b$. For circles, what is the perimeter called?

4. Evaluate $20 \div 10 \div 2$

5. Find three consecutive even whole numbers that add to 2112.

6. Using set builder notation, write the set of all students at KITTENS.

7. A snail was chasing a worm. The worm (at 2"/sec) had a 75 second head start. The snail was really full of caffeine and was really moving at 5"/sec. How long until the snail catches the worm?

8. Joe was out on a hike in the woods. He had a medical emergency (hangnail) and used his cell phone to call Darlene and ask her to come and rescue him (i.e., bring her nail clipper to him). With great difficulty, Joe was moving along toward Darlene at 2 mph, whereas Darlene was going to his rescue at 4 mph. They were two miles apart when he made the call. How long before Joe was rescued?

9. To pass the time as Fred was being hauled off to somewhere in Texas, he listened to the speech of Sergeant Snow. For every grammar error that Snow made, he uttered five cuss words and made two errors in pronunciation for a total of 312 such occurrences. How many errors in pronunciation did he make?

10. Solve $x - 12 = 6x + 8$

11. A rectangular prison cell (designed for holding 12 prisoners) has an area of 357 square feet. The width of the cell is 17 feet. What is its length?

12. Find three consecutive natural numbers that add to 1236.

13. Evaluate $20 \div (10 \div 2)$

14. Name a number that is not a rational number.

15. Betty is running down the hall to tell Alexander the bad news about

Fred's disappearance. She is running toward him at 7 ft/sec. Alexander, not knowing that something important was happening, was walking toward her at 3 ft/sec. How long before they meet? The original distance between them was 120 ft.

16. A mailman with really tired feet is moving at 10 ft/min. He drops a letter off at Darlene's house. Forty minutes later Darlene looks at her mail and realizes that the letter addressed to "Darleen Duckworth" is not for her (since her last name isn't Duckworth). She races after the mailman at 60 ft/min. How soon before she catches up with him?

17. Solve $5(y - 2) = 8(y + 7)$

answers

1.   *Let 3x = the number of rock songs Fred heard.*
      Then 2x = the number of country western songs.
      Then 2x = the number of heavy metal songs.

Since the total number of songs was 329 $3x + 2x + 2x = 329$

                        Combine like terms $7x = 329$

                     Divide both sides by 7 $x = 47$

The number of rock songs (3x) is equal to $3 \cdot 47 = 141$ rock songs

2. $8y - 12 = 3y + 3$

       Subtract 3y and add 12 to both sides $5y = 15$

       Divide both sides by 5 $y = 3$

3. The perimeter of a circle is called the circumference.

4. $20 \div 10 \div 2$

               $2 \div 2$

                1

> First, we do whatever is inside parentheses.
> Second, multiplication and division, working left-to-right.
> Last, addition and subtraction.

5.   *Let x = the first number.*
      Then x + 2 = the second consecutive even whole number.
      Then x + 4 = the third consecutive even whole number.

Since the three numbers add to 2112 $x + (x + 2) + (x + 4) = 2112$

                                       $3x + 6 = 2112$

                                       $3x = 2106$

                                       $x = 702$

The three consecutive even numbers (*x, x + 2*, and *x + 4*) are 702, 704, 706.

6. {x | x is a student at KITTENS}

7.   *Let t = the number of seconds the snail ran. (or is it walked?)*
Then t + 75 = the number of seconds the worm ran.
Then 5t = the distance the snail went.
Then 2(t + 75) = the distance the worm went.

Since they traveled the same distance $5t = 2(t + 75)$

Distributive property $5t = 2t + 150$

Subtract 2t from both sides $3t = 150$

Divide both sides by 3 $t = 50$ seconds

8. Instead of using Let statements, let's use "six pretty boxes."

	distance	rate	time
Joe		2 mph	t
Darlene		4 mph	t

Using $d = rt$ to fill in the last two boxes . . .

	distance	rate	time
Joe	2t	2 mph	t
Darlene	4t	4 mph	t

Together they went two miles $2t + 4t = 2$

$6t = 2$

$t = 1/3$ hour or 20 minutes

9.   *Let 2x = the number of errors in pronunciation.*
Then x = the number of grammar errors.
Then 5x = the number of cuss words.

Since the total number was 312 $2x + x + 5x = 312$

$8x = 312$

$x = 39$

The number of pronunciation errors (2x) is $2 \cdot 39 = 78$.

10. $x - 12 = 6x + 8$

$-12 = 5x + 8$

$-20 = 5x$

$-4 = x$

11. The formula for the area of a rectangle is Area = (length)(width) or A = ℓw.

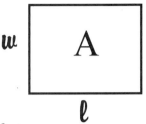

We know that A is 357 square feet and that w is 17 feet.

Putting these numbers into A = ℓw, we have 357 = 17ℓ

Dividing both sides by 17, we have ℓ = 21 feet

12.   *Let x = the first number.*   (I hope this is starting to look familiar by now.)

Then x + 1 = the next consecutive number.

Then x + 2 = the third consecutive number.

Since the total of those three numbers is 1236 . . .

$$x + (x + 1) + (x + 2) = 1236$$
$$3x + 3 = 1236$$
$$3x = 1233$$
$$x = 411$$

The three consecutive numbers (x, x + 1, and x + 2) are 411, 412, and 413.

13. $20 \div (10 \div 2)$

    $20 \div 5$

    4

> First, we do whatever is inside parentheses.
> Second, multiplication and division, working left-to-right.
> Last, addition and subtraction.

14. The number that many students might think of is π. Other possibilities are $\sqrt{2}$ or $\sqrt{3}$. (Not $\sqrt{4}$ since that is equal to 2.)

15. Using the "six pretty boxes" . . .

	distance	rate	time
Betty		7 ft/sec	t
Alexander		3 ft/sec	t

Using d = rt to fill in the last two boxes . . .

	distance	rate	time
Betty	7t	7 ft/sec	t
Alexander	3t	3 ft/sec	t

Since together they went 120 feet

$$7t + 3t = 120$$
$$10t = 120$$
$$t = 12 \text{ seconds}$$

16. Using "six pretty boxes" . . .

	distance	rate	time
mailman			
Darlene			t

	distance	rate	time
mailman			t + 40
Darlene			t

	distance	rate	time
mailman		10 ft/min	t + 40
Darlene		60 ft/min	t

	distance	rate	time
mailman	10(t + 40)	10 ft/min	t + 40
Darlene	60t	60 ft/min	t

Since they travel the same distance

$$10(t + 40) = 60t$$
$$10t + 400 = 60t$$
$$400 = 50t$$
$$8 = t$$

It takes Darlene 8 minutes to catch up with the mailman.

17.
$$5(y - 2) = 8(y + 7)$$
$$5y - 10 = 8y + 56$$
$$-10 = 3y + 56$$
$$-66 = 3y$$
$$-22 = y$$

We know what the trig class was up to. We know what the chancellor of KITTENS and his campus cops were up to. We know what the department secretary Belinda was up to. We know what Betty and Alexander were up to.

Have we forgotten anyone?

Oh, yes. Our hero Fred.

He sat on a wooden bench inside a dark prisoner-transport van. Only one little patch of light came in. He was grateful that they hadn't handcuffed him. *It's not very promising on your sixth birthday to be handcuffed* Fred thought to himself. There was only a little chain around his neck with the other end of the chain attached to the wall of the van.

"Would you look at that window!" Fred exclaimed out loud even though there was no one with him. "That's perfect. Just what I should use in my beginning algebra class."

Now Fred wasn't going to use this paddy wagon window to oppress his students. Fred thought to himself *This fenestration is enlightening* as he giggled at his own little pun. (*Fenestration* refers to how the windows are designed and placed on a building.)

Fred mentally called the height of the window a and the two widths of the window that were separated by the vertical bar b and c.

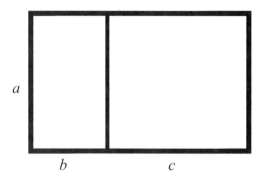

There it is! Isn't that great! he thought as he considered the possibilities for using that diagram in beginning algebra. Unfortunately, some of his students—those with less than a 300 IQ—might not have immediately seen what Fred saw.

In his imagination he asked his students four questions:

① What's the area of the rectangle on the left?   answer: ab
② What's the area of the rectangle on the right?   answer: ac
③ If you take out the vertical bar, what's the length
    of the new large rectangle?   answer: $(b + c)$

④ What's the area of the new large rectangle?   answer: $a(b + c)$

The students still looked at him with blank faces. Fred continued, "The area of the window without the bar is equal to the area of the two smaller original windows."

Seconds passed in silence.

More seconds passed.

Then from the back of the room one student would exclaim, "I get it. You've just proved that $a(b + c)$ equals $ab + ac$. You've proved the distributive law."

So that's how Fred figured that the noon lecture would play itself out, assuming the army figured out that they didn't want this little six-year-old and would send him back to KITTENS.

He stared up at that prison van window and was delighted with his discovery of a proof of the distributive property. He hummed to himself the Happy Birthday song and then started singing quietly "Oh Happy Day" and burst into full voice with "Jerusalem the Golden."

In the front of the van, Sergeant Snow and the driver couldn't figure out what in the world was going on in back. "Another nut case," grunted Sergeant Snow, and he turned up the volume on the country western song playing on the radio:

> ♪ ♫   She ran away and took the dog,
> And left me with the kid.
> I sure do miss my good ol' Blue,
> Now all I got is Sid.   ♪♫

"You know," said the driver, "I sure don't like that there song much that you got a-playing there on the radio."

Sergeant Snow assured the driver, "It's okay, Sidney. That's just how the thing rhymes. They don't mean nothing personal by it."

At 12:15 p.m. Sergeant Snow broke out the doughnuts for their lunch and noticed a car drive up next to them. A man and a woman were waving their hands and shouting. Snow rolled down his window and threw them a doughnut (it was a kind he didn't like) and told the driver to keep on driving.

Snow said, "I'm mighty tired of them protesters."

Betty and Alexander didn't know what to do. They could hear Fred working on the second verse of "Jerusalem the Golden," (written by

Bernard of Cluny in about 1140 A.D.)   He seemed to be in fine spirits as his little voice sang, "They stand, those halls of Zion / All jubilant with song. . . ."

Since they couldn't stop the convoy, the best they could do was to fall in line behind and see what they could do when they got to Lampasas.

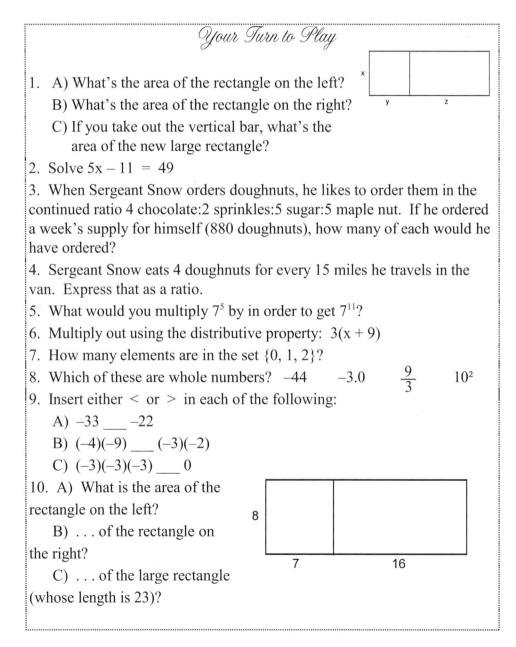

Your Turn to Play

1. A) What's the area of the rectangle on the left?

    B) What's the area of the rectangle on the right?

    C) If you take out the vertical bar, what's the area of the new large rectangle?

2. Solve $5x - 11 = 49$

3. When Sergeant Snow orders doughnuts, he likes to order them in the continued ratio 4 chocolate:2 sprinkles:5 sugar:5 maple nut. If he ordered a week's supply for himself (880 doughnuts), how many of each would he have ordered?

4. Sergeant Snow eats 4 doughnuts for every 15 miles he travels in the van. Express that as a ratio.

5. What would you multiply 7^5 by in order to get 7^{11}?

6. Multiply out using the distributive property: $3(x + 9)$

7. How many elements are in the set $\{0, 1, 2\}$?

8. Which of these are whole numbers? -44 -3.0 $\dfrac{9}{3}$ 10^2

9. Insert either $<$ or $>$ in each of the following:

    A) -33 ___ -22

    B) $(-4)(-9)$ ___ $(-3)(-2)$

    C) $(-3)(-3)(-3)$ ___ 0

10. A) What is the area of the rectangle on the left?

    B) . . . of the rectangle on the right?

    C) . . . of the large rectangle (whose length is 23)?

.......COMPLETE SOLUTIONS.......

1. A) xy

    B) xz

    C) x(y + z)

2.
$$5x - 11 = 49$$

   Add 11 to both sides $\qquad\qquad 5x = 60$

   Divide both sides by 5 $\qquad\qquad x = 12$

3. *Let 4x = the number of chocolate doughnuts ordered.*

   Then 2x = the number of sprinkles doughnuts.

   Then 5x = the number of sugar doughnuts.

   Then 5x = the number of maple doughnuts.

Since the total number of doughnuts is 880, $\quad 4x + 2x + 5x + 5x = 880$

   Combine like terms $\qquad\qquad\qquad\qquad 16x = 880$

   Divide both sides by 16 $\qquad\qquad\qquad x = 55$

He would have ordered (*4x*) 220 chocolate, (*2x*) 110 sprinkles, (*5x*) 275 sugar, and (*5x*) 275 maple nut doughnuts.

4. 4 doughnuts:15 miles

5. $\qquad 7^5(7^?) = 7^{11}$ is a restatement of the question.

   $(7)(7)(7)(7)(7)(7^?) = (7)(7)(7)(7)(7)(7)(7)(7)(7)(7)(7)$

   $\qquad\qquad 7^?$ would have to be $(7)(7)(7)(7)(7)(7)$ or 7^6.

   $\qquad 7^5(7^6) = 7^{11}$

6. $3(x + 9) = 3x + 3(9) = 3x + 27$

7. There are three elements—the numbers zero, one, and two.

8. $\dfrac{9}{3}$ and 10^2, which are better known as 3 and 100

9. A) $-33 < -22$

    B) $(-4)(-9) > (-3)(-2)$ $\qquad$ since $+36 > +6$

    C) $(-3)(-3)(-3) < 0$ $\qquad$ since $-27 < 0$

10. A) 56

    B) 128

    C) $8(7 + 16) = 8(23) = 184$

    Note that $8 \cdot 7 + 8 \cdot 16$ also equals 184.

	56	128
8	7	16

A little while later, Fred overheard the driver say that they had covered half the distance to Lampasas going 45 mph. After uttering a cuss word, Sergeant Snow told the driver, "We gotta get there by five. They're having my favorite beans and hot dogs tonight in the chow hall and I don't want to miss that. Step on it. If you go 65 mph at this point, we'll make it there by five." Snow's vehicle pulled out of the caravan and sped on ahead. Betty and Alexander followed.

It was too dark for Fred to read his watch, but overhearing the conversation in the front, he could tell what time it was now. That's because Fred knows algebra.

First, let t = the amount of time we've been traveling at 45 mph.

Recall: The first step is to assign a letter to the quantity we are trying to find the value of.

Then $45t$ = the distance we've gone thus far.

Since we left KITTENS at 9 a.m. and want to get to the beans and hot dogs by 5 p.m., the whole trip is 8 hours.

Then $8 - t$ = the number of hours left on our journey.

Finding the 8 − t   (important, so I put it in a box)

Stop! "How did we get $8 - t$?" you may ask. It's not really magic. Suppose, for a moment, that the whole trip takes 8 hours and I tell you that we've gone 3 hours already. You would immediately tell me we have 5 hours left to go.

If the whole trip took 8 hours and we had gone 2 hours, you would immediately tell me there were 6 hours left to go.

If I tell you the whole trip takes 8 hours and we've gone t hours, then, using the same process you instinctively used in arithmetic, you would tell me there are $8 - t$ hours left to go.

Here is the trick: if you're not sure how to write it with variables such as t or x, do the problem with some made up numbers and after you've done it, notice what arithmetic operation you used and do the same with the letters.

Then $65(8 - t)$ = number of miles on second part of the trip.

The second part of the trip was at 65 mph and took $8 - t$ hours.

The driver had said that they had covered half the distance already, so the distance for the first part of this trip, which is $45t$, equals the distance for the second half of the trip, which is $65(8 - t)$.

In "six pretty boxes" . . .

	distance	rate	time
first half of the trip	$45t$	45	t
second half	$65(8 - t)$	65	$8 - t$

$$45t = 65(8 - t)$$

Since he didn't have any writing materials, he solved the equation in his head:

$$45t = 520 - 65t \qquad \text{using the distributive property}$$
$$110t = 520 \qquad \text{adding 65t to each side}$$
$$t = \frac{520}{110} = 4\frac{80}{110} = 4\frac{8}{11} \text{ hours.}$$

So it was about 4 hours and 44 minutes* after we started. So it must be about 1:44 p.m. Solitary confinement wasn't a torture for Fred. He had a blackboard in his mind that he could write on, and the hours passed quickly.

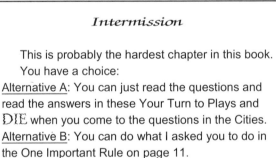

Intermission

This is probably the hardest chapter in this book.
You have a choice:
Alternative A: You can just read the questions and read the answers in these Your Turn to Plays and DIE when you come to the questions in the Cities.
Alternative B: You can do what I asked you to do in the One Important Rule on page 11.

---

* $\frac{8}{11}$ of 60 minutes $= \frac{8}{11} \times \frac{60}{1} = \frac{480}{11} \approx 43.6363 \doteq 44$ minutes

Your Turn to Play

1. The diagram to prove a(b + c) = ab + ac was: a

b   c

What would the diagram look like to prove a(b + c + d) = ab + ac + ad?

2. If it takes t minutes for Darlene to fix her hair and it takes Joe three times as long for him to fix his hair, how long (in terms of t) does it take Joe to fix his hair?

3. If it takes Darlene t minutes to walk from KITTENS to the shopping mall and if Joe can make that walk in four minutes less than Darlene, how long (in terms of t) does it take for Joe to walk from KITTENS to the mall?

4. If Joe and Darlene contribute a total of $16 toward Fred's birthday present and Joe had contributed $x, how much (in terms of x) did Darlene contribute?

5. Darlene and Joe are typing Betty's Ph.D. thesis. For every 4 typing errors that Darlene makes, Joe makes 17. All together they made 147 errors. How many did Darlene make?

6. Joe walks at the rate of 2 mph from KITTENS to the top of a local mountain. (Actually, since we're in Kansas, it's not really a mountain since things are pretty flat there, but Joe liked to think of it as a mountain. Calling it a hill might be more accurate.) He returned at the rate of 4 mph (since it was downhill). The whole trip took three hours. How long did it take him to go from KITTENS to the top of the mountain?

7. On a previous trip, Sergeant Snow ordered his driver to stop at Fort Worth, Texas, so that they could pick up some more doughnuts. "To hold us over till we get to those beans and hot dogs," was the way Snow expressed it. The main convoy continued on while they were in the doughnut store making their selection. An hour later, Snow and his driver got back in their van and continued their trip (at 60 mph). The convoy was going at 45 mph. How long before the doughnut-laden Snow catches up to the convoy?

·······COMPLETE SOLUTIONS·······

1. Instead of two rectangles, we would have three rectangles with widths b, c, and d.

2. $3t$ minutes

3. $t - 4$

4. $\$(16 - x)$

5. *Let 4x = the number of errors that Darlene made.*

   Then 17x = the number of errors that Joe made.

Since they made a total of 147 errors . . .

$$4x + 17x = 147$$
$$21x = 147$$
$$x = 7$$

Darlene made (4x) errors. $4(7) = 28$ errors.

6. *Let t = the number of hours for Joe to go from KITTENS to the top of the mountain.*

   Then 3 - t = the number of hours to go from the mountain back to KITTENS.

   Then 2t = the distance going up the hill.

   Then 4(3 - t) = the distance going back.

These distances are equal.

$$2t = 4(3 - t)$$
$$2t = 12 - 4t$$
$$6t = 12$$
$$t = 2$$

It took Joe t hours to go to the top of the mountain. It took him 2 hours.

7. *Let t = the number of hours till Snow's van catches up to the convoy.*

   Then 60t = the distance Snow drives to overtake the convoy.

   Then t + 1 = the number of hours that the convoy will go until it is overtaken.

   Then 45(t + 1) = the distance the convoy drives until it is overtaken by Snow.

These two distances are equal

$$60t = 45(t + 1)$$
$$60t = 45t + 45$$
$$15t = 45$$
$$t = 3 \text{ hours}$$

T hey cranked the radio a notch higher. Any higher and they would have to stop to buy a new battery for the car.

♪ ♫   She ran away and took the dog,
And left me with the kid.
She came back home and left some puppies
And that is what she did. ♪ ♫

"See, I told yah," commented the driver. "They didn't have ta make it rhyme with Sid."

"It's okay, Sidney. They don't mean nothing personal by it either which way," Snow assured him again.

Highway 35 heads south out of Fort Worth. After they passed through Temple, they turned right on 190 and pulled into Lampasas. (It's all there on any good atlas. You can check for yourself. The fiction in this book is at a bare minimum: just enough to keep things moving along.)

Under a sign that read American Army Amusement Arena, Snow's van pulled in past the guard booth. The two helmeted and armed guards at the booth recognized Sergeant Snow and let him through. He had the driver head directly to the chow hall. They got out and headed toward the beans and hot dogs. Fred sat in his stuffy semi-dark chamber.

Betty and Alexander pulled up to the guard house seconds after Snow had passed through. The guards blocked their entrance and asked for their passes. Betty, without even blinking, shot back to them, "And where do you get these passes?"

One of them pointed, "Right down there at the ticket booth ma'am."

"Ticket booth?" Alexander asked no one in particular as Betty drove in the direction that the guard had indicated. Then he turned to Betty and asked, "And calling you 'ma'am'? You're in your early 20s and unmarried. I don't get it."

Betty couldn't resist. She had to use the line that every faithful Wizard of Oz fan knows by heart, "Honey, I don't think we're in Kansas anymore."

Alexander got out and approached the ticket booth. "Howdy," he said, trying to sound like he was a Texan.

"Howdy back," responded the inhabitant of the booth.

There was a silence. Alexander thought it was obvious why he was standing there in front of the ticket booth. The guy just looked at him. It looked like he could just sit there and look out over the landscape for hours at a time. Maybe, thought Alexander, that's exactly what he does with his days.

Then he realized that the ticket seller's eyes didn't blink and that he was only 18 inches tall. "Ha! Ha! You're only a dummy," said Alexander.

"Ha! Ha! What's that make you?" the dummy responded.

Finally Alexander made eye contact with the guy who was standing behind the dummy ticket seller.

eye contact
achieved

"I want to buy a ticket," Alexander announced.

"To what?"

"To that A.A.A.A. thing over there."

"What part of A.A.A.A. do you want to see?"

Alexander was tempted to start to tell the whole story of how Prof. Gauss had been wrongfully taken by the army, etc., but decided to keep things simple. He asked, "What's it take to get in A.A.A.A. and see the boss there?"

"The best deal is the See-the-Boss Package. In the package, there are some general admission tickets to the American Army Amusement Arena. Those are 50¢ each. Then there are some special get-into-the-officers'-part-of-camp tickets. Those are 75¢ each. Finally, there are the see-the-colonel tickets. Those are 5¢ each. There's the same number of each kind of ticket. The whole package is $9.10."

Alexander quickly figured out that he was buying seven tickets of each kind. (Please take out a piece of paper and see if you can also figure it out. Alexander's work is in the box.) He told the ticket seller that there were only two of them and asked to buy the individual tickets. The dummy said that the tickets are only sold in the See-the-Boss Package.

Alexander forked over $9.10 and received the package of tickets.

> Let x = the number of general admission tickets.
>
> Then x = the number of get-into-the-officers'-part-of-the-camp tickets.
>
> Then x = the number of see-the-colonel tickets.
>
> Then 50x = the value of the general admission tickets. (That's x tickets at 50¢ each.)
>
> Then 75x = the value of the get-into-the-officers'-part-of-the-camp tickets.
>
> Then 5x = the value of the see-the-colonel tickets.
>
> We are given that the value of all the tickets is 910¢. (If we work just in pennies, we don't have to deal with decimals.)
>
> $$50x + 75x + 5x = 910$$
> $$130x = 910$$
> $$x = 7$$

Your Turn to Play

1. You buy *n* copies of Jeanette MacDonald's latest CD (to give to your *n* closest friends). You also buy *n* copies of Gene Audrey's latest CD and *n* copies of Judy Garland's latest CD. Then each friend will receive three wonderful music treats, which they will treasure for a lifetime.

The Jeanette MacDonald CDs cost $8 each.

The Gene Audrey, $7.

The Judy Garland, $5.

The bill came to $680. How many friends will you give these CDs to?

2. The local cleaners charges:

$1 for a shirt

$2 for a suit

$2 for a dress

The Cleaners

You bring in an equal number of shirts, suits, and dresses. The bill is $35. How many of each did you bring in?

3. Your piggy bank has the same number of pennies, nickels, dimes, and quarters. They are worth $9.43. How many of each coin do you have?

4. $(-2)^5$

5. How much would *n* 43¢ stamps cost?

6. Solve $9y = 5y + 36$

.......COMPLETE SOLUTIONS.......

1. Let n = the number of friends who will receive the CDs.

> If one Jeanette MacDonald CD cost $8, then n of them cost $8n$.
>
> If one Gene Audrey CD cost $7, then n of them cost $7n$.
>
> If one Judy Garland CD cost $5, then n of them cost $5n$.

The cost of all the CDs is $8n + 7n + 5n$.

The cost of all the CDs was $680.

$$8n + 7n + 5n = 680$$
$$20n = 680$$
$$n = 34$$

You have 34 closest friends.

2. Let x = the number of shirts you brought in. Then x is also the number of suits you brought in and x is also the number of dresses you brought in.

> The cost of the cleaning of the shirts was $x.
>
> The cost of the cleaning of the suits was $2x.
>
> The cost of the cleaning of the dresses was $2x.

The total cost of the cleaning was x + 2x + 2x.

The total cost of the cleaning was $35.

$$x + 2x + 2x = 35$$
$$5x = 35$$
$$x = 7$$

You brought in seven shirts, seven suits, and seven dresses.

3. Let y = the number of pennies in the bank. Then y is also the number of nickels, dimes, and quarters.

The total value of the coins is $1y + 5y + 10y + 25y$ cents.

The total value of the coins is 943¢.

$$1y + 5y + 10y + 25y = 943$$
$$41y = 943$$
$$y = 23$$

You have 23 pennies, 23 nickels, 23 dimes, and 23 quarters.

4. $(-2)^5 = (-2)(-2)(-2)(-2)(-2)$
    $$= +4\,(-2)(-2)(-2)$$
    $$= \;-8\,(-2)(-2)$$
    $$= \quad +16\,(-2)$$
    $$= \qquad -32$$

5. If I had seven 43¢ stamps, they would cost 7(43)¢.

    If I had eight 43¢ stamps, they would cost 8(43)¢.

    I have *n* 43¢ stamps, and they cost 43*n*¢.

6. $9y = 5y + 36$

    $4y = 36$ Subtract 5y from both sides

    $y = 9$ Divide both sides by 4

B ack at the guard booth was a sign.

Betty couldn't believe it. "An army base and they close it at night!"

Alexander consoled her with the thought that even if they had gotten in, seeing the colonel at this time of day might not be possible.

They checked into the campground, which turned out to be an old converted drive-in movie theater. For $55 they were allowed to park in one of the spaces for the night.

They were getting hungry. The old candy concession building at the drive-in had been replaced with a King of French Fries franchise. Their speciality was the Giant French Fry made from a single potato, deep fried in the best beef fat available. The sign at the counter with their motto, "One GFF from KFF and you'll be full" had been edited by a vandal. He had crossed out "full" and replaced it with dead.

Alexander looked over the rest of the menu. The Giant French Fry ($8) was out of the question.

The small fries were sold separately for 60¢ each.

The tiny fries were 45¢ each.

Each was dipped in butter and rolled in sugar to make an extra special treat.

Alexander bought some of the small fries and took them back to the car where Betty was getting blankets out of the trunk to shield them

against the January Texas night. After they snuggled in and Alexander offered her the small fries, Betty rolled down the window and held one of the small fries out the window. She squeezed it to get rid of some of the oil.

Alexander commented, "If I had bought the

tiny fries instead of the small fries, I would have gotten one more for the same total price."

Betty figured this in her head. She loved the challenge. She let n = the number of small fries that he had purchased. Then the value of those small fries was $60n$ since they each cost 60¢.

If he had purchased tiny fries instead of small fries he would have gotten one more, so $n + 1$ is the number of tiny fries and the value of those tiny fries would be $45(n + 1)$ since each tiny fry cost 45¢.

Since the values of the small fries and tiny fries are equal,

$$60n = 45(n + 1)$$
$$60n = 45n + 45$$
$$15n = 45$$
$$n = 3$$

Alexander had bought three small fries and could have bought four tiny fries for the same price.

Wait! Did you notice something? Bells should be going off in your head right now. Lights should be lighting. This should seem somehow very familiar. About eight pages ago (in problem 7), Sergeant Snow, traveling 60 mph, was trying to catch up with the convoy, going 45 mph, that had a one-hour head start.

In the solution, we arrived at the equation $60t = 45(t + 1)$. Here's a side-by-side comparison:

60 miles per hour	60¢ per small fry
for t hours	bought n of them
45 miles per hour	45¢ per tiny fry
for one more hour	buys one more of them
miles are equal	costs are equal
$60t = 45(t + 1)$	$60n = 45(n + 1)$

Good News and Bad News

The Good News is that there aren't *that* many different things to learn in beginning algebra. Once you've solved $60t = 45(t + 1)$, it isn't that much more difficult to tackle $60n = 45(n + 1)$. The problems may come in different guises, but underneath there are many similarities. Once you learned the distributive law, $45(t + 1)$ and $45(n + 1)$ both are done with the same ease. The powerful thing about algebra is that these fundamental techniques that you're learning have so many different applications. You're handed one hammer and the world turns out to be full of nail opportunities.

The Bad News is that it takes work to learn algebra. That four-letter word strikes terror in the hearts of many: **WORK**. If you are reading *Life of Fred* in order to learn algebra (instead of just for the story), then it will take time, effort, pain, and patience.

My daughters own cats. And none of them (the cats) do any work. They sit around all day eating, napping, and watching TV. You can't get a lick of work out of a cat. The bottom line is that none of my daughters' cats will ever *achieve* anything. There are people who try to live like cats. Your choice is whether you want to be one of those people.

As Friday comes to a close, Sergeant Snow is finishing up his beans and hot dogs. Betty and Alexander are heading to the bathroom to wipe the grease off their hands.

And Fred?

Happy Birthday little guy.

Before Saturday dawns, please get out a pencil and paper and work through these questions:

Your Turn to Play

1. Outside the drive-in theater bathroom Betty noticed a stamp machine. She figured that she'd better let everyone back at KITTENS know what was happening. It offered a packet of stamps containing 34¢, 21¢, and 5¢ stamps. There were twice as many 21¢ stamps as there were 34¢ stamps, and there was one more 5¢ stamp than there were 21¢ stamps. The whole packet cost $5.21. How many 34¢ stamps were there in each packet?

2. Using set builder notation, describe the set of all people in this book.

3. If Sergeant Snow eats four pounds of his favorite beans and hot dogs and it's 27% fat by weight, how much fat has he eaten?

4. The doctor has advised Sergeant Snow to cut down a little bit on the fat intake. Looking at Snow's height, weight, and cholesterol count, Snow shouldn't eat more than 2.16 pounds of fat per day. One day he ate only candy bars, Waddle doughnuts, and fried rice. For every pound of candy bars, he had two pounds of doughnuts, and he had one more pound of rice than he had of doughnuts. The amount of fat in these three items as a percentage of their total weights is 14%, 27%, and 16%, respectively. How many pounds of candy bars did he have, assuming he stayed on the doctor's diet? (Hint: Let x = the pounds of candy bars Snow ate.)

5. The doctor should have been looking at Sergeant Snow's alcohol intake. Last week Snow drank several quarts of Chipper Suds (which is 5% alcohol by volume) and a lot more of his favorite Army Ale (which is 6% alcohol by volume). He had 100 quarts more of the Army Ale than he did of the Chipper Suds. The total amount of (pure) alcohol he consumed last week was 11.94 quarts. How many quarts of Chipper Suds did he have?

6. Because of an unsurprising interaction between grease, alcohol, and his brain, Sergeant Snow was found unconscious in an alley behind Waddle Doughnuts. After they hauled him into Army General Hospital of Texas, they did a fat scan and found that the continued ratio of the fat in his body to the lean muscle to the bone was 30:20:5. How much of his 385 pounds were fat? (Let 30x = pounds of fat in his body.)

7. **[This is the classic mixture problem.** For many students, mixture problems are the hardest beginning algebra word problems. They really aren't that hard once you have done a couple of them. If you like, tear a one-inch long piece of scratch paper and make a bookmark for this page. Write on the top of the bookmark: ᴍɪxᴛᴜʀᴇ.]

Snow wanted 24 pounds of pancake flour that was 50% sugar. The cook had some that was 30% sugar and some that was 60% sugar. How much low-sugar flour should he use in the mixture?

8. [A second mixture problem]

Snow stopped breathing, and the nurse indicated to the doctor that there might be something wrong. The doctor concurred and ordered that Snow be given some medicine. What he said to the nurse was, "30 cc of 25% cough medicine," figuring that if Snow would cough, he would have to start breathing again.

The nurse ran to the cabinet and found two bottles. One of them was too weak. It only contained 20% of the cough medicine by volume. The other bottle was too strong. It was 35% of the cough medicine by volume. The nurse had to mix the two together to get 30 cc of 25% medicine. (A "cc" is a cubic centimeter. That's about the volume of a small sugar cube.) How much of each were used? (Let x = the number of cc of the 20% medicine that is used.)

bookmark

........COMPLETE SOLUTIONS........

1. *Let n = the number of 34¢ stamps.*   ←⋘ That's what you want to find out.

   Then 2n = the number of 21¢ stamps.

   Then 2n + 1 = number of the 5¢ stamps.

   Then 34n is the value of the 34¢ stamps.

   Then 21(2n) is the value of the 21¢ stamps.

   Then 5(2n + 1) is the value of the 5¢ stamps.

   Adding these all up, we get the value of the whole packet, which is 521¢

$$34n + 21(2n) + 5(2n + 1) = 521$$
$$34n + 42n + 10n + 5 = 521$$
$$86n + 5 = 521$$
$$86n = 516$$
$$n = 6$$

   There were six 34¢ stamps in the packet.

2. $\{x \mid x \text{ is a person in this book}\}$

3. 27% of 4 pounds is the same as (0.27)(4), which is 1.08 pounds of fat.

4. *Let x = the number of pounds of candy bars he ate.*

   Then 2x = the number of pounds of Waddle doughnuts he ate.

   The 2x + 1 = the number of pounds of rice he ate.

   The amount of fat in each of these three items is 0.14x, (0.27)(2x), and (0.16)(2x + 1). This has to equal 2.16 pounds.

$$0.14x + (0.27)(2x) + (0.16)(2x + 1) = 2.16$$
$$0.14x + 0.54x + 0.32x + 0.16 = 2.16$$
$$x + 0.16 = 2.16$$
$$x = 2$$

   He ate two pounds of candy bars.

5. *Let y = the number of quarts of Chipper Suds he drank.*

   Then y + 100 is the number of quarts of Army Ale he drank.

   Then the alcohol in the Chipper Suds was 0.05y.

   Then the alcohol in the Army Ale was 0.06(y + 100).

   The total amount of alcohol was 11.94.

$$0.05y + 0.06(y + 100) = 11.94$$

Distributive property	$0.05y + 0.06y + 6 = 11.94$
Combine like terms	$0.11y + 6 = 11.94$
Subtract 6 from both sides	$0.11y = 5.94$
Divide both sides by 0.11	$y = \dfrac{5.94}{0.11}$
Arithmetic	$y = 54$ quarts of Chipper Suds*

6. *Let 30x = the pounds of fat.*

   Then 20x = the pounds of lean muscle.

   Then 5x = the pounds of bone.

Since the total fat plus lean muscle plus bone is equal to 385 . . .

$$30x + 20x + 5x = 385$$
$$55x = 385$$
$$x = 7$$

Then the weight of his fat (30x) is 30(7) = 210 pounds.

7. *Let x = the pounds of the low-sugar (30%) pancake flour used.*

   Then 24 – x = the pounds of the high-sugar (60%) pancake flour used.

The trick with mixture problems is to concentrate on the ingredient in the mixture. In this case it is sugar.

   Then the low-sugar flour will contribute 0.3x pounds of sugar. (30% of x)

   Then the high-sugar flour will contribute 0.6(24 – x) pounds of sugar.

How much sugar do we want in the final mixture? 12 pounds of sugar
(We want 24 pounds of flour that is 50% sugar.)

   The amount of sugar contributed by the low-sugar flour plus the amount of sugar contributed by the high-sugar flour must equal the amount of sugar in the final mixture.

$0.3x + 0.6(24 - x) = 12$	
$0.3x + 14.4 - 0.6x = 12$	distributive property
$-0.3x + 14.4 = 12$	$0.3x - 0.6x$ equals $-0.3x$
$2.4 = 0.3\,x$	add 0.3x and subtract 12 from each side
$8 = x$	divide both sides by 0.3

The cook should use 8 pounds of low-sugar flour.

---

* Alcohol in very large quantities can act as a poison. It's a stupid way to die.

8. *Let x = the number of cc of the 20% medicine that is used.*

    Then 30 – x = the number of cc of the 35% medicine that is used. The nurse has to make a total of 30 cc.

    Then 0.20x is the amount of medicine from the 20% bottle.

    Then 0.35(30 – x) is the amount of medicine from the 35% bottle.

The total amount of medicine used is supposed to be 25% of 30. That's 7.5 cc.

$$0.20x + 0.35(30 - x) = 7.5$$
$$0.20x + 10.5 - 0.35x = 7.5$$
$$-0.15x + 10.5 = 7.5$$
$$-0.15x = -3$$
$$x = \frac{-3}{-0.15}$$

Using pencil and paper or a calculator and the fact that a negative divided by a negative is positive   x = 20 cc of the 20% medicine

The amount of the 35% medicine (30 – x) is 10 cc of the 35% medicine.

The nurse (whose parents only had sons) mixed up the shot and gave Sergeant Snow the injection. Snow began coughing.

Let's do this previous problem using "six pretty boxes."

The ingredient in this case is the cough medicine.

	the ingredient	concentration	amount used
weak mixture			x
strong mixture			30 – x

	the ingredient	concentration	amount used
weak mixture		20%	x
strong mixture		35%	30 – x

	the ingredient	concentration	amount used
weak mixture	0.20x	20%	x
strong mixture	0.35(30 – x)	35%	30 – x

Since we want 7.5 cc of ingredient (30 cc of 25%) . . .

$$0.20x + 0.35(30 - x) = 7.5$$

The sun came up early over A.A.A.A. The rooster had been crowing for about an hour when the loudspeaker announced that it was time to "rise and shine." Over the loudspeakers (like they do at some youth camps), they played popular songs:

♪ ♫   She ran away and took the dog,
And left me with the kid.
She wrote me, "Sold the dog for food.
Could sell for more our Sid."   ♪ ♫

Betty and Alexander stretched and folded up the blankets and headed toward the main gate of A.A.A.A. The guards took the seven general admission tickets, the seven special get-into-the-officers'-part-of-camp tickets, and the seven see-the-colonel tickets that Alexander had proffered . . . and threw them in the garbage.

"Those tickets are expired. They were dated yesterday," the guard announced to the stricken couple. Neither anger nor tears were of any use at this point, and they headed back to the ticket booth.

Sergeant Snow was waking up. He scratched his tummy, climbed into his fatigues, and headed to the chow hall for some coffee, bacon, and eggs. His head hurt, but that was nothing unusual.

Alexander shelled out another $9.10 to the dummy for some new tickets and they headed back to the guard station. After a couple of hours of delay (finger printing, blood testing, oath of allegiance), they arrived at the colonel's door.

Alexander and Betty looked at each other. He squeezed her hand and then knocked.

The door opened in front of Alexander and a man with a helmet and gun asked, "Yes?"

"Colonel Coalback?" Alexander began.

"No," the guard replied. The guard took their see-the-colonel tickets and ushered them in. The room was all decorated in emerald green, and in front of them, at the end of a large marble hallway, were puffs of smoke. A thundering loudspeaker voice uttered, "I am Coalback. The mighty Coalback. You clinking couple of cringing calla lilies."

Betty giggled, "It's just like the Wizard of Oz movie. This is the hall of the mighty wizard. No wonder they only charge five cents to see this."

"Silence!" bellowed the voice behind the smoke. "Come forward you callipygian kid from Kansas to confer with Colonel Coalback." The colonel was apparently in love with *alliteration*—repeating the same sound at the beginning of several words in a series.

"He's referring to me," Betty whispered to Alexander as she stepped forward with a smile on her face.

"I am called 'The Old One' here at A.A.A.A.," the voice bragged. "I am 20 years older than this camp. In 14 years from now, the headlines of the camp newspaper will proclaim that I'm twice as old as this camp."

"So what's the big deal? So you're 26," Betty retorted. "We used to do those age problems in beginning algebra."

Betty was feeling really confident. The loudspeakers and puffs of smoke hadn't scared her. His use of fancy words hadn't intimidated her. She read a lot and had a large vocabulary. She had a good dictionary beside the chair she usually did her reading in and would look up words that she didn't know. And when Colonel Coalback threw beginning algebra at her, she almost laughed out loud. After all, she was a graduate student in math and had taught several of Fred's classes over the years when Fred couldn't make it to class.

Almost on automatic pilot, she continued, "Yeah. Let x equal the age of this camp now. And since you're 20 years older than this camp, your age is x + 20. Big deal.

"In 14 years, the camp will be x + 14 and you'll be x + 20 + 14 or x + 34.

"And you tell me that at that time you'll be twice as old as the camp, so if I double the camp's age it'll equal your age:

$$2(x + 14) = x + 34.$$

"The rest is trivial. $2x + 28 = x + 34$ by the distributive property. Then I subtract x from each side and get x + 28 = 34. Subtract 28 from each side and we have x = 6. So your age, which is x + 20, is 26.

"So let's cease all the smoke and scary stuff," she said, using a bit of alliteration herself, "and tell us where you're holding Fred." She walked back to stand next to Alexander.

Coalback was stunned. If the smoke and the fancy English didn't work, the algebra almost always seemed to work. As a last resort, he used the most primitive method of the animal world. He used force.

Six helmeted and armed guards came in and took Alexander and Betty to the front gate of the camp and threw them out. They instructed the guards at the gate to never let them in again.

Sergeant Snow sat as close as he could to his breakfast, his tummy touching the edge of the table. He looked at the black coffee and the fried eggs lying in their pool of bacon fat. The coffee was several days old. It always was. He thought to himself that the cooks must keep it in the back room somewhere for days after they make it so that it will develop the right amount of staleness. The fried eggs were two days fresher than the coffee. The cook had commented to Snow that if he had "rushed things and served this stuff yesterday," the coffee would have been three times older than the eggs.

This all meant nothing to Sergeant Snow as he slurped his coffee and scarfed down his chicken fruit, but if algebra students from KITTENS had been listening, they might have written down:

Let x = age of coffee.
Let x − 2 = age of eggs. (since they're two days fresher)
The age of the coffee yesterday is x − 1.
The age of the eggs yesterday is x − 2 − 1, which is x − 3.

Yesterday the coffee was three times older than the eggs. So if we triple the smaller number, we'll get the larger.

$$3(x - 3) = x - 1$$
$$3x - 9 = x - 1$$
$$2x - 9 = -1$$
$$2x = 8$$
$$x = 4 \quad \text{The coffee is four days old.}$$

That never bothered Sergeant Snow, because he always put in four spoonfuls of sugar to "help things along."

Your Turn to Play

1. If I am x years old now, how old will I be in 37 years?

2. If I am y years old right now, how old will I be when I'm three times my current age?

3. If I am z years old right now, how old was I five years ago?

4. Joe is one year younger than Darlene. Eighteen years ago she was twice as old as he was. How old are they now?

Intermission

Sometimes algebra students say that they don't know where to start when faced with a word problem.

For some people, it helps to draw some pictures or diagrams in order to visualize the situation. We'll do that in the problem below.

---

5. This semester 60% of the freshmen and 70% of the sophomores at KITTENS University are taking a class from Fred. There are 40 fewer sophomores than freshmen at the university. There are a total of 518 freshmen and sophomores enrolled in Fred's classes. How many freshmen are at the university?

.......COMPLETE SOLUTIONS.......

1. $x + 37$ 2. $3y$ 3. $z - 5$

4. *Let x = Darlene's age now.*

   Since Joe is one year younger than Darlene, he is x - 1 years old.

   Eighteen years ago, Darlene was x - 18.

   Eighteen years ago, Joe was x - 1 - 18, which is x - 19.

Eighteen years ago, Darlene $(x - 18)$ was twice as old as Joe $(x - 19)$.

$x - 18$ is twice as large as $x - 19$.

$$x - 18 = 2(x - 19)$$ Notice that the *is* is changed to "=".

$$x - 18 = 2x - 38$$ Distributive property

$$-18 = x - 38$$ Subtract x from both sides

$$20 = x$$ Add $+38$ to both sides

Darlene is 20 years old.

Joe is $x - 1$ which is 19 years old.

Intermission

Now let's check to see if we got the correct answers. This is a good thing to do with word problems. The original problem said that Joe is one year younger than Darlene. Since we have Darlene's age as 20 and Joe's as 19, this checks. Next, the problem said that eighteen years ago, she was twice his age. Eighteen years ago she would have been two years old and he would have been one. She would have been twice his age.

---

5. *Let x = the number of freshmen at the university.*

   Then x - 40 = the number of sophomores.

   Then number of freshmen taking a class from Fred is 0.6x.

   Then number of sophomores taking a class from Fred is 0.7(x - 40).

Freshmen plus sophomores taking a class from Fred is 518.

$$0.6x + 0.7(x - 40) = 518$$

$$0.6x + 0.7x - 28 = 518$$ Using the distributive property

$$1.3x - 28 = 518$$ Combine like terms

$$1.3x = 546$$ Adding 28 to both sides

$$x = 420$$ Dividing both sides by 1.3

There are fewer math problems in *Life of Fred: Beginning Algebra Expanded Edition* than in most beginning algebra books. In this first city for example, there are only a dozen problems. That's not bad for a day's work.

Do not skip any of the problems. Do not skip any of the cities. Later you can take your vacations.

Larned, Kansas   named for Colonel Benjamin F. Larned

1. It takes 175 seconds for Sergeant Snow to walk from his room to the Waddles Doughnut franchise store located on the base and back to his room. Going there, he walks at 3 ft/sec, and on the way back he zooms along at 4 ft/sec since he's so delighted about carrying eight dozen doughnuts. How long did it take him to walk to the store? (Start by letting t = the time he took to walk to the store. This problem is almost identical to Lesson 26, problem 6.)

2. Alexander and Joe were going to have a speed cooking contest. Since there was only one pan in the student recreation kitchen, Joe said that he would use the pan to make flapjacks and Alexander could use the waffle iron to make waffles. Joe can make 60 pancakes/hour and Alexander 45 waffles/hour. Since waffles take longer, Joe gave Alexander an hour's head start. How long will it take Joe to catch up with Alexander and make as many flapjacks as Alexander makes waffles?

3. On Snow's dinner plate were beans and hot dogs. The beans were one calorie each, and the hot dogs were 80 calories apiece. He ate 1192 more beans than hot dogs, for a total intake of 1840 calories. How many hot dogs did he eat?

4. Solve: $8n + 6(2n) + 5 = 55$

5. Is the number of elements in the set { x | x is a soldier living at A.A.A.A. right now} a rational number?

6. Betty and Alexander are the same age. In fact, they were born on the same date. Betty announced that if she were to get younger at the same rate that Alexander were getting older, in eleven years, he would be three times her age. Alexander didn't like that idea for several reasons. How old are they now?

163

7. Alexander and Betty walk at 4 mph to their car after being ejected from A.A.A.A. It takes them almost an hour, because their car had been towed to the south parking lot by the A.A.A.A. traffic squad. In their car, they travel (at 16 mph) the same distance as they had walked. All together, their walking and traveling by car takes exactly an hour. How long did they walk?

8. Sergeant Snow takes an unattended army jeep and heads out to a local bar at 73 mph. He zooms past a waiting motorcycle cop who notes Snow's speed on his radar gun, puts his radar gun away, starts his engine, and goes after Snow one minute (1/60 hour) after Snow has passed him. With his colored lights on and his siren screaming, the cop chases him at 69 mph. How long before the cop catches up with Snow?

9. Sergeant Snow bought several large french fries at King of French Fries. They were 140¢ each. Sidney bought two more fries than Snow, but his french fries were the medium ones, which only cost 90¢ each. Together they spent $11. How many medium fries did Sidney buy?

10. Solve: $2x + 5(3x) + 17 = 21$

11. A student was asked to name a rational number that is *not* an integer. That student answered $\frac{30}{5}$ Was that answer correct?

12. Col. Coalback is 24 years older than his robot maid. In 4 years he will be five times as old as his robot maid will be. How old is the maid?

answers

1. Since we did problem 6 in Lesson 26 using Let x = statements, let's do this one using "six pretty boxes."

	distance	rate	time
to the store			t
back to his room			

Since it took him a total of 175 for the whole round trip . . .

	distance	rate	time
to the store			t
back to his room			175 − t

	distance	rate	time
to the store		3 ft/sec	t
back to his room		4 ft/sec	175 − t

Using d = rt . . .

	distance	rate	time
to the store	3t	3 ft/sec	t
back to his room	4(175 − t)	4 ft/sec	175 − t

Since the distance to the store equals the distance back to his room . . .

$$3t = 4(175 - t)$$
$$3t = 700 - 4t$$
$$7t = 700$$
$$t = 100 \text{ seconds to walk to the store}$$

2.   *Let t = the time in hours for Joe to catch up with Alexander.*

Then t + 1 = the time Alexander spent making waffles.   (He had an hour head start.)

Then 60t = the number of flapjacks Joe made.

Then 45(t + 1) = the number of waffles Alexander made.

Since they made the same number . . .

$$60t = 45(t + 1)$$
$$60t = 45t + 45$$
$$15t = 45$$
$$t = 3 \text{ hours}$$

3. Using "six pretty boxes" . . .

	total calories	calories/item	number of items
beans		1	x + 1192
hot dogs		80	x

	total calories	calories/item	number of items
beans	x + 1192	1	x + 1192
hot dogs	80x	80	x

Since we are told that the total number of calories eaten was 1840 . . .

$$x + 1192 + 80x = 1840$$
$$81x + 1192 = 1840$$
$$81x = 648$$
$$x = 8 \text{ hot dogs consumed}$$

4.
$$8n + 6(2n) + 5 = 55$$
$$8n + 12n + 5 = 55$$
$$20n + 5 = 55$$
$$20n = 50$$
$$n = \frac{50}{20} \quad \text{which is } 2\tfrac{1}{2} \text{ or } 2.5$$

5. The number of elements in {x | x is a soldier living at A.A.A.A. right now} must be a whole number (which is {0, 1, 2, 3, . . . }), and since every whole number is a rational number, the answer is yes.

6. *Let x = their current age.*

Then x – 11 = Betty's age eleven years from now.

Then x + 11 = Alexander's age eleven years from now.

His age 11 years from now is three times Betty's age 11 years from now.

$$x + 11 = 3(x - 11)$$

Look how nicely the English translates to the equation!

Distributive property	$x + 11 = 3x - 33$
Subtract x from both sides	$11 = 2x - 33$
Add 33 to both sides	$44 = 2x$
Divide both sides by 2	$22 = x$ They are both 22 right now.

7. I can choose either Let x = statements or "six pretty boxes." Since I used Let x = statements for the previous problem, I use "six pretty boxes" for this one.

	distance	rate	time
walking			t
driving			

Since the total time was one hour and they spent t walking, they must have spent the rest (1 – t) driving.

	distance	rate	time
walking			t
driving			1 – t

Filling in the rates . . .

	distance	rate	time
walking		4 mph	t
driving		16 mph	1 – t

Using d = rt . . .

	distance	rate	time
walking	4t	4 mph	t
driving	16(1 – t)	16 mph	1 – t

Since they walked the same distance as they drove . . .

$$4t = 16(1 - t)$$
$$4t = 16 - 16t$$
$$20t = 16$$
$$t = \frac{16}{20} \text{ hours } \left(= \frac{4}{5} \text{ hours} = 48 \text{ minutes}\right)$$

8. Vigilant* readers laugh when they read this problem. Snow is going faster than the cop. Unless Snow runs out of gas or the cop speeds up a bit, you'll finish this book and advanced algebra, geometry, trig, and calculus first. You will get married, have children, get old, and die and the cop will still be chasing Snow.

Let's pretend you didn't notice that silliness of mine. We will do the algebra and see what happens.

Let t = the hours that the cop chases Snow.
Then t + $\frac{1}{60}$ = the hours that Snow drives.

---

* VIJ-eh-lent   Vigilant = alert and ever awake

Then 69t = the distance the cop drives until he catches up with Snow.

Then $73\left(t + \frac{1}{60}\right)$ = the distance Snow drives.

Since they will drive the same distance . . . $69t = 73\left(t + \frac{1}{60}\right)$

It looks fine so far.

Distributive property $69t = 73t + \dfrac{73}{60}$

Subtract 69t and subtract $\dfrac{73}{60}$ from both sides $-\dfrac{73}{60} = 4t$

Divide both sides by 4 $-\dfrac{73}{240} = t$

Wait a minute! The cop would have to travel a negative amount of time to catch up with Snow. The algebra is screaming: This is nuts!

9.   Let x = number of fries Sidney bought.

Then x – 2 = number of fries Snow bought.

Then 90x = cost (in cents) of Sidney's fries.

Then 140(x – 2) = cost (in cents) of Snow's fries.

They spent a total of 1100¢ . . . $90x + 140(x - 2) = 1100$

Distributive property $90x + 140x - 280 = 1100$

Combine like terms $230x - 280 = 1100$

Add 280 to both sides $230x = 1380$

Divide both sides by 230 $x = 6$

Sidney had 6 medium fries.

10. $x = \dfrac{4}{17}$

11. $\dfrac{30}{5}$ is the same as 6 and that is an integer, so the student's answer was not correct.

12.   *Let x = the age of the maid today.*

Then x + 24 = the age of Coalback today.

Then x + 4 = age of the maid four years from now.

Then x + 28 = age of Coalback four years from now.

In four years he will be five times older than the robot maid.

$$x + 28 = 5(x + 4)$$
$$x + 28 = 5x + 20$$
$$8 = 4x$$
$$2 = x$$

The robot maid is two years old.

Dugger, Indiana home to the Dugger Coal Museum

1. Joe makes four times as many faults in tennis as Darlene. In one set, they made a total of 60 faults. How many did Joe make?

2. Inside the army prison van, Fred started counting the rivets on the wall (at 22 rivets/minute) and then switched after a while to counting them on the ceiling. They were harder to count on the ceiling since the chain around his neck made it harder to look up, so he could only count the ceiling rivets at 18/minute. After counting a total of four minutes, he stopped. He had counted the same number of wall rivets as ceiling rivets. How long had he counted wall rivets?

3. Solve: $3w + 7(4w) - 20 = 12$

4. The quartermaster at A.A.A.A. bought a bunch of helmets at $60 each. (The old ones were wearing out.) He bought ten more pistols than helmets (since some of the men liked to carry two pistols). The pistols cost $200 each. The total bill was $9800. How many helmets did he buy?

5. [a mixture problem] Many people suspect that there are coffee grounds in the coffee that's served at A.A.A.A. What they don't know is that army requirements call for 55 coffee grounds per quart of coffee. The cooks have one barrel of coffee with 67.5 grounds per quart in it and another barrel with 45 grounds per quart. How many quarts of each should they mix together to make up 18 quarts?

6. I am 5 years older than my pet cat who's now 10. When will I be twice as old as the cat?

7. Joe and Darlene hired a fellow to chase the tennis balls that went over the fence when they played. During one tennis set, Joe and Darlene each ran an equal distance. The fellow who retrieved the balls ran six times as far as Darlene ran. Together all three ran a total of ten miles. How far did Joe run?

8. Sergeant Snow and Sidney each drove to Fort Worth. Sidney was in an army jeep that could do 60 mph. Snow was in a Lincoln Continental (which he had "borrowed" from Colonel Coalback), that could do 120 mph. Driving at their maximum speeds they spent a combined total of six hours on the road. How long did Snow drive?

9. The librarian at A.A.A.A. took a survey of which magazines were the favorites of the men at the camp. They liked American Mercury

169

(subscription price of $18) and Look magazine (subscription price $22). The librarian ordered three more subscriptions of Look than of American Mercury. The total bill came to $306. How many American Mercury subscriptions did he order?

10. Solve: $5y + 6(6y) - 100 = -80$

11. [a mixture problem] Army regulations require that their pickle relish use 10 pickles per pound of relish. In one giant bowl, the cooks have relish that only uses 8 pickles per pound (too weak). In another bowl, they have relish that has 16 pickles per pound (too strong). They need to make up 200 pounds of relish. How many pounds of pickle relish from each bowl should they use?

12. You're four years older than your 18-year-old sister. How many years ago were you three times her age?

answers

1.   *Let 4x = the number of faults Joe made.*

       Then x = the number of faults Darlene made.

Since the total number of faults was 60 . . .

$$4x + x = 60$$
$$5x = 60$$
$$x = 12$$

Joe made 4x faults. He made 48 faults.

2.   *Let x = the number of minutes he counted wall rivets.* You almost always start by letting x equal the thing you are trying to find out.

       Then 4 – x = the number of minutes he counted ceiling rivets. Look at the box on the first page of Lesson 26.

       Then 22x = the number of wall rivets he counted. He counted at the rate of 22 rivets per minute for x minutes.

       Then 18(4 – x) = the number of ceiling rivets he counted. He counted at the rate of 18 rivets per minute for 4 – x minutes.

Since the number of wall rivets counted equals the number of ceiling rivets he counted . . .

$$22x = 18(4 - x)$$
$$22x = 72 - 18x$$
$$40x = 72$$
$$x = \frac{72}{40} \text{ minutes, which is 1.8 minutes}$$

0.8 minute $\times \dfrac{60 \text{ seconds}}{1 \text{ minute}} = 48$ seconds. He counted for 1 minute and 48 seconds. Conversion factors were taught in Chapter 2 of Life of Fred: Pre-Algebra 2 with Economics.

On the next page, this problem will be redone using "six pretty boxes."

	rivets counted	rate	time
wall rivets			x
ceiling rivets			

Using the box on the first page of Lesson 26 . . .

	rivets counted	rate	time
wall rivets			x
ceiling rivets			4 − x

The rates were given in the problem.

	rivets counted	rate	time
wall rivets		22	x
ceiling rivets		18	4 − x

The number of rivets counted equals the rate of counting times the number of minutes. If you count at the rate of 100 rivets per minute for 7 minutes, you will count 700 rivets.

	rivets counted	rate	time
wall rivets	22x	22	x
ceiling rivets	18(4 − x)	18	4 − x

He counted the same number of wall rivets as ceiling rivets.

$$22x \ = \ 18(4 - x)$$

Finding the equation is usually the hard part of doing a word problem. We solved this equation on the previous page.

3. $3w + 7(4w) - 20 \ = \ 12$
$3w + 28w - 20 \ = \ 12$
$31w - 20 \ = \ 12$
$31w \ = \ 32$
$w \ = \ \dfrac{32}{31}$

4.   *Let x = the number of helmets he bought.* You almost always start by letting x equal the thing you are trying to find out.

    Then x + 10 = the number of pistols he bought.

    Then 60x = the cost of the helmets.

    Then 200(x + 10) = the cost of the pistols.

The total cost of both was $9800 . . .

$$60x + 200(x + 10) = 9800$$

Distributive property $\qquad\qquad 60x + 200x + 2000 = 9800$

Combine like terms $\qquad\qquad\qquad 260x + 2000 = 9800$

Subtract 2000 from both sides $\qquad\qquad 260x = 7800$

Divide both sides by 260 $\qquad\qquad\qquad x = 30$ helmets

5.   *Let x = quarts of the strong coffee used.*

    Then 18 – x = the number of quarts of the weak coffee used. Look at the box on the first page of Lesson 26.

    Then 67.5x = the number of grounds contributed by the strong coffee.

    Then 45(18 – x) = the number of grounds contributed by the weak coffee.

We want (55 grounds/quart)(18 quarts) = 990 grounds of coffee in the final mixture.

$$67.5x + 45(18 - x) = 990$$
$$67.5x + 810 - 45x = 990$$
$$22.5x + 810 = 990$$
$$22.5x = 180$$
$$x = \frac{180}{22.5}$$

$$22.5 \overline{)\,180.}$$

$$\begin{array}{r} 8 \\ 225. \overline{)\,1800.} \\ \underline{1800} \\ 0 \end{array}$$

         8 quarts of the 67.5 grounds/quart and

         10 (which is 18 – x) quarts of the 45 grounds/quart

6.   My cat is now 10. I am now 15. (from the first sentence of the problem)

    Let x = the number of years until I will be twice as old as my cat.

    In x years my cat will be x + 10 years old, and I will be x + 15 years old.

In x years I will be twice as old as my cat. $\qquad x + 15 = 2(x + 10)$

Let's look at this equation: $x + 15 = 2(x + 10)$

The English: In x years I will be twice as old as my cat.

Solving $x + 15 = 2(x + 10)$

$$x + 15 = 2x + 20$$
$$15 = x + 20$$
$$-5 = x$$

The algebra tells me that I will be twice as old as my cat −5 years from now. Translation: I *was* twice as old as my cat five years ago.

7.   *Let x = how many miles Joe ran.*

Then x = how many miles Darlene ran.

Then 6x = how many miles the ball retriever ran.

Since they ran a total of 10 miles . . . $x + x + 6x = 10$

$$8x = 10$$
$$x = \frac{10}{8}$$

$$\begin{array}{r} 1.25 \\ 8\overline{)10.00} \\ \underline{8} \\ 20 \\ \underline{16} \\ 40 \\ \underline{40} \\ 0 \end{array}$$

Joe ran 1.25 miles (or 1¼ miles)

8.   *Let t = the number of hours Snow drove.* You almost always start by letting x equal the thing you are trying to find out.

Then 6 – t = the number of hours Sidney drove. The box on the first page of Lesson 26

Then 120t miles = the distance Snow drove.

Then 60(6 – t) = the distance Sidney drove.

Since they drove the same distance . . . $120t = 60(6 - t)$

$$120t = 360 - 60t$$

Add 60t and subtract 120 from both sides $180t = 360$

$$t = 2$$

Sergeant Snow drove for two hours.

9. Using "six pretty boxes" . . .

	total cost of subscriptions	subscription rate	number of subscriptions
American Mercury			x
Look			

Since there were three more subscriptions to Look than to American Mercury . . .

	total cost of subscriptions	subscription rate	number of subscriptions
American Mercury			x
Look			x + 3

Filling in the subscription rates . . .

	total cost of subscriptions	subscription rate	number of subscriptions
American Mercury		$18	x
Look		$22	x + 3

	total cost of subscriptions	subscription rate	number of subscriptions
American Mercury	$18x	$18	x
Look	$22(x + 3)	$22	x + 3

The total cost of both subscriptions was $306.

$$18x + 22(x + 3) = 306$$
$$18x + 22x + 66 = 306$$
$$40x + 66 = 306$$
$$40x = 240$$
$$x = 6$$

The librarian ordered six subscriptions to American Mercury.

10.
$$5y + 6(6y) - 100 = -80$$
$$5y + 36y - 100 = -80$$
$$41y - 100 = -80$$
$$41y = 20$$
$$y = \frac{20}{41}$$

11. *Let x = the number of pounds of weak relish used.*

Then 200 − x = the number of pounds of strong relish used. Look at the box on the first page of Lesson 26.

Then 8x = the number of pickles contributed by the weak relish.

Then 16(200 − x) = the number of pickles contributed by the strong relish.

Since we want 2000 pickles (200 pounds × 10 pickles/pound) in the final mixture . . .

$$8x + 16(200 - x) = 2000$$
$$8x + 3200 - 16x = 2000$$
$$-8x + 3200 = 2000$$

Subtract 3200 from both sides $-8x = -1200$

Divide both sides by −8 $x = 150$

The question asks how many pounds *of each* should be used. For the weak relish use 150 pounds (which is x), and for the strong relish use 50 pounds (which is 200 − x).

12. Your sister is 18 years old and you are 22. (That was given in the first sentence of the problem.)

Let x = how many years ago that you were three times her age.

Then
$$22 - x = 3(18 - x)$$
$$22 - x = 54 - 3x$$
$$22 + 2x = 54$$
$$2x = 32$$
$$x = 16$$

It was 16 years ago when you were 6 and she was 2.

Seward, Alaska located at mile 0 of the Iditarod trail

1. At the phone booth, Betty and Alexander had a lot of phone calls to make to let everyone at KITTENS know what had happened so far. Alexander made a bunch of calls (18 calls/hour), and then Betty took over and made an equal number of calls (12 calls/hour). In five hours the pair had completed all the calls they needed to make. How long was Alexander on the phone?

2. There are two doctors who do the induction physicals at A.A.A.A. Each one does half of the exams. One can do 4 exams/hour and the other, 5/hour. It took them a combined total of 36 hours to do the exams for all the prisoners in this week's convoy. How long did it take the slower doctor to do his half of the exams?

3. The cooks counted the knives, forks, and spoons in the kitchen. There were ten more forks than knives and ten more spoons than forks. All together there were 363 utensils. How many forks were there?

4. Solve: $12x + 3(7x) - 6 = -8$

5. At Waddles Doughnuts, where you work part time (to put yourself through school), they ask you to make up ten gallons of doughnut dough, which is 44% sugar. There's some dough in the backroom that's 20% sugar, and there's some out in the trunk of the car that's 60% sugar. How many gallons of the stuff in the car trunk should you bring in?

6. You were 22 when your son was born. He's grown up a bunch since then. In fact, last week he asked you for the keys to the family car. You remarked, "Do you realize that in four years, I'll be twice your age?" How old is he now?

7. Betty and Darlene made cookies for the Math Fair. Each contributed the same number of cookies. In the small student recreation kitchen where only one could work at a time, Betty started first, making her coconut cookies at the rate of 180/hour.

Then Darlene took over the kitchen and made her pizza dumpling cookies at the rate of 140/hour. All together they were in the kitchen for six hours. How long did Betty work?

8. Betty and Alexander were grading homework papers for Fred's classes. Alexander corrected 15 papers/hour. Betty started two hours after Alexander, but the assignments she was correcting were longer so she could only correct 10 papers/hour. Together they corrected 130 papers. How long did Betty work?

9. In the army kitchen there are four pots for every three pans for every 21 glasses. All together there are 476 of these items. How many pans are there?

10. Solve: $2y + 4(5y) - 18 = -15$

11. Army jeeps require gasoline that's exactly 50.6% ethanol. You need to put ten gallons in the tank, but the gas station only carries 24% and 62% ethanol gas. How much of each should you put in?

12. Your daughter was just born. In 24 years you'll be three times as old as she will be. How old are you now?

answers

1. Using "six pretty boxes" . . .

	calls made	rate	time
Alexander			t
Betty			

Together they spent 5 hours phoning people. See the box on first page of Lesson 26.

	calls made	rate	time
Alexander			t
Betty			5 − t

	calls made	rate	time
Alexander		18 calls/hr	t
Betty		12 calls/hr	5 − t

	calls made	rate	time
Alexander	18t	18 calls/hr	t
Betty	12(5 − t)	12 calls/hr	5 − t

Since they made the same number of calls . . .

$$18t = 12(5 - t)$$
$$18t = 60 - 12t$$
$$30t = 60$$
$$t = 2$$

Alexander was on the phone for two hours.

2.   *Let x = the number of hours the slower doctor took to do his half of the exams.*

Then 36 – x = the number of hours the faster doctor took.

Then 4x = the number of exams the slower doctor did.

Then 5(36 – x) = the number of exams the faster doctor did.

Since they did the same number of exams . . .

$$4x = 5(36 - x)$$
$$4x = 180 - 5x$$
$$9x = 180$$
$$x = 20$$

The slower doctor took 20 hours.

3.   *Let x = the number of forks.* In almost every word problem, you let x equal the thing you are trying to find.

Then x – 10 = the number of knives.

Then x + 10 = the number of spoons.

Since there is a total of 363 utensils . . .

$$x + (x - 10) + (x + 10) = 363$$
$$x + x - 10 + x + 10 = 363$$
$$3x = 363$$
$$x = 121$$

There are 121 forks.

4. $$12x + 3(7x) - 6 = -8$$

$$12x + 21x - 6 = -8$$
$$33x - 6 = -8$$
$$33x = -2$$
$$x = \frac{-2}{33}$$

5. This is a mixture problem.

Let x = the number of gallons of 60% dough from the trunk.

Then 10 - x = the number of gallons of 20% dough from the backroom.

Then 0.6x = sugar contributed by the 60% dough.

Then 0.2(10 - x) = the sugar contributed by the 20% dough.

Since we want 4.4 (10 gallons × 44%) sugar . . .

$$0.6x + 0.2(10 - x) = 4.4$$
$$0.6x + 2 - 0.2x = 4.4$$
$$0.4x + 2 = 4.4$$
$$0.4x = 2.4$$
$$x = 6$$

We need 6 gallons of dough from the trunk.

6. *Let x = the current age of your son.*

Then x + 22 = your current age.   You were 22 years old when your son was born. You will always be 22 years older than him (while you are both alive).

Then x + 4 = his age four years from now.

Then x + 26 = your age four years from now.

In four years you will be twice his age.

$$x + 26 = 2(x + 4)$$

$$x + 26 = 2x + 8$$
$$18 = x \qquad \text{Subtract x and subtract 8 from both sides of the equation}$$

He is now 18 years old.

7.

	cookies made	rate	time
Betty			t
Darlene			

Using the box on the first page of Lesson 26 . . .

	cookies made	rate	time
Betty			t
Darlene			6 – t

	cookies made	rate	time
Betty		180 cookies/hr	t
Darlene		140 cookies/hr	6 – t

	cookies made	rate	time
Betty	180t	180 cookies/hr	t
Darlene	140(6 – t)	140 cookies/hr	6 – t

Since they made the same number of cookies . . .

$$180t = 140(6 - t)$$
$$180t = 840 - 140t$$
$$320t = 840$$
$$t = \frac{840}{320}$$

$\frac{840}{320} = \frac{84}{32} = \frac{21}{8} = 2\ 5/8$ Betty worked for $2\frac{5}{8}$ hours.

8.   *Let x = the number of hours that Betty worked.*

Then x + 2 = the number of hours Alexander worked since Betty started two hours after Alexander.

Then 10x = the number of papers Betty corrected.

Then 15(x + 2) = the number of papers that Alexander corrected.

Since they corrected a total of 130 papers . . .

$$10x + 15(x + 2) = 130$$
$$10x + 15x + 30 = 130$$
$$25x + 30 = 130$$
$$25x = 100$$

Betty worked for four hours. $x = 4$

9.   *Let 3x = the number of pans in the army kitchen.*

   Then 4x = the number of pots.

   Then 21x = the number of glasses.

Since there are 476 items . . .

$$3x + 4x + 21x = 476$$
$$28x = 476$$
$$x = 17$$

There are (3x =) 51 pans.

10.
$$2y + 4(5y) - 18 = -15$$
$$2y + 20y - 18 = -15$$
$$22y - 18 = -15$$
$$22y = 3$$
$$y = \frac{3}{22}$$

11. A mixture problem

	gallons of ethanol	percent of ethanol	gallons
24% gas	0.24x	24%	x
62% gas	0.62(10 – x)	62%	10 – x

We want 5.06 gallons of ethanol (50.6% × 10 gallons) in the final mixture.

$$0.24x + 0.62(10 - x) = 5.06$$
$$0.24x + 6.2 - 0.62x = 5.06$$
$$-0.38x + 6.2 = 5.06$$
$$-0.38x = -1.14$$
$$x = 3$$

We should use 3 gallons of the 24% and (10 – x), which is 7, gallons of the 62%.

12. *Let x = your current age.*

   Then x + 24 = your age in 24 years.

   Then 24 = your daughter's age 24 years from now. She was just born.

Since in 24 years you will be three times her age . . .

$$x + 24 = 3(24)$$
$$x + 24 = 72$$
$$x = 48$$

T he convoy of prisoner vans had been parked out on the south flat parking lot since it had arrived at A.A.A.A. on Friday afternoon.

As each of the vans had been driven onto the asphalt of the five-acre parking lot, the military police opened each of them and escorted the draft evaders to the chow hall. They were fatigued after their trip, and many of them complained loudly when they found that their evening meal was beans and hot dogs.

Now it was just past noon on Saturday. A detail of soldiers was opening up each van and cleaning it out. It was 105° out on the asphalt parking lot, and they were tired and hot. Opening up the back doors of each transport vehicle subjected the men to a blast of hot air from the inside.

When they came to the 84th vehicle, they were shocked. Unconscious, Fred was still chained to the wall. Had he been conscious, he would have been blinded by the white light streaming in but would have welcomed the 20° drop in temperature.

One of the men unlocked the handcuff that had been placed around his neck and muttered, "This guy's so small that handcuffs wouldn't stay on his wrists." He tossed Fred over his shoulder and took him off to the camp hospital.

And small he was. Fred had lost four pounds and now weighed only 33 pounds (when they put him on a postal scale at the infirmary). In his delirium, he imagined a blackboard in his mind upon which was written:

$$\text{Let } x = \text{my original weight.}$$
$$\text{Then } x - 4 = \text{my current weight.}$$
$$\text{Since my current weight is 33 lbs,}$$
$$x - 4 = 33$$

This is really beginning, beginning algebra. We all know how to solve $x - 4 = 33$. You just add four to each side.

You start with: $\quad x - 4 = 33$
and you turn it into: $\quad x \quad = 33 + 4$

If you look at the "before" line and then the "after" line and go back and forth between the two lines—before/after/before/after/before/after/before/after/before/after/before/after/before/after/before/after/before/after/before/after/before/after/before/after/before/after/before/after/before/after it will start to look as if the – 4 jumped from the left side of the equation to the right side and became + 4.

$$
\begin{aligned}
\text{Before:} \qquad & x - 4 = 33 \\
\text{After:} \qquad & x \quad\ = 33 + 4
\end{aligned}
$$

When a "– 4" jumps over the equal sign, it turns into a "+ 4". Now you and I know that the "– 4" really didn't jump over the equal sign. It just looks like it did. And we know that on the way over the equal sign it didn't trip and change its sign. It just looks like it did.

This is called **transposing**. People will say that they transposed the – 4 to the other side of the equation. However, we will not be mentioning the idea of transposing in this book.*

The nurse came in to feed Fred. He (the nurse) propped up Fred, placing three pillows behind him. "How about some chow for the little fellow?" he asked Fred, trying to rouse him.

Fred's mouth fell open a bit. He wanted to say, "Water!" but his mouth was dry as talcum powder and he couldn't get the word out.

Just as the nurse was about to shovel some beans and hot dogs into Fred's mouth, one of the doctors who does induction physicals walked in and moved the nurse out of the way.

"Time for this runt's physical," he announced.

He got out his clipboard.

He borrowed a pencil from the nurse.

✶ Virtually all great writers use irony in their writing. Irony means saying one thing, but implying something quite different. If you are in the middle of a hurricane and say, "Isn't this a lovely day!" you are using irony. Your ironic statement about the weather really means quite the opposite of the literal words. Readers often get into trouble when they take an author's words literally and miss the irony. You miss a lot in Shakespeare if you fail to see his irony. In the Bible (2 Samuel 6:20), Michal says to David, "How glorious was the king of Israel today. . . ." She is really being quite nasty to him. She means the opposite of what she is literally saying. Pure irony.

He sat down on the bed so that he could be more comfortable. His weight on the bed tipped Fred over. The nurse set Fred up again.

He then filled out the form:

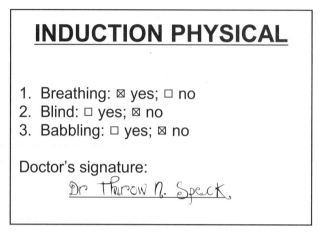

INDUCTION PHYSICAL

1. Breathing: ☒ yes; ☐ no
2. Blind: ☐ yes; ☒ no
3. Babbling: ☐ yes; ☒ no

Doctor's signature:
Dr Throw N. Speck.

The nurse noticed that the doctor had hardly looked at Fred and asked the doctor, "Have you looked at his eyes? They look kinda funny."

The doctor, who had already gotten up and was heading out the door, glanced back and uttered, "Two."

From across the room you couldn't see much—just a kid who

 barely filled half the bed. But if you sat on the bed and really looked at the patient as the nurse had done, then you would detect the burned-in image of the prison van window that had been Fred's only source of light. The Texas sun had done its work.

It would probably be a week before that image of the proof of the distributive property would fade from his eyeballs.

Your Turn to Play

1. Transpose the -7 in the equation: $x - 7 = 50$.

2. Solve by transposing: $y - 100 = 2$.

3. To solve $z + 8 = 19$ by transposing, your first step would be $z = 19 - 8$. What did you do to both sides of the equation?

4. Solve $6w = 2w + 44$

5. Solve $5x = 7x - 18$

6. Solve $2y + 10 = 7y - 16$

7. Solve $2(w - 5) = 28$ (Get rid of the parentheses before transposing.)

8. Sometimes you get parentheses within parentheses—nested parentheses. The best place to start is with the innermost pair.

   For example: $3(x + 7(2w - 5))$

   $3(x + 14w - 35)$

   $3x + 42w - 105$

Your turn: $8(z + 12(5 + 7x))$

9. Solve $6(5x + 2(7 - x)) = 101$

Intermission

There are two kinds of algebra problems:

① simplifying expressions, which we did in problem 8 and

② solving equations, which we did in problem 9.

---

10. If I am x years old now, how old will I be when I am twice as old as I am now?

11. Solve $8(y - 3) = 5 - 21y$

12. The formula for the volume of a cylinder is $V = \pi r^2 h$ where V is the volume, r is a radius, and h is the height.

   If the volume is 252 cubic inches and the radius is 6 inches, what is the height of the cylinder?

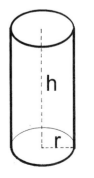

· · · · · · · COMPLETE SOLUTIONS · · · · · · ·

1. $x - 7 = 50$ becomes

$x = 50 + 7$ It's the same as adding +7 to both sides.

2. $y - 100 = 2$

$y = 2 + 100$ Transpose the -100

$y = 102$

3. To go from $z + 8 = 19$

to $z = 19 - 8$ Subtract 8 from both sides of the equation

or add -8 to both sides

4. $6w = 2w + 44$

$6w - 2w = 44$ Transpose the 2w

$4w = 44$

$w = 11$

5. $5x = 7x - 18$

$18 = 7x - 5x$ Transpose both the 5x and the -18

$18 = 2x$

$9 = x$

6. $2y + 10 = 7y - 16$

$26 = 5y$ Transpose both the 2y and the -16

$\dfrac{26}{5} = y$

7. $2(w - 5) = 28$

$2w - 10 = 28$ Distributive property

$2w = 38$ Transpose

$w = 19$

8. $8(z + 12(5 + 7x))$

$= 8(z + 60 + 84x)$ Do the innermost parentheses first

$= 8z + 480 + 672x$

9. $6(5x + 2(7 - x)) = 101$

$6(5x + 14 - 2x) = 101$

$6(3x + 14) = 101$

$18x + 84 = 101$

$18x = 17$

$x = \dfrac{17}{18}$

10. 2x years old

11.

$8(y - 3) = 5 - 21y$

$8y - 24 = 5 - 21y$

$8y + 21y = 5 + 24$ **Transposing −24 and 21y**

$29y = 29$

$y = 1$

12. Start with the formula $V = \pi r^2 h$

We are given V = 252 and r = 6.

Put those values into the formula. $252 = \pi(6^2)h$

$252 = 36\pi h$

Divide both sides by 36π $\dfrac{252}{36\pi} = h$

Reducing the fraction $\dfrac{252}{36\pi} = \dfrac{126}{18\pi} = \dfrac{63}{9\pi} = \dfrac{7}{\pi}$ inches he beans kept falling out of Fred's mouth, so the nurse wet his mouth down a bit so the beans would stick. The water helped. Between mouthfuls, which in Fred's case consisted of about three beans, the nurse was looking at a problem in his Army Algebra for Medical Personnel Manual, which is known affectionately as AAMPM. The nurse scratched his chin and turned to Fred and asked, "Do you know anything about math?"

Fred coughed and responded, "A bit, I guess." He didn't mention that he had become a math professor at KITTENS University at the age of nine months.

"You see here," the nurse said to Fred as he pointed to the AAMPM, "there's this Rat and this Mouse. The Rat is 15 days older than the Mouse."

On the blackboard in Fred's head immediately clicked:

> Let x = age of Rat
> Then $x - 15$ = age of Mouse

"I think I understand," Fred meekly stated.

"Okay, then," the nurse continued. "Now things get really complicated. We're told that the sum of the ages of the two animals is 55 days."

Fred's mind was on automatic pilot at this point. The lines just wrote themselves in his mind:

> Let x = age of Rat
> Then $x - 15$ = age of Mouse
> Then since the sum of the ages of the Rat and the Mouse is 55
> we have
> $$x + (x - 15) = 55$$
> $$2x - 15 = 55$$
> $$2x = 70$$
> $$x = 35$$
> So the age of the Rat is 35 days and the age of the Mouse is 20 days.

Fred kept quiet at this point wondering where the nurse had gotten stuck. He had seen a problem like this in an ordinary banal algebra book, but the problem in that book dealt with a mother and a daughter. He figured, since this is the army, they had to be tough, aggressive, and mean, and therefore, had to use examples with a Rat and a Mouse in them instead.

"So you see," the nurse explained to Fred as he wrote on the bed sheet, "since the Rat is 15 days older than the Mouse, we have

$$R = M + 15$$

Two unknowns! thought Fred. We did this with just one unknown which we called x.

"And since the sum of their ages is 55, we get. . . ."

$$R + M = 55$$

Two equations and two unknowns noted Fred to himself. The nurse seemed to be stuck at that point as he stared at this system of equations:

$$\begin{cases} R & = M + 15 \\ R + M = 55 \end{cases}$$

After a moment of silence, Fred suggested that things get organized a bit by putting all the terms with unknowns in them on the left side and the numbers on the right side of each equation.

The nurse transposed the M in the first equation to the left side:

$$\begin{cases} R - M = 15 \\ R + M = 55 \end{cases}$$

"But how does that help?" the nurse frowned. "We've still got two equations with two unknowns."

"Wouldn't it be nicer if we had one equation with one unknown?" Fred asked.

Fred took the pen from the nurse and added the equations. He added the left sides together and added the right sides together.

$$\begin{cases} R - M = 15 \\ R + M = 55 \end{cases}$$
$$\overline{\qquad\qquad\qquad}$$
$$2R \qquad = 70 \qquad \text{☜ (Fred's handwriting)}$$

"Look! One equation and one unknown. Divide both sides by 2

and we get R = 35. We substitute that value back into any of the earlier equations and we find M's value. If we put R = 35 into the equation R + M = 55, we get 35 + M = 55. So M = 20."

Fred handed the pen back to the nurse and waited. The nurse shook his head a bit. The only thing he couldn't understand is why you were allowed to add two equations together. (He *did* like the fact that the M's disappeared when you added the equations.)

Fred explained it this way. He held out his two little fists and said, "Suppose I've got equal weights in each of my hands. Now you hold out your hands."

The nurse held out his two fists.

Fred continued, "Now suppose your hands also contain equal weights. They might contain more weight than my hands since yours are larger, but your left hand and your right hand would be equally heavy.

"Now suppose that we combine what we each have in our left hands and put that on the table next to my bed.

"Then suppose we combine what we have in our right hands and put that on the bed. Which would weigh more: the stuff on the table or the stuff on the bed?"

"That's easy," replied the nurse. "They'd have to weigh the same."

"That's the whole point," said Fred. "If equals are added to equals, then the results have to be equal."

He asked for the pen back and wrote on the bed sheet:

$$\begin{array}{r} \alpha = \beta \\ \underline{\gamma = \delta} \\ \alpha + \gamma = \beta + \delta \end{array}$$

"What's that stuff?" the nurse exclaimed. "I can't read your writing."

"Oops. That's alpha (α), beta (β), gamma (γ) and delta (δ)—the first four letters of the Greek alphabet. We use them all the time in trig and calculus."

The nurse took the pen back and, mumbling something about "foreign stuff," crossed out what Fred had written and wrote "American":

$$a = b$$
$$\underline{c = d}$$
$$a + c = b + d$$

The nurse felt he was serving his country if he supported "the letters that were invented in this country." (irony alert)

Fred's head was starting to nod. It had been a long 27 hours in the back of the van and he was full of beans. While he takes a rest, let's turn to . . .

Your Turn to Play

1. Let's practice a little transposing:
   $$x - 3 = 22$$
   $$y + 59 = 88$$
   $$87z = 4 - 7z$$

2. The **union** of two sets is the set containing all the elements in either or both sets. For example, the union of {3, #, %} and {4, #, San Diego} is {3, #, %, 4, San Diego}. Remember, you don't list an element twice, so # isn't listed twice in the answer.

The symbol for union is "∪" so, for example,
   $$\{7, 8\} \cup \{7, 11\} = \{7, 8, 11\}.$$

Now it's your turn to play: {rose, star} ∪ {2, star} = ?

3. What do we call the following set? {x | x is a natural number} ∪ {0}

4. I have two books on my shelf. The one about Jeanette is 50 years older than the one about Nelson. Forty years from now, the one about Jeanette will be twice as old as the one about Nelson. Using one unknown (letting x equal the current age of the book about Nelson) find the current age of the book about Nelson.

5. Do the previous problem using two unknowns. Let N = current age of the Nelson book and let J = the current age of the Jeanette book.

6. Solve the system of equations $\begin{cases} 43x + 6y = 87 \\ 20x - 2y = 74 \end{cases}$

☞ "Solve" means find x and y.

☞ Multiply the second equation by 3, and then add them. Once you find the value of x, substitute it back into any of the equations that contain both x and y, and then solve for y.

7. And now for the brain-buster: What would be *the first step* in solving
$$\begin{cases} 80x + 2y = 9 \\ 39x - 3y = 11 \end{cases}$$

8. Let's save a little work.
If you had $\begin{cases} 4x + 37y = 5 \\ 3x - 39y = 371 \end{cases}$

it would be a lot of work to multiply the top equation by 39 and the bottom equation by 37 (although this would work if you elected to do so).

What's the easier first step?

9. In words, how would you describe the set {–x | x is a natural number}?

10. What's wrong with describing {–x | x is a natural number} as "the set of negative natural numbers"?

. . . . **COMPLETE SOLUTIONS** . . . .

1. x = 22 + 3   y = 88 – 59   87z + 7z = 4

2. {rose, star, 2}   (The order in which you list the elements does not matter. You could have written your answer as { rose, 2, star}.

3. That's the set {0, 1, 2, 3, . . .}, which is the whole numbers.

4. *Let x = the current age of the book about Nelson.*

Then x + 40 will be the age of the Nelson book 40 years from now.

Then x + 50 is the current age of the book about Jeanette (since it's 50 years older than the Nelson book).

Then x + 50 + 40 = the age of the Jeanette book 40 years from now.

In 40 years, the Jeanette book will be twice as old as the one about Nelson.

$$x + 50 + 40 = 2(x + 40)$$
$$x + 90 = 2x + 80$$
$$10 = x$$

The Nelson book is now ten years old.

5. *Let N = current age of the Nelson book.*   } two Let statements!
   Let J = current age of the Jeanette book.

   Then N + 40 is the age of the Nelson book 40 years from now.
   Then J + 40 is the age of the Jeanette book 40 years from now.

Here are the two equations:

First equation $J = N + 50$ since the Jeanette book is 50 years older than the Nelson book

Second equation $J + 40 = 2(N + 40)$ since in 40 years the Jeanette book will be twice as old as the Nelson book

From the first equation $J - N = 50$
From the second equation $J - 2N = 40$ (I skipped a couple of steps.)

What to do? If we add these two equations neither unknown disappears. I'm going to multiply the second equation by –1:

$$\begin{cases} J - N = 50 \\ -J + 2N = -40 \end{cases}$$

Now if we add the equations, the J's will disappear and we'll get N $= 10$.

    For people who like to name things . . . This is called **solving systems of equations by elimination**.

6. Multiply the second equation by +3 to obtain $\begin{cases} 43x + 6y = 87 \\ 60x - 6y = 222 \end{cases}$

and then when you add them, you will get $103x = 309$

Divide both sides by 103	$x = 3$

To get the value of y, take $x = 3$ and put that value in *any* equation containing both variables.

Putting $x = 3$ into the second original equation ⇨ $20x - 2y = 74$

$$20(3) - 2y = 74$$

Transpose $\qquad\qquad -2y = 74 - 20(3)$

Arithmetic $\qquad\qquad -2y = 14$

Divide both sides by -2 $\qquad\qquad y = -7$

7. In this system, $\begin{cases} 80x + 2y = 9 \\ 39x - 3y = 11 \end{cases}$

the easiest way is to multiply the top equation by 3 and the bottom equation by 2

$$\begin{cases} 240x + 6y = 27 \\ 78x - 6y = 22 \end{cases}$$

Add and the y-terms disappear leaving $318x = 49$

8. Take $\begin{cases} 4x + 37y = 5 \\ 3x - 39y = 371 \end{cases}$ and eliminate the x-terms instead of the y-terms by multiplying the top equation by 3 and the bottom equation by -4.

9. $\{-x \mid x \text{ is a natural number}\}$ in symbols could also be written as $\{-1, -2, -3, \ldots\}$, which could be described as the negative integers.

10. How many negative natural numbers can you name? There aren't any. All the natural numbers I can think of are positive. The set of negative natural numbers is the same as the set of all twelve-foot-tall Girl Scouts. It's the set that contains no elements which is also known as the empty set.

   $\{-x \mid x \text{ is a natural number}\}$ is the set of the *negatives of* the natural numbers. Little changes in English can make a big difference in meaning.

   When you say that Dan went to the store, that's different than Dan went to this door, although the two sentences sound alike.

When Fred was waking from his nap, he was thinking about why anyone would use two unknowns in two equations when one equation with one unknown would work. At first he couldn't think of a good answer to that.

Then he thought of times where he actually needed two equations and two unknowns—where he couldn't use just one equation and one unknown. Like the time that the department secretary Belinda had put her Diary of Work on the blackboard in her office. (It was something she'd learned in her **Work Speedier and Get Richer** seminar.)

She had written:

> 20 job offer letters answered
> 11 interview request letters answered
> Time spent: 830 minutes

Later she added to her Diary of Work:

> 20 job offer letters answered
> 11 interview request letters answered
> Time spent: 830 minutes
> ==
> 28 job offer letters answered
> 33 interview request letters answered
> Time spent: 1690 minutes

When Fred had seen her public diary, he had thought to himself

Twenty-five minutes to answer a job offer and all she's doing is copying the return address onto an envelope and stuffing in a form letter! She doesn't even read the job offer—just tosses the unopened envelope into the garbage. That is not very fast.

The algebra that Fred had done in his head really called for two equations and two unknowns:

Let x = time to answer a job offer.

Let y = time to answer an interview request letter.

Then $\begin{cases} 20x + 11y = 830 \\ 28x + 33y = 1690 \end{cases}$

There wasn't a natural way to set this up with just one equation and one unknown. With two unknowns it was straightforward. (Imagine, for a moment, how long it would take you to figure out that it was taking Belinda 25 minutes to answer each job offer letter if you had never had any algebra.)

Fred multiplied the top equation by –3: $\begin{cases} -60x - 33y = -2490 \\ 28x + 33y = 1690 \end{cases}$

He then added the two equations: $-32x = -800$

and divided by –32: $x = 25$

And then it hit Fred. He exclaimed, "Oh my!" He had figured out what had happened, why the big army guys had arrested him and taken him to the army camp.

A letter from A.A.A.A. had arrived in the pile of marriage proposals, job offers, and interview requests.

Belinda smelled it and checked for "S.W.A.K." It wasn't a marriage proposal. She had copied the return address onto KITTENS stationery and enclosed the form letter, "Sorry, he's happy where he is."

This must have enraged the army guys when they received it. Probably they had sent several letters, each one more threatening than the previous one, and each had been dutifully processed by Belinda. Her 𝔚𝔬𝔯𝔨 𝔖𝔭𝔢𝔢𝔡𝔦𝔢𝔯 𝔞𝔫𝔡 𝔊𝔢𝔱 �𝔦𝔠𝔥𝔢𝔯 had become Fred's 𝔚𝔬𝔯𝔨 𝔖𝔭𝔢𝔢𝔡𝔦𝔢𝔯 𝔞𝔫𝔡 𝔊𝔢𝔱 𝔇𝔯𝔞𝔣𝔱𝔢𝔡.

Relieved to know how it had happened, Fred had only one problem: It had happened. Physicians encounter this all the time. They announce to their patient, "You've got metastatic carcinomatous melanoma," and the patient appears consoled and responds, "I'm so glad you know what it is."

(A rough translation of *metastatic carcinomatous melanoma* is a thingy on your skin that's cancer, and it's spreading.)

Your Turn to Play

1. One day Belinda sent 5 faxes and unwrapped 8 pieces of gum. It took her 118 minutes. On another day it took her 80 minutes to send 4 faxes and unwrap 4 pieces of gum. Since all the faxes were each one page long, the length of time it took to send a fax didn't vary. It was constant.

How long does it take Belinda to send a fax? How long to unwrap a piece of gum?

2. It took Belinda 264 minutes to type three pages and read four chapters from her 𝔚𝔬𝔯𝔨 𝔖𝔭𝔢𝔢𝔡𝔦𝔢𝔯 & 𝔊𝔢𝔱 𝔕𝔦𝔠𝔥 seminar book.

It took her 342 minutes to type five pages and read two chapters from her seminar book.

How long does it take Belinda to type a page? To read a chapter?

3. This diagram was used to prove what property?

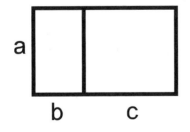

4. Solve $-12 = 4(y + 2(y - 9))$

. . . . COMPLETE SOLUTIONS . . . .

1.   *Let x = minutes it takes to send a fax.*

Let y = minutes it takes to unwrap a piece of gum.

Then it takes 5x minutes to send 5 faxes.

Then it takes 8y minutes to unwrap 8 pieces of gum. } $5x + 8y = 118$

Similarly, $4x + 4y = 136$

$$\begin{cases} 5x + 8y = 118 \\ 4x + 4y = 80 \end{cases}$$ ← This is called a system of equations.

Copy the first equation and multiply the second equation by –2 to obtain

$$\begin{cases} 5x + 8y = 118 \\ -8x - 8y = -160 \end{cases}$$

Add the two equations together (to eliminate the y-terms) $-3x = -42$

Dividing both sides by –3 $x = 14$

Now we know the value of one of the two unknowns. We have x = 14. It takes 14 minutes for Belinda to send a fax. Put this value of x into *any* equation that contains both x and y. Then solve for y.

I choose the equation $4x + 4y = 80$ (because it looks the simplest) and get $4(14) + 4y = 80$. After a couple of steps: y = 6.

2. *Let x = minutes for Belinda to type a page. Let y = minutes to read a chapter.*

$$\begin{cases} 3x + 4y = 264 \\ 5x + 2y = 342 \end{cases}$$

Copy first equation $\begin{cases} 3x + 4y = 264 \\ -10x - 4y = -684 \end{cases}$

Second equation times –2

Add the equations $-7x = -420$

Divide both sides by –7 $x = 60$

Back substitute x = 60 into $3x + 4y = 264$ and obtain y = 21 minutes.

3. The distributive property $a(b + c) = ab + ac$

4. $-12 = 4(y + 2(y - 9))$

$-12 = 4(y + 2y - 18)$ Do the innermost parentheses first

$-12 = 4y + 8y - 72$

$60 = 12y$

$5 = y$

꧁꧂꧁꧂꧁꧂꧁꧂꧁꧂꧁꧂꧁꧂꧁꧂꧁꧂

F red looked at the wall clock. It was 3:30. He had been in the infirmary for a little over three hours. Still woozy, he felt his face. It was all rough. Since he was a bit young to be getting a beard, he figured it was the dried bean sauce. He climbed out of bed and went looking for the bathroom so he could wash his face. He was surprised to find that his room was right next to the hospital administrator's office. The door was clearly marked, HEAD and he knocked on it.* When no one answered, he slowly opened the door and found a room with a wash basin in it.

Sergeant Snow, who had dropped by the hospital to pick up his stomach medicine, spotted Fred and yelled out, "Hey runt! Feeling all better now with some good army chow in your gut?" Fred, in fact, was feeling a bit nauseated from the hot sauce on the beans. He wasn't used to much spiciness in the pizza or the vending machine fare that he had consumed in the first six years of his life.

After a couple of randomly chosen cuss words, Sergeant Snow ordered Fred to report to the quartermaster's to get some decent clothes and then report to the parade ground at 4 o'clock (actually he said "sixteen hundred hours") for general quarters assignments. Fred had still been wearing his bow tie that he had purchased on the bus when he had first gone to KITTENS. It was a polka dot bow tie.

"And get rid of that bow tie!" Sergeant Snow shouted as he left the room.

At the quartermaster's shack, Fred stood in front of the counter but couldn't be seen, because the counter was four feet tall and Fred wasn't. Finally, when the camp chaplain came by, he gave Fred a lift so that he could stand on the counter.

"What's you selling kid?" the quartermaster barked. "Anyway, I don't want any."

"Sir," Fred began in the correct military manner. "I need some clothes."

∗ On ships, a toilet or lavatory is often called a *head*.

"Sure you do, but we don't do charity here, kid. Get lost."

"Sir, but sir," Fred continued, "Sergeant Snow said I was to report here for military clothing—or whatever it's called."

"Look at this graph," exclaimed the quartermaster pointing to the poster in back of him. "It shows the number of nut cases we've been experiencing in the last week or so. And now you. Are you even three feet tall?"

"Yes, I am."

"Put a lid on it kid. If I want you to talk, I'll ask you. And I suppose you want a general's outfit?"

"A private's would be more appropriate," Fred answered.

The quartermaster was about to knock Fred off the counter top when the chaplain intervened. He asked, "Son, how old are you?"

"Six, sir."

"And why," the chaplain continued, "do you want to be in the army at such a young age?" He carefully avoided references to Fred's—how shall we say it—lack of height.

Fred explained, "But I don't. They took me." He pointed to the red welt around his neck. "It was my sixth birthday yesterday, and two big men came and took me here."

He thought about mentioning:

① that he was a college professor at KITTENS University

② that the problem may have occurred because the department secretary had inadvertently failed to read the draft notice since

③ she thought it was one of the many job offers that along with

④ the hundreds of marriage proposals that he had received

⑤ along with the requests for interviews . . .

but decided that that much explanation might be a little too much to give a man he had just met.

While Fred was talking with the chaplain, the quartermaster was on his cell phone talking with Sergeant Snow, whose message to the quartermaster was clear: "Get that runt some military clothes. Sew them if you have to and get him on the parade grounds by sixteen hundred hours. And get rid of that bow tie he has. It drives me crazy."

As Fred was explaining to the chaplain about the graph that the quartermaster had on the wall, the quartermaster came up to Fred and pulled off his bow tie and threw it in the garbage.

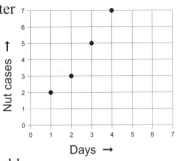

Fred was explaining what was wrong with the graph on the previous page. (There shouldn't be the overly dramatic arrow, the separate days shouldn't have been connected with a line, etc.)

Fred drew a graph showing how it should have been plotted. He drew on his undershirt using the chaplain's pen. Along the horizontal (↔) line are the days, starting with day number one, day number two, etc.

On the first day, suppose there had been two nut cases. (The nut cases are listed on the vertical axis ↕.) So to graph the fact that day number one had two nut cases he placed a dot that matched "1" on the horizontal axis and matched "2" on the vertical axis. It's called plotting the **ordered pair** (1, 2).

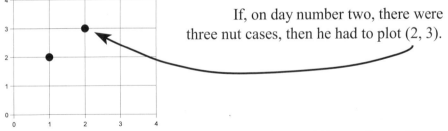

If, on day number two, there were three nut cases, then he had to plot (2, 3).

While the quartermaster is making the world's smallest military uniform, we'll take some time for . . .

Your Turn to Play

1. Looking at the graph at the top of this page, two of the points are (1, 2) and (2, 3). Name the other two points.

2. Draw a graph plotting the three points: (1, 1), (2, 2), and (3, 2).

3. Explain why (2, 4) and (4, 2) have different meanings.

4. Want to get real fancy? Plot the point (–2, 4). That will be the point that corresponds to –2 on the horizontal axis and corresponds to 4 on the vertical axis.

5. The horizontal and vertical axes (*axes* is the plural of axis) divide a piece of paper into four **quadrants**. The upper-right-hand quadrant is where Fred plotted all of the points on the graph on the previous page. That's called the **first quadrant** or **quadrant I** or **QI**.

  The point (–2, 4) that you plotted in the previous question (which is in the upper-left-hand corner) is called the **second quadrant** or **QII**. As we name the quadrants, we go around counter-clockwise. For those of you who own digital watches, that means going the way that people/cars/horses usually run around race tracks. The point (–3, –4) is in the third quadrant. Name a point that's in the fourth quadrant.

6. You are told that the abscissa of a point is positive. Which two quadrants might that point be in? (If you don't know what an abscissa is, read the answer to question 3.)

Quadrant II	Quadrant I
Quadrant III	Quadrant IV

╌╌╌

. . . . COMPLETE SOLUTIONS . . . .

1. (3, 5) and (4, 7)

2.

3. On Fred's graph (2, 4) would mean that on day number 2 there were 4 nut cases. On the other hand, (4, 2) would mean that on day number 4 there were 2 nut cases. So (2, 4) and (4, 2) are not the same. The order in which the **coordinates** are listed makes a difference, and that is why it is called an *ordered* pair.

The coordinates of (2, 4) are 2 and 4. The **first coordinate** (also known as the **x-coordinate**) is 2 and the **second coordinate** (also known as the **y-coordinate**) is 4.

If this were a real fancy book, we'd inform you that the first coordinate is also called the **abscissa** (ab-SIS-ah). The second coordinate, a.k.a. the y-coordinate, is also called the **ordinate**.

4.

5. You're correct if the first coordinate you named was positive and the second coordinate was negative. For example, (3897293, –38797999992).

6. QI or QIV. It's traditional to use Roman numerals to name the quadrants. Luckily, the quadrants only go up to QIV. You might hate to encounter QLV, which would be quadrant fifty-five.

╌╌╌

In the thirty-five years that the quartermaster had been handing out uniforms, he had seen tall men, skinny men, and men with one arm shorter than the other. He thought he had seen it all, but now he had to furnish an infantry uniform for some squirt who was three feet tall. "This has gotta be a joke," he thought to himself. "Anyone under five feet tall would automatically be exempt from the draft."

He went into the backroom and found an old camouflage T-shirt and sewed up the bottom into legs. He didn't have any square helmets, so he popped a coffee cup on Fred's head instead.

By 4 p.m. Fred had made it to the parade ground. The men were all lined up in a single row stretching out to the left and to the right as far as the eye could see.

Somehow this reminded him of a long line of roses (and the integers), but these guys in their olive drab looked more like leaves than roses.

Sergeant Snow stood pompously in front of them and intoned, "Men, this is the modern army! You should feel proud to have been drafted into the best army in Texas."

Fred couldn't figure out that last sentence. How many armies were there in Texas?

"When we're done with you," Snow promised, "you'll be better than any fighting force in the world."

Fred blinked and thought to himself That would mean that we won't be a fighting force. Then he realized that what he was listening to wasn't a well thought out piece of communication. It was like his father who once yelled at his mother, "I could care less what happens to you." When Fred heard that as a small child, he thought that those were words of love. What his father had actually meant was, "I couldn't care less. . . ."

"We'll make you strong," Sergeant Snow predicted and flipped on the overhead projector, which showed against the chow hall wall. "On the first day, the average draftee can do zero pushups."

a very small graph

Fred could see that Snow had plotted the point (1, 0).

"But on day two, you'll be able to do two pushups and on day three, four pushups and by day four, you'll be able to do six pushups."

He had plotted four points:

(1, 0)
(2, 2)
(3, 4)
(4, 6)

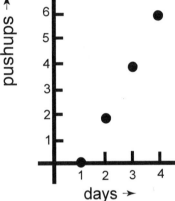

The guy next to Fred mumbled, "Hey, they're all in a line."

Fred thought Let x = the first coordinate, and let y = the second coordinate. Playing with the numbers a little, he came up with the equation: $y = 2x - 2$. (Later on in math, you'll learn how to find that equation. It's called curve fitting.)

The four points all satisfied the equation. For example, the point (3, 4) makes $y = 2x - 2$ true since $4 = 2(3) - 2$.

Sometimes on a graph, the horizontal axis is marked with an "X" since it measures the x-coordinate, and the vertical axis, with a "Y" since it measures the y-coordinate.

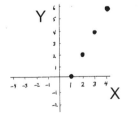

Once you have the equation, you can find other points for the graph. For example, how many pushups might you expect to be doing on the 8th day? That's not hard. We have x = 8 for the equation $y = 2x - 2$. So $y = 2(8) - 2$, which is $y = 14$ pushups.

"Okay, you slimy worms," bellowed the sergeant. "Get down and give me some pushups. Keep your back straight and make sure your chest touches the ground."

Fred had a problem with this order. It wasn't that he was out of shape physically. His muscles, heart, and lungs were in great condition for a six-year-old. For years he had jogged in the morning, and in the privacy of his office, he worked out with weights (using books since they were the only thing available).

When he was in the down position, his facial anatomy offered some difficulties for him.

average recruit   Fred

Fred's nose got stuck in the ground like a tent peg. He got on his hands and knees, pulled his head out of the ground, dug a little hole in the appropriate spot, and then he was okay.

His only other difficulty was that his body was only six years old. Even with his good conditioning, the testosterone-induced musculature of a male in his late teens was not something that Fred possessed.

He could only do six pushups. Snow went up and down the line recording how many pushups each draftee could do. Five of them couldn't do any pushups. When it came to doing one pushup, there were two in that category. Here's Snow's chart:

Fred looked at the chart. The bottom line raised his eyebrows: 473 pushups. Snow had to stop him at 473 because Snow was getting hungry, and it didn't seem like this guy was going to stop doing pushups before the moon came up.

Pushups	How many guys
0	ᵗᴴᴴ
1	II
2	III
3	
4	I
5	
6	I
7	
8	
9	
⋮	⋮
473	I

Dear Reader,

From the very first *Your Turn to Play*, I have asked you to, "Take out a piece of paper and write out the answers."

I know that it is easier just to read the questions and then read the answers—a real temptation.

If you work through the questions in the *Your Turn to Play*, the questions in the cities should be challenging but fun.

For the rest of you, read Aesop's fable of the grasshopper and the ant. (Search the Internet under "Aesop grasshopper ant.")

—Your author *Stan*

Your Turn to Play

1. Plot the points (2, 3) and (4, 1) on a graph.

2. Which of these points satisfies the equation y = 5x – 3?
   (1, 2), (3, 3), (6, 27), (–3, – 18), and (–2, –7)

3. Name the point that is on the x axis and has an x coordinate equal to 5.

4. Name the coordinates of any point that lies on the y axis.

5. Name the point that lies on both axes. That point is called the **origin**.

6. Plot the point (1.5, 1).

7. Does the origin satisfy the equation 3x + y + 6 = 2(x + 3)?

8. Sergeant Snow was giving his lecture to the assembled troops. He spoke at the rate of 60 words/minute. Four minutes after Snow began his oration, Col. Coalback began his speech to the same soldiers. The men didn't know who to listen to. It was very disconcerting.

   Coalback spoke a little faster than Snow (70 words/minute) and they both finished their speeches at the same time.

   When they were done, they had spoken a total of 1020 words. How long did Snow speak? How long did Coalback speak?

9. What is the abscissa of the point (–5, 4)?

. . . . COMPLETE SOLUTIONS . . . .

1.

2. (1, 2), (6, 27), and (–3, –18) satisfy y = 5x – 3.

3. If it's on the x axis, then the y coordinate must be zero.
The point is (5, 0).

4. The answers will vary. You might have written (0, 4) or (0, –3) or (0, 0) or (0, π). The important thing is that the first coordinate will have to be equal to zero.

5. (0, 0). This point is called the origin.

6.

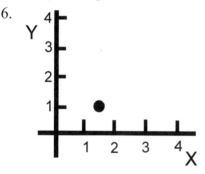

7. In order to find out if the origin, (0, 0), satisfies the equation
$$3x + y + 6 = 2(x + 3)$$
we replace every x by zero and every y by zero
and see if the equation is true.
$$3(0) + 0 + 6 \overset{?}{=} 2(0 + 3)$$
$$0 + 0 + 6 \overset{?}{=} 2(3) \qquad \text{Yes.}$$

8. *Let t = the number of minutes that Snow spoke.*

  Then t – 4 = the number of minutes that Coalback spoke. He started four minutes after Snow began speaking.

Then 60t = the number of words Snow spoke.

Then 70(t – 4) = the number of words that Coalback spoke.

Since they spoke a total of 1020 words . . .

$$60t + 70(t - 4) = 1020$$
$$60t + 70t - 280 = 1020$$
$$130t - 280 = 1020$$
$$130t = 1300$$
$$t = 10$$

Snow spoke for t minutes. He spoke for 10 minutes.

Coalback spoke for t – 4 minutes. He spoke for 6 minutes.

The equation could also have been found using "six pretty boxes."

	words spoken	rate	time
Snow			t
Coalback			

Since Coalback spoke for four minutes less than Snow . . .

	words spoken	rate	time
Snow			t
Coalback			t – 4

	words spoken	rate	time
Snow		60 words/minute	t
Coalback		70 words/minute	t – 4

	words spoken	rate	time
Snow	60t	60 words/minute	t
Coalback	70(t – 4)	70 words/minute	t – 4

Since they spoke a total of 1020 words . . . $60t + 70(t - 4) = 1020$

9. The abscissa is the x coordinate. It is –5.

S now asked the guy who could really do pushups, "What's your name, boy? And what's your secret?"

"Jack LaRoad, sir. My secret? I don't sit around and watch much television, and I do pay attention to what I eat." The muscles on Jack's tummy were all visible: all 320 of them.

Snow's chart could easily be made into a graph. The points of the graph would be (0, 5), (1, 2), (2, 3), (4, 1), (6, 1), and (473, 1).

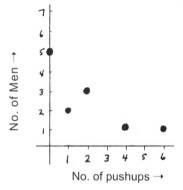

We had to leave out the (473, 1) point on the graph or otherwise the graph might have to look like:

The other item that sharp-eyed readers may have noticed was the statement by Sergeant Snow, "On the first day, the average draftee can do zero pushups."

As you were taught years ago, to find the average of a bunch of numbers, you add them up and divide by the number of numbers that you added together.

In this case there were thirteen numbers that we could add up:

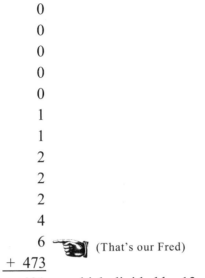

$$0$$
$$0$$
$$0$$
$$0$$
$$0$$
$$1$$
$$1$$
$$2$$
$$2$$
$$2$$
$$4$$
$$6 \quad \text{(That's our Fred)}$$
$$+ \ 473$$

491   which divided by 13 yields approximately 38 as the average number of pushups that a draftee can do on the first day of training.

This kind of average is called the **mean average**. But in this case, reporting the mean average would be very deceiving. Twelve of the thirteen guys on the parade ground would feel very bad, thinking that they couldn't do anywhere near the average number of pushups.

What Snow was explaining to the draftees was the **mode average** number of pushups done by guys on the first day of training. Taking the mode average is a lot easier than taking a mean average.

The mode average is simply the most common number in the bunch of numbers that you're trying to average. The most frequently reported number of pushups was zero, so the mode average was zero.

There's a third kind of average called the **median** and that's also really simple to find. If you want to average the numbers 3, 8, 4, 7, 7, you line them up in order of size—3, 4, 7, 7, 8—and the middle one is the median average.

7 is the median average

Mentioning two new kinds of averages—the mode and the median—in a beginning algebra book might be a little too much, and if it is, we won't talk about the mode or median average at all. We won't even mention them. (irony alert)

What's surprising is that these two maybe-new-to-you kinds of averages are each much simpler to find than the traditional mean average.

Why do they often teach the mean average first?

Multiple choice answers:

(A) The teacher was mean.

(B) They were trying to weed out the slow students in the seventh grade and what better way than to present the hardest kind of average first.

(C) It's the only kind of average that your teacher knew about.

(D) Since, for centuries, they had always taught the mean average to unsuspecting kids, they decided to keep doing it that way when you came along. It's the answer proclaimed in "Fiddler on the Roof," namely, TRADITION!

When you take your first course in statistics, the first thing that your teacher will present is the mean, median, and mode averages. If that course is in college, they will try and make it all sound much more complicated than it is, saying something like (read the following aloud using the most highfalutin, pompous tone you can muster): **"To determine the median average of data, one must first with due diligence order those data in ascending degree of size and then, secondly, ascertain the number of data thus employed so that one may select the middlemost such item—those above and below of equinumerousity. Said item we shall hereinafter designate as the median average of the data."** Gag! Cough! Spit!

Of course, if you get a teacher like Fred, he'll say, "Line 'em up by size and pick the middle one."

Your Turn to Play

Fred received lots of letters offering him jobs at various universities around the country.

According to Belinda's records, during the first ten days of March he received 5, 0, 8, 6, 7, 9, 2, 4, 0, 3 offers.

1. What was the mean average number of offers he received?

2. The mode average?

3. The median average?

4. [a fun question] On March 1 he had received offers from Yale, Harvard, University of California at Berkeley, Stanford, and City College of San Francisco. On March 2 and March 9 Fred received no job offers in the mail. Why wasn't he surprised?

5. Name five integers whose mean average is equal to zero.

6. Is it possible to name five whole numbers whose mean average is equal to zero?

7. Is it possible to name five whole numbers—not all of which are equal to zero—whose median average is equal to zero?

8. Is it possible to name five natural numbers whose median average is equal to 3 and whose mean average is more than a thousand?

9. In what quadrant is the point (–6, 7)?

10. If the abscissa of a point is negative, what two quadrants might the point be in?

11. There were two big stacks of *Modern Trends in Smocking* magazine. Betty read one stack and Darlene read the other.

Let x = the number of hours Betty read. Let y = the number of hours Darlene read. There were the same number of magazines in each stack.

Betty could read 12 magazines/hour.

Darlene could read 10 magazines/hour.

They spent a total of 11 hours going through the magazines. How long did each read?

........COMPLETE SOLUTIONS.......

1. The mean average of 5, 0, 8, 6, 7, 9, 2, 4, 0, 3 is found by adding up the numbers (= 44) and dividing by the number of items (10). The mean average is 4.4 job-offer letters per day.

2. To find the mode average of 5, 0, 8, 6, 7, 9, 2, 4, 0, 3 you pick out the most frequent entry which is zero.

Intermission

They were having a meeting at KITTENS University to determine whether they should increase Fred's salary. (He currently makes $600/month as a full professor.)

The chief financial officer said, "We don't need to increase Fred's salary since he has no other job options. The average number of job-offer letters that he receives each day is zero."

Would you call that lying or telling the truth?

It is a true statement that the mode average of those ten numbers 5, 0, 8, 6, 7, 9, 2, 4, 0, 3 is equal to zero. But part of telling the truth is telling the *whole* truth.

When there are very few items of data, the mode average will often give a misleading summary of the data.

(Averages are summaries of the data.)

---

3. In order to find the median average of 5, 0, 8, 6, 7, 9, 2, 4, 0, 3, you first have to arrange the data from smallest to largest: 0, 0, 2, 3, 4, 5, 6, 7, 8, 9. Then you pick the one in the middle. Oops. The 4 and the 5 are both in the middle. When this happens, you (mean) average those two numbers: $\frac{4+5}{2}$ = 4½. This will be the median average.

4. ". . . no job offers *in the mail* " or anything else in the mail is something that occurs hebdomadally.* March 2 was a Sunday.

---

* *hebdomadal* (heb-DOM-eh-dl) is a word one can hardly live without. It was created sometime between 1605 and 1615. It is an adjective (a word that modifies a person, place, or thing). The adverbial form is a mouthful to pronounce. Try saying, "I enjoy pizza hebdomadally." For the one or two readers who don't know what it means: *hebdomadal* = weekly.

5. Answers will vary. You might have written –33, – 22, 0, 22, 33, or you might have written –7, –2, –1, 4, 6.

6. Yes, if they're all zero.

7. It is possible, if the first three numbers are zero.

8. How about 1, 3, 3, 4, 979230923490236420394902323924096?

9. QII

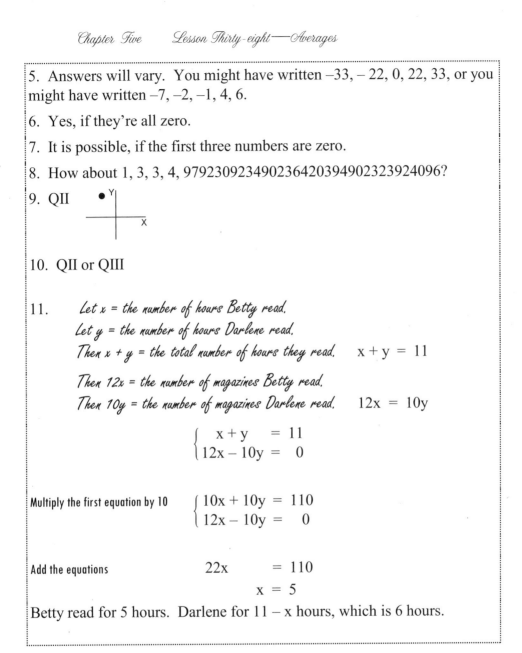

10. QII or QIII

11.   *Let x = the number of hours Betty read.*

     Let y = the number of hours Darlene read.

     Then x + y = the total number of hours they read. $x + y = 11$

     Then 12x = the number of magazines Betty read.

     Then 10y = the number of magazines Darlene read. $12x = 10y$

$$\begin{cases} x + y & = 11 \\ 12x - 10y & = \ 0 \end{cases}$$

Multiply the first equation by 10 $\begin{cases} 10x + 10y = 110 \\ 12x - 10y = \ \ 0 \end{cases}$

Add the equations $\begin{aligned} 22x \quad &= 110 \\ x &= 5 \end{aligned}$

Betty read for 5 hours. Darlene for 11 – x hours, which is 6 hours.

Now it was getting late (4:15 p.m.) and Sergeant Snow was thinking about dinner. He figured that he could assign the men to their bunks tomorrow, which was Sunday. No use rushing things like that.

"Line up in front of the chow hall," he ordered. It was now 4:17 p.m. The hall would open at 5.

They lined up.

Everyone smelled of new canvas clothes. Almost everyone. Fred smelled like an old camouflage T-shirt.

All the other men looked at him. (It wasn't because of his smell, but his stature, but Fred didn't know that.) Feeling self-conscious, he left the line and headed to the bathroom (where he knew they kept wash basins) to launder his shirt so that it wouldn't smell.

As he walked across the parade grounds toward the bathroom, the chaplain met him and said, "I was talking with one of the nurses in the infirmary, and he said that you know something about math."

He didn't want to respond as he had done with the nurse and say that he knew "a bit." His conscience had bothered him after he had said that to the nurse. Telling the truth, as the Ninth Commandment requires, was important to Fred.

So Fred responded, "What do you have in mind?"

"I have to talk tomorrow about the ten young girls who are waiting at night to meet an important dignitary," the chaplain began. "Some of them have fresh batteries for their flashlights and some do not." (He was changing the story a bit, since when he had preached on that passage before, in its more original form, he had trouble explaining why ten virgins were waiting to meet a bridegroom at midnight.)

"The real point of the story," the chaplain continued, "is that either you're ready or you're not."

Fred thought to himself Let x = the number of girls with fresh batteries and let y = the unprepared ones. Then $x + y = 10$.

"I want to put this all in visual form," explained the chaplain. "So the only thing I could think of is a bunch of pictures. For example, if there

216

are nine wise ones, then there must be one foolish one. So I made this drawing that I can use during the sermon."

Fred looked at that and recognized that if x = 9, then y = 1.

"But I also had to make another drawing to show what happens when there are only eight wise ones."

Fred and the chaplain both realized that if he were to illustrate all the different possibilities it would take up most of the thirteen minutes that he had allotted to the sermon time.

Fred suggested that they make one graph to cover all the eleven possibilities from (0, 10), which is zero wise ones and ten foolish ones all the way to (10, 0).

"Those dots are all in a line," the chaplain noted. "Does that happen all the time?"

"Depends on the equation you're graphing," said Fred. "The **linear equations** give you lines, but the other ones can give you all kinds of curves like circles and parabolas."

"Are you joking with me? *Line*ar gives you lines?"

"No," answered Fred. "That's what they call those equations. Would I kid you? Linear equations are any equations that can be put in the form: $\alpha x + \beta y = \gamma$, where alpha ($\alpha$), beta ($\beta$), and gamma ($\gamma$) are any numbers. Oops! I should have written it as $ax + by = c$ where a, b, and c are any numbers. I tend to forget that many people—except for mathematicians and Greeks—don't know the Greek alphabet."

The chaplain laughed and started to recite the Greek alphabet,

"Alpha, beta, gamma, delta, epsilon. . . . I studied Greek in seminary, since it's the language of the New Testament."

Fred hadn't learned that in Sunday School.

"So these are all linear equations," the chaplain said as he took a stick and wrote in the dust:

$5x + 7y = 38$ since it's in the form
$$ax + by = c$$

$32x - 8y = 2$ since this can be written as
$$32x + (-8)y = 2$$

$4x = y - 9$ since this can be written as
$$4x + (-1)y = -9$$

$3x + 99y = 2x + 4$ since this can be written as
$$x + 99y = 4$$

Your Turn to Play

1. Let x = the number of girls with fresh batteries and y = the number of unprepared ones and suppose x + y = 10. What point on the graph would be represented by the case in which there are six girls with fresh batteries?

2. Suppose x represents the number of back issues of *Modern Trends in Smocking* that you order on their website. If you order x back issues, they will charge you y dollars, where y = 4x + 3. [This means that each magazine costs $4 and there is a $3 shipping and handling charge.]

    A) Is y = 4x + 3 a linear equation?

    B) If you order 3 copies, how much do they charge?

    C) If x = 2, what is the value of y?

    D) Find two more points on the graph of y = 4x + 3

3. What is this set called? { x | x = *a/b*, where *a* and *b* are integers and *b* ≠ 0}

4. Name three points that are on the graph of the equation of y = 5x – 7. (First, name any number you like for x, and second, figure out what the corresponding value of y will be.)

5. Is the origin one of the points on the graph of $y = 5x - 7$?

6. Is $4x - 7y = 3x + 88$ a linear equation?

7. Is $9x + 2y = 0$ a linear equation?

8. Is $5y - 4x = \pi$ a linear equation?

Intermission

In case you're curious, here are all twenty-four letters of the Greek alphabet (in lower case and capitals). They only have 24 letters which makes their alphabet easier to learn than ours.

alpha αA

beta βB

gamma γΓ

delta δΔ

epsilon εE

zeta ζZ

eta ηH

theta θΘ

iota ιI

kappa κK

lambda λΛ

mu μM

nu νN

xi ξΞ

omicron oO

pi πΠ

rho ρP

sigma σΣ

tau τT

upsilon υY

phi φΦ

chi χX

psi ψΨ

omega ωΩ

. . . . COMPLETE SOLUTIONS . . . .

1. $(6, 4)$

2. A) $y = 4x + 3$ is a linear equation because it can be put in the form $ax + by = c$, where a, b, and c are any numbers. By transposing, $y = 4x + 3$ can be written $-4x + y = 3$.

    B) If $x = 3$, the equation $y = 4x + 3$ becomes $y = 4(3) + 3$. $y = 15$. They will charge $15.

    C) If $x = 2$, the equation $y = 4x + 3$ becomes $y = 4(2) + 3$. $y = 11$.

    D) Answers will vary depending on what values of x you choose. For example, if you let $x = 7$, then $y = 4x + 3$ becomes $y = 4(7) + 3$. $y = 31$. The point on the graph will be $(7, 31)$. If you let $x = 100$, then $y = 403$ and the point on the graph will be $(100, 403)$. If you let $x = 20$, then $y = 83$ and the point on the graph will be $(20, 83)$.

3. The set of all numbers which can be written as $\dfrac{a}{b}$ where a and b are integers and b isn't equal to zero is called the rational numbers.

4. Answers will vary. If you said that x was equal to 4, then $y = 5x - 7$ would become $y = 5(4) - 7$, and y would be 13. You would have the point $(4, 13)$. If you said that x was equal to 30, then $y = 5x - 7$ would become $y = 5(30) - 7$, and y would be 143. The point would be $(30, 143)$.

5. The origin is the point $(0, 0)$. If we let x equal 0, then $y = 5x - 7$ becomes $y = 5(0) - 7$, and y would be -7. No, the origin isn't a point on the graph of the equation $y = 5x - 7$.

6. By transposing, $4x - 7y = 3x + 88$ becomes $x - 7y = 88$. It is a linear equation.

7. $9x + 2y = 0$ does fit the definition of a linear equation: *Any equation that can be put in the form ax + by = c, where a, b and c are any numbers.* Zero is a number.

8. $5y - 4x = \pi$ can be written as $-4x + 5y = \pi$. Since π is a number, it is a linear equation. Recall that π is approximately equal to

3.141592653589793238462643383279502884197169399375105820974944592307816406286208998628034825342117067982148086513282306647093844609550582231725359408128481117450284102701938521105559644622948954930381964428810975665933446128475648233786783165271201909145648566923460348610454326648213393607260249141273724587006606315588174881520920962829254091715364367892590360011330530548820466521384146951941511609433057270365759591953092186117381932611793105118548074462379962749567351885752724891227938183011949129833673362440656643086021394946395224737190702179860943702770539217176293176752384674818467669405132000568127145263560827785771342757789609173637178721468440901224953430146549585371050792279689258923542019956112129021960864034418159813629774771309960518707211349999998372978049951059731732816096318595024459455346908302642522308253344685035261931188171010003137838752886587533208381420617177669147303598253490428755468731159562863882353787593751957781857780532171226806613001927876611195909216420198938095257201065485863278865936153381827968230301952035301852968995773622599413.

They continued walking and the chaplain said, "When I was studying calculus back in college, I was thinking of going into computer programming and making a lot of money. They made us buy graphing calculators for about $80, which could graph all kinds of curves. Someone stole mine a couple of weeks into the semester, and I had to borrow my friend's calculator several times during the course. I didn't have enough money at that time to buy another one."

At this point it was Fred's turn to laugh a little. "You never needed a graphing calculator. Graphing *any equation* is so simple. Any beginning algebra student can learn to graph anything in three easy steps. How many times did you have to borrow your friend's graphing calculator?"

"Oh, maybe ten times during the semester," the chaplain responded.

"An $80 purchase and it gets used ten times. That works out to eight dollars for each graph."

By this time they had reached the bathroom, and the chaplain, with a frown on his face, asked, "Three easy steps?"

"Sure. All you have to do is plot a bunch of points and connect the dots. It's called **point plotting**."

"Excuse me. Isn't that just two steps?" asked the chaplain.

"You're right. The first step is to find some points." They stepped in front of one of the wash basins. Fred ran the hot water and then began to write on the steamy mirror above the basin:

"Is finding points difficult?" asked the chaplain. "Certainly plotting points and connecting the dots isn't hard."

"No. It's child's play." Fred wiped the mirror with his T-shirt and threw the shirt into the basin to let it soak. He added a couple of squirts of hand soap into the shirt soup and turned his attention back to the mirror, which was steaming over again. "To find points, all you do is name a value for x and then find out the corresponding value of y." He drew a vertical line down the mirror and indicated that the right half of the mirror was the chaplain's. "I'll write out the steps on the left and you graph it on the right. What equation—any equation—would you like?"

"How about y = 2x – 1," the chaplain suggested.

They wrote with Fred on the left and the chaplain on the right:

This is the THEORY side	I'm going to graph y = 2x – 1
Name any value of x. Just name a number.	Okay. I'll say "0"
Find out what y equals when x is 0.	y = 2(0) – 1 so y = –1

"Now you have a point. It's the point (0, –1)," Fred continued. "Plot the point and repeat until you have enough points to figure out what the curve looks like and then connect the dots."

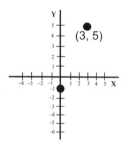

The chaplain said, "I don't think one point is enough to figure out what the curve looks like. Let me try another value for x. Let's say that x is equal to 3. Putting that value in the equation y = 2x – 1, I get y = 2(3) – 1, so y is 5 when x is 3. That's easy enough." He put a second point on the graph and having done that he smiled. He knew that he was virtually done. Since the original equation is linear (since it can be written in the form ax + by = c), all he needed was two points, so he drew a line through those points and he had graphed y = 2x – 1.

Find a steamy bathroom mirror (or use paper). It's . . .

Your Turn to Play

1. Sergeant Snow has had a tendency to gain weight. His weight gain for any week depends on how much lard he's consumed during that week. If x = the number of pounds of lard consumed and y = his weight gain, and we know that y = 2x + 1, plot the graph of his weight gain vs. the amount of lard he's eaten. Since he can't eat less than zero, nor has he ever eaten more than five pounds in a week, we'll limit the graph to the region where 0 ≤ x ≤ 5.   (≤ means *less than or equal to.*)

2. Sometimes when you're plotting, your graph will consist of separate dots: for example, when you're graphing pushups. It wouldn't make sense to talk about 4.78 pushups. Other times your graph will be a line. This happens when the x and y values vary continuously. Temperature or weight or length are good examples of things that can vary continuously so that their graphs would be lines instead of just separated points. Which of the following would be items that varied continuously? A) the number of pigeons that have landed on the chow hall roof today; B) phone calls received; C) area of lawn I could plant; D) time it takes to mow my lawn; E) Col. Coalback's bowling score.

3. Plot 2x + 3y = 18. You do this equation like all the others that we've done so far. You name values for x and compute the corresponding y values. The fact that the equation isn't in the form y = ax + b (where a and b are numbers) doesn't affect the procedure. Let me do one point to get things started. I'll let x = 2. Then the equation becomes 2(2) + 3y = 18.
Then   4 + 3y = 18

3y = 14

y = 14/3 or 4⅔. So I plot the point (2, 4⅔).

4. Find the mean, median, and mode averages for the following numbers:
3, 4, 5, 5, 9

5. Now let's graph something that isn't linear. $xy = 12$ has the shape of a hyperbola ["hi-PERB-eh-lah"], which is studied in calculus. In this case it will be important for you to try negative numbers for x also. Plot values of x in the interval $-12 \leq x \leq 12$.

6. Since we're plotting fancy curves from calculus, let's do a parabola ["pah-RAH-buh-lah"]. Plot $y = x^2$ for values of x in the interval $-5 \leq x \leq 5$. Oops! You haven't had exponents yet: x^2 means "x times x." Let's just say: Plot the curve $y = xx$.

7. The parabola that you graphed in the previous question has no points in which quadrants? (Note: A point on either axis is not contained in any quadrant.)

8. When Fred said that you could graph any equation, he wasn't stretching the truth much at all. You can plot—right now!—trigonometric equations like $y = \sin x$ if you wanted to. And you probably don't even know what sin x means! After this beginning algebra course will normally come, advanced algebra, geometry, and then you'll get to trig. At that point you'll learn what sin x means. But you can still plot $y = \sin x$ now if you've got access to a "scientific" hand calculator that has buttons on it like "sin," "cos," and "tan." If it's also got buttons on it that read "log" or "ln" then you've got all the calculator you need to get through all of *Life of Fred: Calculus Expanded Edition* as well as all of advanced algebra and trig. Those calculators are not too hard to find for under $15—a lot less than those $80 graphing things!

    Now $y = \sin x$ has nothing to do with theology. "Sin" is an abbreviation of "sine" (pronounced "sign"). And $y = \tan x$ has nothing to do with getting a suntan. They both deal with triangles, but that's about all we can say right now. If you must really know what sin x means right now, you'll need to ask your older sister or read faster.

    On your calculator, if you've got one, type in 30 and then press the "sin" button. If things are working right the answer 0.5 will appear. So sin 30 = 0.5 and we have the point (30, 0.5) to plot. Plot the graph of $y = \sin x$ in the interval $0 \leq x \leq 90$.

    If you don't have one of these calculators (and your older sister won't let you borrow hers), then use these values off my calculator:

if x = 0 then y = sin x = sin 0 = 0

if x = 10 then y = sin x = sin 10 = 0.17

if x = 20 then y = sin x = sin 20 = 0.34

and the points (30, 0.5), (40, 0.64), (50, 0.77), (60, 0.87), (70, 0.94), (80, 0.98), (90, 1).

. . . . COMPLETE SOLUTIONS . . . .

1. First we need to find some points to plot for y = 2x + 1. If x = 0, then y = 2(0) + 1.   We have the point (0, 1). That means that if he eats no lard during a week, we can expect a weight gain of one pound. (That's because he's a heavy beer drinker, and that also can contribute to a beer belly.)

   If we let x equal, for example, three, then y = 2(3) + 1 = 7, so we have the point (3, 7). If x = 4, then y = 2(4) + 1 = 9 and we have the point (4, 9). Plotting these three points and drawing a line segment for all x between zero and five we get

2. A) The number of pigeons is not continuous; it is a whole number. It's hard to imagine having a third of a pigeon land on the roof. B) The number of phone calls must be also a whole number. C) I could plant any (reasonable) number of square feet of lawn. Area can be measured as a continuous quantity. One way to look at continuous quantities is to ask the question, "Between any two numbers is there another number?" Between, for example, 34.7 and 34.8 square feet of lawn, there is 34.75 square feet of lawn. This wouldn't work with phone calls. Between three and four phone calls, there isn't 3.7 phone calls. D) Time varies continuously. E) Bowling scores can be any whole number between 0 and 300 depending on how many pins you knock over. Since it's tough to knock a pin slightly over, it would be equally as tough (impossible) to obtain a bowling score of 237½.

3. If I let x = 0, then 2x + 3y = 18 becomes 2(0) + 3y = 18, so that y = 6. The point will be (0, 6)

If I let x = 1, then 2x + 3y = 18 becomes 2(1) + 3y = 18.

Then 2 + 3y = 18

$$3y = 16$$

y = 16/3 or 5⅓. The point will be (1, 5⅓).

Other possible points are (3, 4), (4, 3⅓), (5, 2⅔), (6, 2). You may have named different x values than these.

4. For the mean average: 3 + 4 + 5 + 5 + 9 = 26, then divide by five, which gives 5.2.

For the median average the numbers are already arranged in order so all we need to do is pick out the middle one 3 4 5 5 9, which is 5.

For the mode average we pick out the most common item, which is 5 since it's mentioned twice.

5. If I let x = 3, for example, in xy = 12, we get 3y = 12 so that y = 4. This gives the point (3, 4).

Other points that you may have obtained are: (1, 12), (2, 6), (4, 3), (6, 2), (12, 1), (–1, –12), (–2, –6), (–3, –4).

The only time the value of x that gives us any "excitement" is when we let x equal zero.

Then xy = 12 becomes 0y = 12, which is the same as 0 = 12. Since this is never true, there are no values of y that make 0y = 12 true. Hence there are no points whose first coordinate is zero.

6. Possible points are (0, 0), (1, 1), (2, 4), (3, 9), (–1, 1), (–4, 16). Your selection of x values may differ from mine.

This is the first time that we've plotted the point (0, 0). That point has a special name. It's called the **origin**. It's the place where the x- and y-axes meet.

7. There are no points in QIII or in QIV.

8.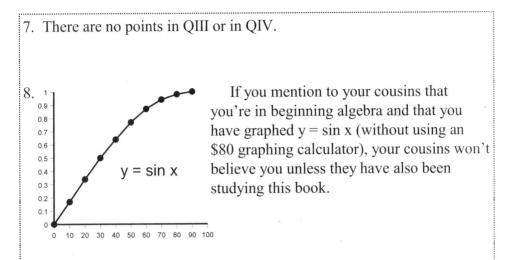

If you mention to your cousins that you're in beginning algebra and that you have graphed y = sin x (without using an $80 graphing calculator), your cousins won't believe you unless they have also been studying this book.

Note that we have different scales on the x- and y-axes. The horizontal axis goes from zero to 90 and the vertical axis from zero to 1.0. It's okay to do that.

Fred drained out the yellow-brown sudsy water and rinsed out his shirt in clear water. He wrung it out and put it on. Saying goodbye to the chaplain, he hurried back to get in the line in front of the chow hall.

Stigler, Oklahoma known for a statue of the Ten Commandments on the courthouse lawn

1. Which of the following would be plotted as a bunch of separate points rather than as a continuous line?

     ✦ The number of teeth in pet cats.
     ✦ The height of pet dogs.
     ✦ The number of gallons of ice cream eaten.

2. Solve
$$\begin{cases} 2x + 5y = 25 \\ 7x - 5y = 20 \end{cases}$$

3. $\{\pi, \text{my aunt}\} \cup \{5, \text{my aunt}\} = ?$

4. Name the coordinates of a point in quadrant four.

5. The point (α, β) is directly below the point $(4, 9)$. What can you say about α? What can you say about β?

6. How many points are there that are *not* in any quadrant?

7. Graph $y = \log x$. You may not know what log x is, but that doesn't matter right now. (Logarithms are covered in *Life of Fred: Advanced Algebra Expanded Edition*.) If you have a scientific calculator with a "log" button on it, here's how you would find, for example, log 20: you enter 20 and then press the "log" button. You'll get an answer of approximately 1.3. If you don't have one of those calculators, here are some points that I got from my calculator: $(0.1, -1)$, $(1, 0)$, $(2, 0.3)$, $(3, 0.48)$, $(4, 0.6)$, $(10, 1)$.

The graph will be a continuous curve. As was noted in the answer to question 8 on the previous page, you might want to use different scales for the x- and y-axes.

8. Which of these equations have graphs that pass through the origin?

     $2x + 3y = 10$
     $y = x + 4$
     $33x = 8y$

9. Solve
$$\begin{cases} 2x + 3y = 7 \\ -10x - 3y = 25 \end{cases}$$

10.   Four chocolate Waddle doughnuts and a jelly doughnut weigh 20 pounds.

     Nine chocolate Waddle doughnuts and two jelly doughnuts weigh 43 pounds. How much does a jelly doughnut weigh?

11. A graph can be considered a set of ordered pairs. For example, the graph on the second page of Lesson 37 could be expressed as $\{(1, 0), (2, 2), (3, 4), (4, 6)\}$.

      Using set-builder notation, how might you write the set of all points on $77x + 32y = 4$?

12. Find the mean, median, and mode averages for: 3, 6, 6, 7, 9.

13. Which of these equations are linear?

      $3xy + 6y = 3xy + x + 14$

      $\pi x + 9y = 14$

      $y = 6$

14. Graph $y = x - 6$ for all abscissa from zero to 7.

15. You are told that the abscissa of a point is negative. Is it possible that that point is neither in QII nor QIII?

answers

1. The number of teeth in pet cats would be plotted as a bunch of separate points.

2.
$$\begin{cases} 2x + 5y = 25 \\ 7x - 5y = 20 \end{cases}$$

Add the equations $9x = 45$

$$x = 5$$

Back substituting $x = 5$ into the first equation

$$2(5) + 5y = 25$$
$$10 + 5y = 25$$
$$5y = 15$$
$$y = 3$$

3. $\{\pi, \text{my aunt}, 5\}$

4. Points like $(4, -5)$ or $(\pi, -2)$ are in QIV. The first coordinate must be positive and the ordinate must be negative.

5. If one point is directly below the other, then their first coordinates must be the same. So $\alpha = 4$. Points that are below other points have smaller y-coordinates. So $\beta < 9$.

6. An infinite number of points can be plotted that aren't in any quadrant. The first of these is the origin, $(0, 0)$, but also, any point on either of the two axes isn't in any quadrant.

7.

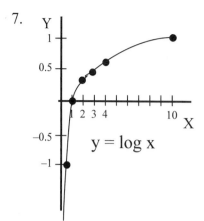

$$y = \log x$$

8. $33x = 8y$ passes through the origin since $(0, 0)$ satisfies the equation.

9.
$$\begin{cases} 2x + 3y = 7 \\ -10x - 3y = 25 \end{cases}$$

Add the equations
$$-8x = 32$$
$$x = -4$$

Back substituting $x = -4$ into the first equation . . .
$$2(-4) + 3y = 7$$
$$-8 + 3y = 7$$
$$3y = 15$$
$$y = 5$$

10. *Let x = the weight of a chocolate doughnut.*
Let y = the weight of a jelly doughnut.
$$\begin{cases} 4x + y = 20 \\ 9x + 2y = 43 \end{cases}$$

Multiply the first equation by -2
$$\begin{cases} -8x - 2y = -40 \\ 9x + 2y = 43 \end{cases}$$

Add the equations
$$x = 3$$

Back substituting $x = 3$ into the first equation . . .
$$4(3) + y = 20$$
$$12 + y = 20$$
$$y = 8$$

A Waddle jelly doughnut weighs eight pounds. They also make great chairs—just pull up a doughnut and sit down.

11. $\{(x, y) \mid 77x + 32y = 4\}$ Once you have seen how to do this, you can do twenty of them per minute.

12. mean $= 6.2$; median $= 6$; mode $= 6$.

13. They all are.

     The first equation can be put in the form $-x + 6y = 14$ by subtracting $3xy$ and x from each side.

     In the second equation, since π is a number, $\pi x + 9y = 14$ is already in the form $ax + by = c$.

     The third equation can be written as $0x + y = 6$.

14.

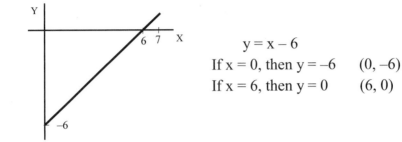

$$y = x - 6$$
If $x = 0$, then $y = -6$ $(0, -6)$
If $x = 6$, then $y = 0$ $(6, 0)$

15. Yes! If the point were, for example, $(-4, 0)$ it would not be in any quadrant (since points on the axes are not in any quadrant).

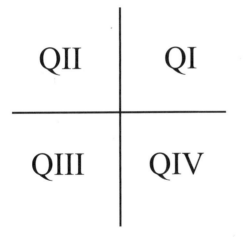

Wyoming, Pennsylvania in the Wyoming Valley

1. Using two equations and two unknowns, set up the equations for: three cans of antacid and twenty kettles of fish that cost $55. Five cans of antacid and eighteen kettles of fish that cost $49. (You need not solve them.)

2. Solve
$$\begin{cases} 5x - 2y = 18 \\ x + 2y = -6 \end{cases}$$

3. $\{\#, 17\} \cup \{\ \} = ?$

4. Name a point in QIII.

5. Five bow ties and a bus ticket cost a total of $22. Two bow ties and three bus tickets cost $40. How much is a bus ticket?

6. Find the mean, median, and mode averages for: 4, 4, 5, 8, 10.

7. Graph $y = 4$. (Hint: for every value of x that you name, the value of y will be 4. So the graph will be the set of all points of the form (a, 4) where a is any number.)

8. Which of these equations have graphs that pass through the origin?

$$7y + 4x = 0 \qquad 2.2x = \pi y \qquad y = 3x - 7$$

9. Which of these are not continuous?
   - ✦ The time it takes to get a haircut.
   - ✦ The number of roads leading out of a city.
   - ✦ The cars on the road.

10. Solve
$$\begin{cases} 7x + 4y = -29 \\ -7x + 3y = 64 \end{cases}$$

(Hint: There is an easier way than multiplying the top equation by 3 and the bottom equation by −4.

11. At the infirmary, four aspirin and five bandages cost a total of $49. Four aspirin and eight bandages cost $64. How much does an aspirin cost?

12. Graph $3x + 2y = 6$.

13. To graph x = y^4 (which is the same as x = yyyy), it is much easier to first name values for y and then find corresponding values for x. For example, if y were equal to 3, then x would be 81 and the point to plot would be (81, 3). Name four other points on the curve x = y^4.

14. Graph x = y^2
(which is the same as x = yy).

answers

1. 3x + 20y = 55   5x + 18y = 49 (You may have used different letters than x and y.)

2.
$$\begin{cases} 5x - 2y = 18 \\ x + 2y = -6 \end{cases}$$

Add the equations $6x \qquad = 12$

$$x = 2$$

Back substituting x = 2 into the second equation $2 + 2y = -6$

$$2y = -8$$
$$y = -4$$

3. {#, 17} (The union of the empty set with any set A always yields A.)

4. Any point of the form (a, b) where a and b are both negative.

5.   *Let x = the price of a bow tie.*
        Let y = the price of a bus ticket.

$$\begin{cases} 5x + y = 22 \\ 2x + 3y = 40 \end{cases}$$

Multiply the first equation by −3 $\begin{cases} -15x - 3y = -66 \\ 2x + 3y = 40 \end{cases}$

Add the equations $-13x \qquad = -26$

$$x = 2$$

Back substituting x = 2 into the first equation $5(2) + y = 22$

$$10 + y = 22$$
$$y = 12$$

A bus ticket cost $12.

6. mean = 6.2   median = 5   mode = 4

7.

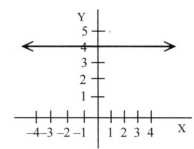

8. $7y + 4x = 0$ and $2.2x = \pi y$ pass through the origin.

9. The number of roads leading out of a city. The cars on the road.

10.
$$\begin{cases} 7x + 4y = -29 \\ -7x + 3y = 64 \end{cases}$$

Add the equations
$$7y = 35$$
$$y = 5$$

Back substituting $y = 5$ into the second equation
$$-7x + 3(5) = 64$$
$$-7x + 15 = 64$$
$$-7x = 49$$
$$x = -7$$

11. *Let x = the cost of an aspirin.*
Let y = the cost of a bandage.
$$\begin{cases} 4x + 5y = 49 \\ 4x + 8y = 64 \end{cases}$$

Multiply the first equation by -1
$$\begin{cases} -4x - 5y = -49 \\ 4x + 8y = 64 \end{cases}$$

Add the equations together
$$3y = 15$$
$$y = 5$$

Back substituting $y = 5$ into the first equation
$$4x + 5(5) = 49$$
$$4x + 25 = 49$$
$$4x = 24$$
$$x = 6$$

An aspirin costs $6.

12. $3x + 2y = 6$

If x = 0, then y = 3. The point (0, 3).
If x = 1, then y = 1½. The point (1, 1½).
If x = 2, then y = 0. The point (2, 0).

13. The answers will vary.
The most common answers might be (0, 0), (1, 1), (1, –1), (16, 2), (16, –2).

14. $x = y^2$

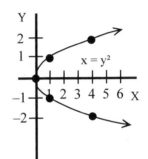

If y = 0, then x = 0. The point (0, 0).
If y = 1, then x = 1. The point (1, 1).
If y = –1, then x = 1. The point (1, –1).
If y = 2, then x = 4. The point (4, 2).
If y = –2, then x = 4. The point (4, –2).

Roswell, New Mexico near the supposed 1947 sighting of a UFO

1. Using two equations and two unknowns, set up the equations for: six pies and five cakes that cost a total of $27. Seven pies and three cakes that cost $25. (You need not solve them.)

2. Solve
$$\begin{cases} 2x + 5y = 7 \\ 8x - 5y = 53 \end{cases}$$

3. {May, June} $\cup$ {May, #} = ?

4. The point (a, b) is directly to the left of $(5, 7)$. What can you say about a? What can you say about b?

5. In 50 minutes a nurse at the infirmary can inject four patients and clean two bowls. It takes that nurse 58 minutes to inject five patients and clean one bowl. How long does it take to clean a bowl?

6. Graph x = –3. (Hint: There are many points on this curve and all of them have an abscissa equal to –3.)

7. What is the equation of the vertical line that passes through the points $(2, 3), (2, 6), (2, 55), (2, \pi)$, and $(2, ½)$?

8. Which of these equations have graphs that pass through the origin?
$$4x + 9y = 12$$
$$x + 3.8y = 0$$
$$3x - 9y = 2x + 1$$

9. Solve
$$\begin{cases} 3x - 13y = -59 \\ -5x + 13y = 81 \end{cases}$$

10. In what quadrant is the point $(-3, 7)$?

11. It takes an army cook 59 minutes to wash seven pans and clean three ovens. It takes 103 minutes to wash fourteen pans and clean five ovens. How long does it take to wash a pan?

12. Find the mean, median, and mode averages for: 1, 1, 3, 7, 8.

13. Is there a point that is both in QII and in QIII?

14. Graph –x + y = 2

15. Points (a, b) and (c, d) lie on the same vertical line. What, if anything, can be said about the numbers a, b, c, and d?

answers

1. $\begin{cases} 6x + 5y = 27 \\ 7x + 3y = 25. \end{cases}$

Or you could have used other letters such as $6p + 5c = 27$; $7p + 3c = 25$.

2. $\begin{cases} 2x + 5y = 7 \\ 8x - 5y = 53 \end{cases}$

Add the equations $10x \quad\quad = 60$

$x = 6$

Back substituting $x = 6$ into the first equation $2(6) + 5y = 7$

$12 + 5y = 7$

$5y = -5$

$y = -1$

3. {May, June, #}

4. $a < 5$ and $b = 7$

5.   *Let x = the number of minutes to inject a patient.*

Let y = the number of minutes to clean a bowl.

$\begin{cases} 4x + 2y = 50 \\ 5x + y = 58 \end{cases}$

Multiply the second equation by −2 and add the equations $-6x \quad\quad = -66$

$x = 11$

Back substituting $x = 11$ into the second equation $5(11) + y = 58$

$55 + y = 58$

$y = 3$

It takes three minutes to clean a bowl.

6. $x = -3$

If $y = 0$, then $x = -3$. The point $(-3, 0)$.
If $y = 1$, then $x = -3$. The point $(-3, 1)$.
If $y = 2$, then $x = -3$. The point $(-3, 2)$.
If $y = 3$, then $x = -3$. The point $(-3, 3)$.
If $y = 398{,}925{,}996$, then $x = -3$.

The point $(-3, 398{,}925{,}996)$.   Can you tell when I get bored?

7. $x = 2$ This is very similar to the previous problem.

8. $x + 3.8y = 0$ has a graph that passes through the origin.

9.
$$\begin{cases} 3x - 13y = -59 \\ -5x + 13y = 81 \end{cases}$$

Add the equations $-2x \qquad = 22$

$$x = -11$$

Back substitute $x = -11$ into the first equation $3(-11) - 13y = -59$

$$-33 - 13y = -59$$
$$-13y = -26$$
$$y = 2$$

10. QII

11.   *Let x = the number of minutes to wash a pan.*

     Let y = the number of minutes to clean an oven.

$$\begin{cases} 7x + 3y = 59 \\ 14x + 5y = 103 \end{cases}$$

Multiply the first equation by -2 $\begin{cases} -14x - 6y = -118 \\ 14x + 5y = 103 \end{cases}$

Add the equations $-y = -15$

$$y = 15$$

Back substituting $y = 15$ into the first equation $7x + 3(15) = 59$

$$7x + 45 = 59$$
$$7x = 14$$
$$x = 2$$

It takes two minutes to wash a pan.

12. mean = 4; median = 3; mode = 1

13. No. To be in QII the ordinate must be positive and in order to be in QIII the ordinate must be negative.

14.

$-x + y = 2$

If $x = 0$, then $y = 2$. The point $(0, 2)$.
If $x = 1$, then $y = 3$. The point $(1, 3)$.
If $x = -2$, then $y = 0$. The point $(-2, 0)$.

15. If (a, b) and (c, d) lie on the same vertical line then they have the same x-coordinates so it must be true that $a = c$. If we assume that we're talking about two different points, then $b \neq d$.

Fred arrived back at the chow hall line with a clean, old camouflage T-shirt. The rest of the guys in the line were still staring at him. (He was still short.) It was 5 p.m. and they all filed in to take their seats.

While Fred had been away, the Sergeant had carefully explained to the recruits how the seating arrangements worked in the chow hall. Each table could seat two men. Snow had been worried about this since he had 13 men in his group, but when he counted them as they stood in line there were only 12 of them, so pairing them up had been easy.

They marched into the hall and sat down in pairs leaving Fred standing there alone. He had not been paired with anyone and didn't know what to do.

"Hey!" yelled one of the recruits. "Look at him standing there. He must be the chaplain. Aren't you going to say grace for us?"

Not realizing that they were making fun of him, Fred said, "Okay. Let's bow our heads. Thank you Lord. . . ." He was cut off by the racket of a dozen mouths inhaling their chow. They had bowed their heads, not to pray, but to eat.

> As Brillat-Savarin wrote in 1825, "Gluttony is mankind's exclusive prerogative."

As Fred stood there, he looked at the tables. At each table the cooks placed a bowl of ten hot dogs between a pair of men.

In his mind Fred thought Let x = the number of hot dogs the first man eats and let y = the number that the second man eats. Then x + y = 10. The graph would look like————→

> The graph is continuous because one man might eat 2⅔ hot dogs and the other 7⅓.

Nothing much new here. Then he noted that each man was getting the same number of hot dogs as his partner.

If that's the case Fred continued in his thoughts it must be true that x = y.

He then thought of a new way to solve the system of equations:

$$\begin{cases} x + y = 10 \\ x = y \end{cases}$$

His new way: On top of the graph of $x + y = 10$, he also graphed the line $x = y$ and obtained:

The two lines intersected at (5, 5). That would be the **solution by graphing** to

$$\begin{cases} x + y = 10 \\ x = y \end{cases}$$

Why does that work? Answer: Every point on the line $x + y = 10$ satisfies $x + y = 10$. Every point on $x = y$ satisfies $x = y$. The point that lies on both lines is the point that satisfies both equations.

small essay
Solving by Graphing

Objection #1: Why learn how to solve pairs of equations by graphing when we already know how to solve them by algebra (the elimination method)?

Objection #2: When we graph the two lines, we have to sometimes estimate the point of intersection. With the elimination method, we could always get the exact answer.

Answers to the objections: Sometimes we have a pair of equations that cannot be solved by elimination. For example, it is pretty tough to eliminate either the x or the y from $\begin{cases} xy = y^x \\ 7 + \sqrt{x+y} = \log x \end{cases}$

However, you can graph any equation by point-plotting. (You graphed $y = \log x$ in problem 7 of the first city of Chapter 5.)

Solving by graphing is sometimes the only way to go.

end of small essay

Your Turn to Play

1. Solve by graphing: $\begin{cases} x + y = 6 \\ x = y \end{cases}$

2. Solve by graphing: $\begin{cases} y = 2x - 1 \\ y = -x + 5 \end{cases}$

3. Solve by graphing $\begin{cases} 2x + y = 6 \\ y - x = 3 \end{cases}$

4. $\xi^4 \xi^7 = \xi^?$ (ξ is the letter xi. It's my favorite Greek letter.)

5. If the diameter of a circle is equal to 6 miles, what is the exact circumference?

6. In Fred's trigonometry class, for every five students who like the sine function best, there are two for whom the sine function is not their favorite. There are 91 students in that class. How many of them like the sine function best?

· · · · · · · **COMPLETE SOLUTIONS** · · · · · · ·

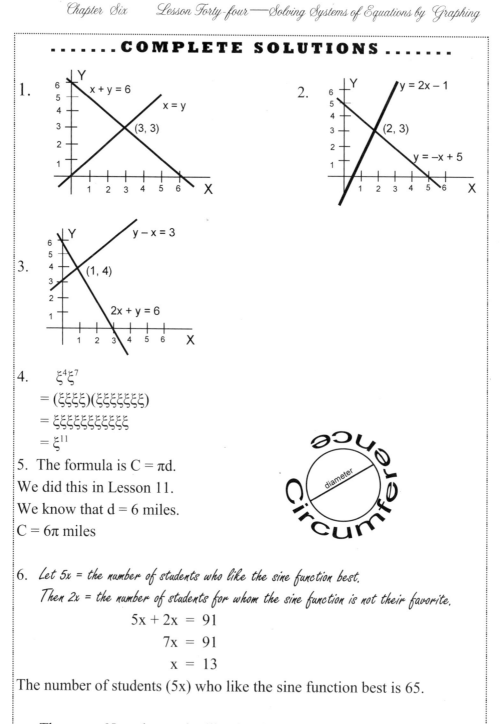

1. [graph with lines x + y = 6 and x = y intersecting at (3, 3)]

2. [graph with lines y = 2x – 1 and y = –x + 5 intersecting at (2, 3)]

3. [graph with lines y – x = 3 and 2x + y = 6 intersecting at (1, 4)]

4. $\xi^4\xi^7$

    $= (\xi\xi\xi\xi)(\xi\xi\xi\xi\xi\xi\xi)$

    $= \xi\xi\xi\xi\xi\xi\xi\xi\xi\xi\xi$

    $= \xi^{11}$

5. The formula is $C = \pi d$.

    We did this in Lesson 11.

    We know that d = 6 miles.

    $C = 6\pi$ miles

6. *Let 5x = the number of students who like the sine function best.*

    Then 2x = the number of students for whom the sine function is not their favorite.

    $$5x + 2x = 91$$
    $$7x = 91$$
    $$x = 13$$

    The number of students (5x) who like the sine function best is 65.

    There are 65 students who like the sine function best and 26 for whom the sine function is not their favorite.

Solving by **substitution** is a third way that Fred thought of. If the system of equations looks like $\begin{cases} 53x + 20y = 2114 \\ x = 7y + 3 \end{cases}$ using substitution can be the easiest way to go.

Here is how it is done.

Step one: Solve one of the equations for one of the variables.

That has already been done. We have $x = 7y + 3$.

Step two: Substitute a "$7y + 3$" for every "x" in the other equation.

$$53x + 20y = 2114$$

becomes $53(7y + 3) + 20y = 2114$

We now have one equation in one unknown.

Step three: Solve that equation. We find that $y = 5$.

Step four: Back substitute $y = 5$ into any equation containing both x and y.

$x = 7y + 3$ looks like the easiest one. $\quad x = 7(5) + 3$

$$x = 38$$

We now have three ways of solving two equations, two unknowns:

☆ by the elimination method of the previous chapter,

☆ by graphing the two lines and seeing where they intersect, and

☆ by substitution.

The method we use depends on the equations.

This system $\begin{cases} 34x + 92y = 4 \\ 26x - 92y = 39 \end{cases}$ cries out "Use elimination!"

This system $\begin{cases} 2x + y = 22 \\ y = 6x - 50 \end{cases}$ calls for substitution.

The second equation is already solved for y, so we substitute the value of y (which is $6x - 50$) into the first equation and obtain

$$2x + (6x - 50) = 22$$
$$2x + 6x - 50 = 22$$
$$8x - 50 = 22$$
$$x = 9$$

Back substituting this value of $x = 9$ into any of the original equations will give us the value of y. If we substitute x = 9 into y = 6x – 50, we obtain

$$y = 6(9) – 50$$
$$y = 54 – 50$$
$$y = 4.$$

Your Turn to Play

1. Solve by substitution: $\begin{cases} 3x + y = 14 \\ y = 2x - 6 \end{cases}$

2. Solve by substitution: $\begin{cases} 5x - 2y = 1 \\ y = x + 4 \end{cases}$

3. Solve by substitution: $\begin{cases} x - 3y = 14 \\ x = 6y + 23 \end{cases}$

4. Solve: $\begin{cases} 15x + 3y = 27 \\ 7x - 3y = -5 \end{cases}$

.......COMPLETE SOLUTIONS.......

1. $\begin{cases} 3x + y = 14 \\ y = 2x - 6 \end{cases}$

Substituting $2x - 6$ in place of y in the first equation:

$$3x + (2x - 6) = 14$$
$$3x + 2x - 6 = 14$$
$$5x - 6 = 14$$
$$5x = 20$$
$$x = 4$$

Back substituting x = 4 into the second equation:
$$y = 2(4) - 6$$
$$y = 2$$

2. $\begin{cases} 5x - 2y = 1 \\ y = x + 4 \end{cases}$

Substituting $x + 4$ in place of y in the first equation:

$$5x - 2(x + 4) = 1$$
$$5x - 2x - 8 = 1$$
$$3x - 8 = 1$$
$$3x = 9$$
$$x = 3$$

Back substituting x = 3 into the second equation:
$$y = 3 + 4$$
$$y = 7$$

3. $\begin{cases} x - 3y = 14 \\ x = 6y + 23 \end{cases}$

Substituting $6y + 23$ in place of x in the first equation:

$$(6y + 23) - 3y = 14$$
$$6y + 23 - 3y = 14$$
$$3y + 23 = 14$$
$$3y = -9$$
$$y = -3$$

Back substituting y = −3 into the second equation:
$$x = 6(-3) + 23$$
$$x = -18 + 23$$
$$x = 5$$

4. $\begin{cases} 15x + 3y = 27 \\ 7x - 3y = -5 \end{cases}$

The easiest way to solve this system is by the elimination method.
Adding the two equations together:
$$22x = 22$$
$$x = 1$$

Back substituting x = 1 into the first equation:
$$15(1) + 3y = 27$$
$$y = 4$$

~~~

Fred had three ways of determining how many hot dogs each man at the two-man tables would receive. What he wasn't able to determine was how he could obtain *any* hot dogs.

He noticed one old man (26) was sitting by himself. Fred walked over to him and asked, "Mind if I join you?" Six helmeted and armed guards seized Fred and started to drag him away.

"It's okay boys," the old guy said. "Let him come sit with me." The guard dragged Fred back to the table. One of them obtained three phone books for him to sit on and they dropped him opposite Col. C. C. Coalback. "Have a dog, boy!" Coalback said as he gestured toward the bowl between them.

Fred thanked the man and reached for the silver tongs to serve himself a "dog" out of the Lennox china bowl. He noticed that the hot dogs at this man's table were larger than at any of the other tables.

Fred couldn't lift it out of the bowl. The diameter of the end was about two feet and the dog was about three feet long. If this thing were hollow thought Fred I could crawl inside it. I wonder how much it weighs?

The place to start, Fred figured, would be to find the volume of that hot dog (and then convert it to a weight). The hot dog was shaped like a cylinder and the volume of a cylinder is $V = \pi rrh$. Using exponents, the formula is $V = \pi r^2 h$ where r^2 means $r \cdot r$ (read "r squared").

diameter

Fred had everything he needed now to compute the volume. He had $V = \pi r^2 h$. He knew h was 3 feet since that's the length of the hot dog, and he knew that the radius was one foot since that's half of the diameter.

$$V = \pi r^2 h$$
$$V = \pi(1)^2 3$$
$$V = \pi(1)(1)(3)$$
$$V = 3\pi \text{ cu ft}$$

radius

Since Fred's measurement of the dimensions of the hot dog was only approximate to begin with, he called π approximately equal to three, even though you and I and Fred know π is more precisely equal to

3.14159265358979323846264338327950288419716939937510582097494459230781640628620899862803482534211706798214808651328230664709384460955058223172535940812848111745028410270193852110555964462294895493038196442881097566593344612847564823378678316527120190914564856692346034861045432664821339360726024914127372458700660631558817488152092096282925409171536436789259036001133053054882046652138414695194151160943305727036575959195309218611738193261179310511854807446237996274956735188575272489122793818301194912983367336244065664308602139494639522473719070217986094370277053921717629317675238467481846766940513200056812714526356082778577134275778960917363717872146844090122495343014654959585371050792279689258923542019956112129021960864034418159813629774771309960518707211349999998372978049951059731732816096318595024459455346908302642522308253344685035261931188171010003137838752886587533208381420617177669147303598253490428755468731159562863882353787593751957781857780532171226806613001927876611195909216420198938095257201065485863278865936153381827968230301952035301852968995773622599413891249721775283479131515574857242454150695950829533116861727855889075095838175463746493931925506040092770167113900984882401285836160356370766010471018194295559619894676783744944825537977472684710404753464620804668425906949129331367702898915210475216205696602405803815019351125338243003558764024749647326391419927260426992279678235478163600934172164121992458631503028618297455570674983850549458858692699569092721079750930295532116534498720275596023648066549911988183479775536636980742654252786255181841757467289097777279380008164706001614524919217321721477235014144197356854816136115735255213345757418494684385233239073941433345477624168625189835694855620992192221842725502542568876717904946016534668049886272327917860857843838279679766814541009538837863609506800642251252051173929848960841284886269456042419652850222106611863067442786220391949450471237137869609563643719172874676774657573962413890865832645995813390478027590 . . .

The way that's written is π ≈ 3. A wavy equal sign means "approximately equal to."

$$V = 3\pi \approx 3(3) = 9 \text{ cu ft}$$

And guessing that hot dogs have a density of about 60 lbs/cu ft, he found that the weight of this giant dog was approximately 9(60) = 540 lbs. No wonder he couldn't lift it!

The reason Fred estimated the density of the hot dog at 60 lbs/cu ft was that he knew from his study of physics that water weighs approximately 62.4 lbs/cu ft and he had taken a little chunk out of the hot dog and found that it barely floated in a glass of water he had in front of him.

It gets tiring after a while to keep writing $V = \pi rrh$ instead of $V = \pi r^2 h$. It gets really tiring to keep writing 10,000,000,000,000,000,000,000,000,000,000,000,000,000,000,000,000,000,000,000,000,000,000,000,000,000,000,000,000 atoms in the observable universe (as we did in the first chapter). It's much easier to write 10^{79} atoms. It's time to talk about exponents.

r^2 means rr.   The little ^{2} is called an exponent.

y^5 means yyyyy.

ξ^{14} means ξξξξξξξξξξξξξξ   (where ξ—my favorite Greek letter, xi—rhymes with "eye.")

TiMe oUt FoR ChemISTRY!

The **atomic weight** of an atom is (roughly speaking) the number of protons and neutrons it has.

You can look up the atomic weight of any atom in a chemistry book. For example, the atomic weight of copper is 63.55. Copper always has 29 protons. Some atoms of copper have 34 neutrons and some have 35 neutrons. That's why the atomic weight of copper is an average and not just a whole number such as 63 or 64.

A chem teacher can do the famous "Count the Atoms" trick. One student is told to weigh out 63.55 grams of copper and put it in a beaker. Another student is told to weigh out 107.87 grams of silver and put it in another beaker. (The atomic weight of silver is 107.87.)

Each student can take any element and find its atomic weight in a chemistry book. Then weigh out that many grams in a beaker.

The chem teacher goes to the beaker of copper and pretends to count the atoms. "There are 602,200,000,000,000,000,000,000 atoms in that beaker."

Then to the beaker of silver. "There are 602,200,000,000,000,000, 000,000 atoms of silver in that beaker."

Every beaker will have 602,200,000,000,000,000,000,000 atoms in it. It is a bit of chemistry magic.

The problem comes when students are trying to take notes on what the chem teacher has just said. They are trying to count the number of zeros in that magic number 602,200,000,000,000,000,000,000. One student writes $6.022 \times 100{,}000{,}000{,}000{,}000{,}000{,}000{,}000$. Another student who has had algebra shortens it up to 6.022×10^{23} using exponents. That is a lot easier to look at.

6.022×10^{23} is called Avogadro's number.

Amedeo Avogadro

Your Turn to Play

1. $5^4 = ?$

2. Solve by graphing: $\begin{cases} x + y = 5 \\ x - 2y = -1 \end{cases}$

3. Solve by substitution: $\begin{cases} 8x + 2y = 46 \\ y = 2x - 1 \end{cases}$

4. What's the volume of a cylinder that is five inches tall and has a radius of two inches?

5. The Super-Thick Bell-Pepper-and-Anchovy Pizza™ is a favorite of Sergeant Snow. The pizza is 1" tall and is packed with melted cheese, bell peppers, and you-know-what. Snow normally has a 15" Super-Thick Bell-Pepper-and-Anchovy Pizza™ for lunch. (A 15" pizza has a diameter equal to 15".) Today, feeling a little more hungry, having missed his beer brunch, he ate a 22" instead. Compute the volumes of these two pizzas and compare them. (Hint: these are cylinders with h = 1.)

6. $x^2 x^3 = ?$ (Careful! There isn't going to be a six in your answer.)

7. $y^{40} y^2 = ?$

8. Mathematicians estimate that there are 10^{120} possible games of chess. How many times more games of chess are there than there are atoms in the observable universe? (That means, what would you multiply 10^{79} by in order to obtain 10^{120}?)

As a further hint, that means $(10^{79})(10^?) = 10^{120}$.

9. (Attempt to) solve by graphing: $\begin{cases} -x - 2y = 4 \\ 2x + 4y = 2 \end{cases}$

10. Let's take the same system of equations as in the previous question and attempt to solve them by elimination. $\begin{cases} -x - 2y = 4 \\ 2x + 4y = 2 \end{cases}$

11. When we solve two equations with two unknowns by graphing, we usually get intersecting lines. We just looked at a case where we got parallel lines. There is a third possibility. Can you picture this third case before you graph the following system of equations?

$$\begin{cases} 6x - 2y = 12 \\ 3x - y = 6 \end{cases}$$

. . . . COMPLETE SOLUTIONS . . . .

1. 625

2. x + y = 5 is easy to graph. For x – 2y = –1, I substituted in values for y (since it makes the arithmetic easier) rather than substituting in values for x. When I set y equal to zero, x was equal to –1. This gives the point (–1, 0). When I set y equal to 1, then x was also equal to 1, yielding the point (1, 1).

   The intersection of the two lines appears to be at the point (3, 2). With graphing, in contrast to using the methods of elimination and substitution, you only get an approximate answer. More accurate drawing of the graphs will give more accurate answers.

3. $\begin{cases} 8x + 2y = 46 \\ y = 2x - 1 \end{cases}$

substituting the value of y from the second equation into the first

$$8x + 2(2x - 1) = 46$$

Distributive property $\qquad\qquad 8x + 4x - 2 = 46$

$$12x - 2 = 46$$

$$12x = 48$$

$$x = 4$$

Back substitute x = 4 into y = 2x – 1 $\qquad$ y = 2(4) – 1

$$y = 7$$

4. V = πr²h = π2²(5) = π4·5 = 20π cubic inches.

5. The 15" pizza has a radius of 7.5", so the volume = πr²h = π(7.5)²(1) = 56.25π cubic inches.

   The 22" pizza has a volume of π(11)²(1) = 121π cubic inches.

   Comparing: 121π ÷ 56.25π = 121/56.25. This is approximately equal to 2.15 so a 22" pizza is approximately 215% the size of a 15" pizza. We may need to say a word about how the π's disappeared. Here is the whole thing worked out in more detail:

$$121\pi \div 56.25\pi = \frac{121\pi}{56.25\pi} = \frac{121\cancel{\pi}}{56.25\cancel{\pi}} = \frac{121}{56.25}$$

You learned to do this canceling when you reduced fractions in elementary school. When you had 6/8 and you were told to reduce it, you might have done it in a similar way:

$$\frac{6}{8} = \frac{3 \times 2}{4 \times 2} = \frac{3 \times \cancel{2}}{4 \times \cancel{2}} = \frac{3}{4}$$

When we come to study algebraic fractions later in this book, a lot of the things we learned in arithmetic will be imitated in algebra. For example, watch how this algebraic fraction is simplified:

$$\frac{(x + 7)(x + 23)}{(x - 9)(x + 23)} = \frac{(x + 7)\cancel{(x + 23)}}{(x - 9)\cancel{(x + 23)}} = \frac{(x + 7)}{(x - 9)} = \frac{x + 7}{x - 9}$$

6. $x^2 x^3 = (xx)(xxx) = xxxxx = x^5$

When the bases are the same (that's the number under the exponent), then you *add* the exponents. It may seem weird at first (like the first week of wearing braces), but after that it begins to feel more natural. Let's put it in a box and you are encouraged to stare at the box for ten seconds:

$$\boxed{x^2 x^3 = x^5}$$

7. $y^{40} y^2 = y^{42}$

8. $10^{79}\, 10^? = 10^{120}$ which means, what do you *add* to 79 to get 120? $10^{79}\, 10^{41} = 10^{120}$, which translates into: There are 10^{41} chess games for every atom in the observable universe. There are 100,000,000,000,000,000,000,000,000,000,000,000,000,000 chess games for every atom in the observable universe.

9. The lines don't intersect! They are parallel. That means there is no ordered pair of numbers (a, b) that will make both equations true at the same time. This is one of the advantages of graphing. You can *see* what's happening.

10. $\begin{cases} -x - 2y = 4 \\ 2x + 4y = 2 \end{cases}$

Doubling the first equation:

$\begin{cases} -2x - 4y = 8 \\ 2x + 4y = 2 \end{cases}$

Adding the two equations:

$$0 = 10$$

  Both of the unknowns dropped out. What we have is something that is never true (0 = 10).

  Here is our reasoning at this point. If there were an ordered pair of numbers (a, b) that could make both original equations true, then (a, b) would also make the resulting equation (0 = 10) true. But since 0 = 10 is never true, there will never be an ordered pair of numbers that makes both of the original equations true.

  Such a pair of equations is called **inconsistent**.

11. Instead of intersecting lines or parallel lines, we have the same line drawn twice. Every ordered pair of numbers that satisfies one equation will satisfy the other. We have an infinite number of solutions.

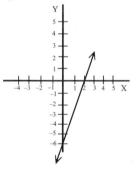

  If you tried to solve this system of equations

$\begin{cases} 6x - 2y = 12 \\ 3x - y = 6 \end{cases}$

by elimination, you'd arrive at 0 = 0. Since 0 = 0 is always true, any ordered pair (a, b) that makes one equation true will make the other one true. Such equations are called **dependent equations**.

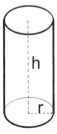

Intermission

  The volume of a cylinder is V = πr²h. The area of a circle is πr². Putting those two formulas together, we can say that the volume of a cylinder is the area of the circular base times its height.

A = πr²

Chapter Six
Lesson Forty-seven—Factorial

❦❦❦❦❦❦❦❦❦❦❦❦

Sergeant Snow bellowed. "Everyone line up single file at the door." Fred hopped off the telephone books on his chair and said to the old guy, "Excuse me, sir. My sergeant is calling me."

The 13 guys were all bunched up at the door. They couldn't figure out which way to line up. One fellow suggested that they do it by height, from the tallest to the shortest. Another said, "How about from the shortest to the tallest?" Another thought it should be by weight. Another claimed that it should be by age.

How many ways are there to line up 13 guys? Fred knew that the answer was approximately 6.227×10^9. That's a lot of ways. He didn't find that out by counting on his fingers. You could do that if you had, say, three items to arrange in a line. To arrange A, B, and C you'd get:

ABC
  ACB
    BAC
      BCA
        CAB
          CBA. There are six ways to arrange three items in a line.

With four items, a, b, c, and d, things start to get out of hand:
abcd, abdc, acdb, ... There are 24 ways to do this.

How did Fred figure out that there were about 6.227×10^9 ways to line up 13 men in a row?

He first asked himself how many possibilities there are for the first position in line. Any one of the 13 men might be first in line.

After that first-in-line fellow is selected, how many possibilities are there for the second-in-line fellow? Any one of the 12 remaining guys could be second in line.

So there are 13×12 ways of selecting the first and second persons in line. (Think of it this way: if you have three shirts and two pants, how many different outfits can you have? It's 3×2.)

After the first two places in line are filled, there are 11 possible guys to be third in line. So there are $13 \times 12 \times 11$ possible ways of selecting the first, second, and third people in line. Think of it this way: if you have two cars, three chauffeurs, and two body guards (one of which you use and the other one stays home and guards the family dog), how many possible ways can you be driven to school? $2 \times 3 \times 2$.

So with 13 men to line up, there are $13 \times 12 \times 11 \times \ldots \times 3 \times 2 \times 1$ ways to do it. Doing the arithmetic, Fred came up with a rough estimate of 6.227×10^9, which is approximately the population of the earth.

Now, here are two things to make your life a little easier:
First: We abbreviate $13 \times 12 \times 11 \times \ldots \times 3 \times 2 \times 1$ as 13! (which is read, "thirteen **factorial**"). Factorials grow quickly. When we looked at arranging A, B and C we had that $3! = 6$. With a, b, c, and d we had that $4! = 24$. Then 13! = about six billion. What about something like 50! That's approximately equal to 30,414,000,000,000,000,000,000,000,000, 000,000,000,000,000,000,000,000,000,000,000,000, or, since we're now using exponents, 3.0414×10^{64}.

Second: If you—or your older sister—has one of those scientific calculators that has sin, cos, and tan buttons on it, it may also have a button that reads, "x!" in which case you can find 13! by just punching in 13 and hitting the x! button. The display will read either ь.227   09 or ь.227 Е09. In either case this means 6.227×10^9. Calculators use ь.227   09 or ь.227 Е09 because they have trouble making little exponents appear on the display.

Factorials grow very fast. On my calculator, $20! = 2.4329 \times 10^{18}$. $50! = 3.0414 \times 10^{64}$. When I try 70!, my calculator quits. 70! is too large for it to compute.

After they had lined up in one of the 13! ways, the men were ordered to go outside and pick up cigarette butts for an hour.

"This'll work off some of the dinner you had," Sergeant Snow informed them as he headed back inside to see if there were any leftover hot dogs for him to snack on.

". . . work off some of the dinner you had," reverberated in Fred's mind. He hadn't had any. In the last two days he'd eaten twice: about 18 beans in the infirmary before he stopped to help the nurse with his Rat-and-Mouse problem and yesterday's breakfast. He wondered if he'd be 33 pounds for the rest of his life.

Your Turn to Play

1. What does 8! equal?

2. Which is larger: 2^{20} or 20! (You can do this without multiplying everything out.)

3. Which is larger: 20^{20} or 20!

4. $y^{44}y^6 = ?$

5. Of the 13 soldiers, Sergeant Snow picked one for guard duty, another to run an errand, and a third to hang some drywall. How many different ways could Snow have made his selections?

6. Classify each system of equations as either inconsistent, dependent or independent. (Independent systems of equations are those that have exactly one answer. They are the ones that are neither inconsistent nor dependent. When you graph two independent equations with two variables, you get two intersecting lines.)

$$\begin{cases} 3x + y = 14 \\ 6x + 2y = 30 \end{cases} \quad \begin{cases} 5x + 2y = 16 \\ 2x - 4y = 16 \end{cases} \quad \begin{cases} -2x + 3y = 10 \\ 4x - 6y = -20 \end{cases}$$

7. Is it possible to name three numbers so that their mode average is different than their median average?

8. What quadrant is the point (4338, −7) in?

9. What is the exact volume of a cylinder with a radius equal to 3 and a height equal to 7?

.......COMPLETE SOLUTIONS.......

1. $8! = 8 \times 7 \times 6 \times 5 \times 4 \times 3 \times 2 \times 1 = 40{,}320$.

2. 2^{20} means $2 \times\ 2 \times\ 2 \times\ 2 \times 2 \times 2 \times 2 \times 2 \times 2 \times 2 \times 2 \times 2 \times 2 \times 2 \times 2 \times 2 \times 2 \times 2 \times 2 \times 2$.

   $20!$ means $20 \times 19 \times 18 \times 17 \times \ldots$ $\times 3 \times 2 \times 1$

   Note that the factors on the bottom line are each larger than the ones on the top line. 20 is larger than 2. 19 is larger than 2. 18 is larger than 2. (The only exception is that 1 isn't larger than 2.) So the product of the bottom line, $20!$, is larger than the product of the top line, 2^{20}.

3. 20^{20} means $20 \times\ 20 \times\ 20 \times\ 20 \times\ 20 \times 20 \times 20 \times 20 \times 20 \times 20 \times 20 \times 20 \times 20 \times 20 \times 20 \times 20 \times 20 \times 20 \times 20 \times 20$.

   $20!$ means $20 \times 19 \times 18 \times 17 \times \ldots$ $\times 3 \times 2 \times 1$.

   Now, each factor on the top line is greater than or equal to the corresponding factor on the bottom line and so $20^{20} > 20!$

4. y^{50} The rule is $y^a y^b = y^{a+b}$.

5. He had 13 possible choices for guard duty. After he chose the soldier for guard duty, he had 12 soldiers to choose among to run an errand. After he had made those two choices, he had 11 soldiers to choose among to hang the drywall. So there are $13 \times 12 \times 11$ ways he could have made the three choices. $(13)(12)(11) = 1716$.

6. When you try to solve the first system of equations, you will get something like $0 = 2$ which is never true. Inconsistent equations.

   When you solve the second system, you will find that $x = 4$ and $y = -2$, which could be written as $(4, -2)$. Independent equations.

   When you solve the third system of equations, you will get something like $0 = 0$ which is always true. Dependent equations.

7. In order to obtain a mode average, at least two of the three numbers will have to be alike. For example, the mode average of 14, 19, 19 is 19 since 19 is the most common number listed.

   If at least two of the three numbers are alike, then when you list them from smallest to largest, the number in the middle will be one of the numbers that is alike. Hence the median average will be the same as the mode average.

8.

9. $V = \pi r^2 h$ $V = \pi(3)^2(7) = 63\pi$

Lesson Forty-eight—Area of a Square, Volumes of Cubes and Spheres,
Like Terms, Commutative Laws of Addition and Multiplication

F red didn't mind picking up cigarette butts. He was a lot closer to the ground than those big ol' guys who were five or six feet tall. He was thankful that, even though he was in the army, he wasn't being shot at or having to shoot other people.

After about twenty minutes in the late-afternoon Texas sun and with very little food, he passed out in the dirt.

After the hour was up, Snow came outside and assigned the recruits to their bunkhouses. Ten of them were assigned to the main bunkhouse and the remaining three—Fred, Pat, and Chris—were put out in the shack around in back of the chow hall. Pat picked up Fred and took him to the shack and laid him on an empty bed.

About nine o'clock that night, Fred began to open his eyes. Pat came over and put a hand on his forehead and said, "How yah doing little fellow? Would you like some orange juice?"

Fred was still disoriented. He looked at the quilt that Pat had put over him. It reminded him of graphing on the **rectangular coordinate system**, which is the system we've been using where (2, 3) means "go over to the right two units and then head up three units."

He looked around. He really wondered whether or not he had died. He knew that his image of heaven at the tender age of six was different

than that of mature Christians who had spent a lifetime getting beyond the Sunday School thoughts of harps and cherubic angels. But in a second he realized that he wasn't in the afterlife. It was simple: he hurt. He hurt all over.

As Pat held up his head and gave him sips of cold orange juice, Fred noticed one other thing. Pat wasn't the nickname of Patrick, but of Patricia. She was a girl-type person instead of a boy-type person. Chris wasn't Christopher, but Christine. Snow had put the women and children in the shack behind the chow hall since it would have been against camp regulations to put them in with the men.

Fred slept through the night. He dreamed of the prisoner van and then of the orange juice.

Fred awoke at 6 a.m. He curled up in his "rectangular coordinate system" quilt and read for a half hour. At 6:30 reveille sounded to awaken the rest of the camp. It was Sunday morning. Our 13 recruits showered. Twelve of them shaved.

In his hebdomadal address, the chaplain was talking about a city that is described on the next-to-the-last page of the Bible. It is laid out in the form of a square—a big square. It is about 1500 miles long and 1500 miles wide. On the blackboard in Fred's mind he wrote:

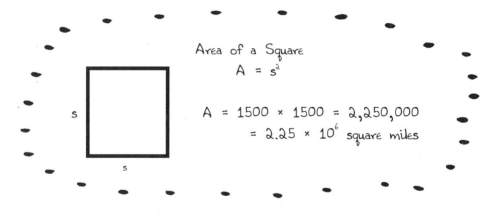

Area of a Square

$$A = s^2$$

$$A = 1500 \times 1500 = 2{,}250{,}000$$
$$= 2.25 \times 10^6 \text{ square miles}$$

When Fred taught beginning algebra at KITTENS it was usually at this point that several of his students would remark, "Now I know why x to the second power is called x squared."

The chaplain finished reading verse 16. It contained a real surprise: ". . . the city was 1500 miles long and was as wide *and as high* as it was long."

"Holy cow!" Chris exclaimed under her breath. "Now that's a skyscraper." Chris had been an architecture student before she had been drafted. She loved to tell everyone about her favorite architect, Frank Lloyd Wright, and how in 1956 he designed a mile-high skyscraper. Wright's skyscraper has never been built (yet).

Fred was thinking about the volume that this city would contain:

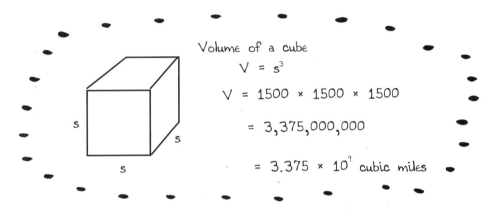

Volume of a cube
$$V = s^3$$
$$V = 1500 \times 1500 \times 1500$$
$$= 3,375,000,000$$
$$= 3.375 \times 10^9 \text{ cubic miles}$$

And that's why x^3 is called x cubed.

We get x *squared* from a square, where the corners consist of two lines meeting at right angles:

    We get x *cubed* from a cube, where the corners consist of three lines all meeting at right angles:

Once we figure out how to get four lines all meeting at right angles to each other, we'll have another name for *x to the fourth power*. If you want to play with this, try taking four pencils and arranging them in your hand so that each pencil is at 90° to the other three.

At lunch Pat and Chris sat with each other as they had done last night. The ten guys sat at their five tables enjoying the spaghetti and meatballs. Sergeant Snow was "helping out in the kitchen" where he could sample the food as it was prepared. Fred looked around and saw that the old guy was still sitting by himself and asked again if he could join him. The old man nodded and Fred turned to the six helmeted and armed guards who were approaching him and called out, "It's okay, boys!" The old guy laughed and the six followed suit, smiled, and backed away.

Fred was learning all kinds of army talk. You say, "It's okay, boys," when you don't want to be attacked, and "head" when you really mean potty.

There were two things that surprised Fred. The biggest meatball in the chow hall was in a crystal bowl at this guy's table. Fred wondered how this man rated so highly with the chow hall staff. He certainly didn't seem to work as hard as any of the other people. Fred had never seen him outside marching with the men or helping out in the kitchen or anything. Maybe, Fred guessed, he was sick or something and needed extra good food.

The meatball was fairly large. It was a sphere (that's the math word for a ball) that had a radius of 1.5 feet. The volume of a sphere is

$V = (4/3)\pi r^3$. (That can be proved in calculus, but for now we'll have to just accept it as true.) Fred tested a tiny bit of the meatball in his water glass and saw it sink pretty quickly. He knew that the density of water was approximately 62.4 lbs/cu ft, so he guessed that the density of the meatball was, say, 70 lbs/cu ft. Fred wondered if he might be injured if this meatball fell out of the bowl and rolled on top of him. He did a quick calculation:

$$V = (4/3)\pi r^3 = (4/3)\pi(1.5)^3 = (4/3)\pi(3.375) \text{ cubic feet.}$$

(He knew that 1.5^3 equaled 3.375, because he had figured out that 1500^3 equals 3,375,000,000 when he was computing the volume of the city during the sermon on the previous page.)

Now, since the 70 lbs/cu ft was only an estimate of the density of the meatball, he started to approximate all the items so that he could easily figure it out in his head:

$(4/3)\pi(3.375)70$

$\approx (4/3)3(3.375)70$ »» letting π equal three as an approximation

$\approx (4)(3.375)70$ »» the threes cancel

$\approx 13(70)$ »» since four times 3.375 is about 13

≈ 910 lbs.   »» Fred could multiply 13 times 70 in his head

At 910 pounds, this could be a very dangerous meatball.

The second thing that surprised Fred was what the old guy was wearing. It was something new. The big bunch of medals were the same ones he wore last evening. He remembered how they had dipped into his coffee cup when he reached for another piece of hot dog. But what was new was his tie—a bow tie! It was polka dotted.

Fred felt he should warn the old guy so that he wouldn't get into trouble. "Sir," Fred began carefully, "I think you should know something about bow ties at this army camp." Fred was going to tell the story of how the quartermaster had ripped Fred's bow tie off and had thrown it into the garbage can, but before he could begin, the old guy said, "It's okay son. Around here, I make the rules."

The colonel had found Fred's polka dot bow tie in the quartermaster's garbage can and had taken a liking to it. He hadn't known that it had been Fred's.

Of course, when Sergeant Snow saw that bow tie around the colonel's neck, he nearly choked on the six meatballs that he had in his mouth. He wanted to say, "Nice tie, sir!" but was afraid he'd lose one of his meatballs as he uttered that lie.

Fred had now eaten two regular meals in a row. His eyes were starting to look a little better. He had gotten a good night's sleep after Patricia had given him the orange juice. He and Jack LaRoad decided to head out for an afternoon jog.

The other eleven decided to watch TV for five hours until they received their get-off-the-base passes, which were good from 6 p.m. until midnight. You have a choice: jogging, television, or . . .

Your Turn to Play

1. $x^4 x^5 = ?$

2. $7x^5y^3z$ and $2x^5y^3z$ are called **like terms**. They have the same exponents and the same **bases** (that's the number or letter under the exponent). The numerical coefficients (that's the number in front) may differ (or be alike)—that doesn't matter.

     There's no way to combine $5x + 9y$. They're not like terms. But $7x^5y^3z + 2x^5y^3z$ combine nicely to give $9x^5y^3z$.

     $3abc^2 + 2abc^2 = ?$

     $8d^2 + 9d = ?$

     $4xy^2 + 5y^2x = ?$

3. State the commutative law of addition.

4. Is the commutative law of addition true?

5. Is the commutative law of subtraction true?

6. $6x^7x^8 = ?$

7. Sergeant Snow walked into his local Globo Ice Cream store and ordered five scoops of ice cream on a cone. He wanted five different flavors: peach, anise, walnut, turkey, and licorice. He was asked in what order he wanted them. How many possible answers could he give?

8. Five students are going to make a movie. One of them will be the director. Another will star. Another will be the editor. Another, the producer, and the last one will operate the camera. How many ways could they make these assignments?

9. $(x^2)^3 = ?$

10. Complete the following general rule: $(x^a)^b = ?$

11. When Fred was picking up cigarette butts, he spotted five butts, all near each other, all of different brands. If he picked them up one-at-a-time, how many different ways could he do that?

12. Is the commutative law of division true?

13. The garbage containers at A.A.A.A. are in the shape of cubes. They come in two sizes. The smaller measures a yard in each dimension (a yard wide, a yard long, a yard deep). The larger one is 1.3 yards in each

dimension. Fred had filled two of the smaller containers with cigarette butts before he had passed out. Jack LaRoad filled one of the larger containers. Who had picked up a greater volume of butts?

14. A Giant Globo scoop of ice cream (known as a "GG") is a sphere with a diameter of 4". What's the volume of a GG?

15. $(5y^{20})^3 = ?$

16. $x^7(x^3y + xy^8) = ?$

17. Name a value of x so that $x^2 < x$.

This is a problem that involves . . . thought!

In many math books, you are shown how to do something and then they give you 30 problems that are all alike. That's called "drill-and-kill"—no thought and lots of work.

The problems that you will get on an SAT exam or in real life are not like that. They are not just reciting what you have memorized.

Paper and pencils were invented so that you do not have to memorize everything.

Would You Care to Memorize 15,693 Lines of Poetry?

We are talking about the *Iliad* (ILL-ee-ed). It is one of the oldest works in Western literature. Every war movie you have ever seen, every superhero, every action/adventure story can be traced back to the *Iliad*. It does a lot of thinking *about what fighting means*—besides describing a lot of blood-and-guts battles.

The background: The kingdom of Troy started around 3000 B.C. By 1200 B.C., it was a walled fortress, famous for its iron and for its horses. The Greeks attack it (the Trojan War) and the *Iliad* tells the story in about 15,000 lines of poetry.

Oops. There was no way to write it down. It wasn't that there was a shortage of paper. There was no written language. The *Iliad* was a part of oral literature. It was passed on for generation after generation after generation after generation after generation after generation after generation for about four or five hundred years. People would memorize it and sing or chant it to audiences. Some younger people in the audience would hear it often enough that they would memorize it and pass it on down the ages.

Finally (see Lesson 39), the Greeks got their alphabet and writing. The last singer/chanter recited it and it was written down.

.......**COMPLETE SOLUTIONS**.......

1. $x^4x^5 = x^9$

2. $3abc^2 + 2abc^2 = 5abc^2$

$8d^2 + 9d$ can't be combined. They must have the same exponents.

$4xy^2 + 5y^2x = \ldots$ Before we start this one, let's recall something that you learned years ago. Does 3×7 give the same answer as 7×3? Does $149 \times 386 = 386 \times 149$?

You know the answer is *yes* even without multiplying them out. In general, for any two numbers *a* and *b*, it's true that *ab = ba*. The fancy name for this is the **commutative law of multiplication**.

Now by the commutative law of multiplication y^2x is equal to xy^2, so we can rewrite the original problem (which was $4xy^2 + 5y^2x$) as $4xy^2 + 5xy^2$. Now they're obviously like terms, and we can add them $4xy^2 + 5xy^2 = 9xy^2$.

3. For any two numbers x and y, it's true that $x + y = y + x$. You may have used different letters than x and y; it doesn't matter. You could even state it with the Greek letters alpha and beta: For any two numbers α and β, it's true that $\alpha + \beta = \beta + \alpha$.

4. Sure it is. $a + b = b + a$ is true for any numbers a and b.

5. No way. You know that $5 - 8$ isn't the same as $8 - 5$. The commutative law for putting on socks and shoes is also not true. It's much easier to put on your socks first.

6. $6x^7x^8 = 6x^{15}$. If you have difficulties with this, I'll work it out in slow motion:

$6x^7x^8 = 6(\text{xxxxxxx})(\text{xxxxxxxx}) = 6\text{xxxxxxxxxxxxxxx} = 6x^{15}$

7. There are five possible answers for the bottom scoop, then four possible answers for the next scoop, then three for the middle scoop, then two for the next-to-last scoop, and whatever was left over would be the top scoop. The number of possibilities would then be $5 \times 4 \times 3 \times 2 \times 1$ or 5!, which is 120 possible ways to line up those five scoops. This means that if he went into Globo's every day and ordered these same five delicious flavors, each time in a different order, it'd take him about a third of a year before he'd have to duplicate the order of these five fine flavors.

8. This is the same problem as the ice cream scoops. Think of lining up the five students where the first one in line will be the director and the second will be the star, etc. The answer is 5!

9. $(x^2)^3 = x^6$ An exponent-on-an-exponent means you multiply! Let's work it out in slow motion: $(x^2)^3 = (x^2)(x^2)(x^2) = (xx)(xx)(xx) = xxxxxx = x^6$

10. $(x^a)^b = x^{ab}$

11. This is the same problem as ice-cream-scoops and students-making-a-movie again. The answer is 5!

12. No. If it were true, it would have to be true for every pair of numbers. Therefore, in order to show that it's not true, all I have to do is exhibit one pair of numbers for which the law fails.

    How about 2 and 3: $2 \div 3$ does not equal $3 \div 2$.

13. Jack did. Fred picked up two cubic yards. The larger container that Jack had filled had a volume of $V = s^3 = (1.3)^3 = 2.197$ cubic yards. For many people, it's surprising that a cube with edges equal to 1.3 has more than twice the volume as one with edges equal to 1.

14. The volume of a sphere is $V = (4/3)\pi r^3$. Since a GG has a diameter of 4", it has a radius of 2", and its volume is $= (4/3)\pi(2)^3 = (32/3)\pi$ cubic inches. For those of you who are interested, that is approximately equal to 33½ cubic inches.

15. $(5y^{20})^3 = 5^3 y^{60} = 125 y^{60}$

16. $x^7(x^3 y + xy^8) = x^{10}y + x^8 y^8$ and these can't be combined since they're not like terms.

17. We want a value of x so that $x^2 < x$. Normally, when you square a number it seems to get larger. Six squared is 36. Eighteen squared is 324.

    What if we try a negative number for x? This won't work. The square of every negative number is positive, so it'll never be true for a negative value of x that $x^2 < x$.

    What's left? Zero doesn't work since 0^2 is not less than 0.

    Would you believe that there is an infinite number of possible values of x that will make $x^2 < x$ true? There are. If you square the number ½ you get ¼. It got smaller! The set of numbers that makes $x^2 < x$ true is $\{x \mid 0 < x < 1\}$.

Out on the road, Jack and Fred were having
a lot of fun. They figured they'd jog for a couple
of hours and then head back to the men's
bunkhouse. Jack had brought a nice set of barbells
and dumbbells to the army camp, and he said
he'd be happy to show Fred how to use them.

A couple of miles into their run, Jack
noticed that Fred wasn't huffing and puffing
very much.

their jogging route
★ = starting and ending spot

He asked, "How old are you? You seem to be in
great shape."

Fred answered that he was six and that he'd been exercising
regularly all of his life.

"Six!" Jack exclaimed. "I thought you were a midget. That
explains why they stuck you in the bunkhouse behind the chow hall."

A couple of miles later Jack had another thing on his mind and said
to Fred, "I hear that you do okay with numbers—like it's your speciality or
something."

Fred replied, "It's one of the things I like. I also like reading good
literature. Years ago I came across Fadiman's *Lifetime
Reading Plan,*★ which lists the books that every well-
educated person might want to read. I've been working
my way through that list, reading for about a half hour
each morning. I like the third edition of Fadiman's book
best."

"Can we talk about math for a moment?" Jack asked. He had
something on his mind and there weren't many people like Fred that he
could talk with. "I was reading in this funny but weird math book called
Life of Fred: Beginning Algebra Expanded Edition about how you add

★ Use your computer browser to search for FADIMAN'S LIFETIME READING PLAN,
THIRD EDITION.

exponents when you're multiplying two numbers with the same base. It showed that $x^4x^5 = x^9$ which easily generalizes into the rule: $x^mx^n = x^{m+n}$."

"That seems to be correct," Fred said reassuringly.

"Well," Jack said proudly, "I think I've come across something. Instead of multiplying, I switched it to dividing. I started out with x^6/x^4, which becomes $\dfrac{x^6}{x^4}$ which is $\dfrac{xxxxxx}{xxxx}$ which after you cancel

$\dfrac{\cancel{xxxxx}x}{\cancel{xxxx}}$ equals x^2. So I've invented the rule: $\dfrac{x^m}{x^n} = x^{m-n}$.

Cool, isn't it?"

Jack was starting to breathe hard as he described this. It wasn't because he was in poor condition. He was just excited about the math he had created.

"I like it," Fred praised. "It's got a nice symmetry with $x^mx^n = x^{m+n}$."

Then Fred led Jack into deeper waters. Ever so gently he asked, "You know, Jack, did you ever consider something like x^{-3}?"

"X to the minus three power?," Jack puffed. "You gotta be kidding. There isn't such a thing. That's nuts. When you write something like x^5 you mean xxxxx, which is a term that has five factors of x in it. How in the world can you talk about x^{-3}? That would mean writing x a negative three times. It can't be done."

[Small historical note: The above attitude is quite common among human beings toward the advancement of any new idea—even in mathematics. When the square root of two was first introduced (we'll see it later in this book) it was called "irrational" which means crazy and nuts. Later when the square root of −1 was introduced, the mobs called it "imaginary." The world has no shortage of bad names to apply to anything new. If you plan to introduce a new idea to the world, be prepared to be insulted, imprisoned or, like Joan of Arc, burned at the stake.]

"You mean," Fred answered, "that x^{-3} has no meaning right now?"

"Yeah."

"Well then," Fred continued, "we are free to assign it any meaning we like. It's only the words or expressions that already have a meaning that it's improper to assign different meanings to."

"Well, what possible meaning could you give x^{-3}?" Jack asked.

"Remember the rule you just discovered?" Fred said. "That rule could be the key to giving a meaning to x^{-3}. We'll take your rule $\frac{x^m}{x^n} = x^{m-n}$ and put in m = 4 and n = 7. Watch what we get." Fred stopped running and knelt down in the dirt and wrote:

$$\frac{x^4}{x^7} = x^{4-7} = x^{-3}$$

"But we know what the left side is," said Jack. He wrote in the dirt:

$$\frac{x^4}{x^7} = \frac{XXXX}{XXXXXXX} = \frac{\cancel{XXXX}}{XXX\cancel{XXX}} = \frac{1}{x^3}$$

And that's how Fred showed Jack that the natural extension of Jack's formula $x^m/x^n = x^{m-n}$ gives us that x^{-3} should equal $\frac{1}{x^3}$

Of course, if m and n are each equal to, say, five, we find that

$$\frac{x^5}{x^5} = x^{5-5} = x^0$$

and since the left side is equal to one, we have that **anything to the zero power is equal to one**.

Jack challenged Fred to an arithmetic game. (Foolish man!) As they ran, Jack called out, "One."

Then Fred was supposed to say, "Two," and Jack would double that to "Four," and so on.

At 131,072 Jack cried, "Uncle!" and Fred suggested that they'd learn a lot if they went the other direction from 131,072. Jack couldn't understand what they'd learn since they had already "been there," but he agreed.

The instructive part came when they got back to 2. Here's a chart of the game in progress:

64	32	16	8	4	2	
2^6	2^5	2^4	2^3	2^2	2^1	

It's the next couple of entries in this chart that are new. Each entry in the top row is half of the previous entry. The pattern in the bottom row is even more obvious.

Here's how the game continued:

64	32	16	8	4	2	1	1/2	1/4	1/8
2^6	2^5	2^4	2^3	2^2	2^1	2^0	2^{-1}	2^{-2}	2^{-3}

$$x^a x^b = x^{a+b} \qquad\qquad (x^a)^b = x^{ab}$$

$$(xy)^a = x^a y^a \qquad\qquad x^{-4} = \frac{1}{x^4}$$

$$\frac{x^a}{x^b} = x^{a-b} \qquad\qquad x^0 = 1$$

Your Turn to Play

1. $\dfrac{y^6}{y^{11}}$

2. $z^5 z^{-7} z^8$

3. Express 10^{-3} as a decimal.

4. Simplify $(\sqrt{3} + 0.38998 + 50! - \log_5 44 + \sin \pi/6 - \xi)^0$.

5. Plot the graph of y = log x.

   Some things to note:

♪#1: "log x" means the logarithm of x. We will get to logarithms in advanced algebra.

♪#2: You don't need to know what logs are in order to do this graph. All you need to know is how to find log x for various values of x. If you have a "scientific" hand calculator that has a **log** button on it, you can graph y = log x.

   Pick a value for x. Say, x = 2. Type in **2** and then hit the **log** button. The display will read 0.30103. So log 2 = 0.30103. You now have one point on the graph: (2, 0.30103).

   Pick another value for x. Say x = 8. Type in **8** and then hit the **log** button. The display will read 0.90305. So log 8 = 0.90305. You can now plot (8, 0.90305).

   If you pick x = –4, your calculator will be unhappy when you hit the **log** button. It will report "E" or "Error" since logs don't exist for negative numbers.

♪#3: If you don't have a scientific calculator and your big sister won't lend you hers, then graph y = 3x instead.

6. Fill in one word: (6 + x) + 3.2 = 6 + (x + 3.2) is an example of the _____ law of addition.

7. Replace the letter *y* by any number that will make $y < -5$ true.

8. $-33 \div -11 = ?$

9. Graph $y = x^{-2}$.

     . . . . **COMPLETE SOLUTIONS** . . . .

1. y^{-5} or $\dfrac{1}{y^5}$

2. z^6

3. $10^{-3} = \dfrac{1}{10^3} = \dfrac{1}{1000} = 0.001$

4. That's duck soup. (The phrase "duck soup" entered the English language in about 1910. It means something that is easily done.) *Anything to the zero power is equal to one.*

Actually, that's not quite true. I'm not sure what I'd get if I raised a duck to the zero power.

5.

6. $(a + b) + c = a + (b + c)$ is the associative law of addition.

7. y could be numbers like –6 or –5.5 or –1200.

8. When the signs are alike, the answer is positive. $-33 \div -11 = +3$

9. $y = x^{-2}$ is the same as $y = \dfrac{1}{x^2}$

If we let x = 1, then y = 1. The point (1, 1) is on the graph.
If we let x = 2, then y = ¼. The point (2, 0.25) is on the graph.
If we let x = ½, then y = 4. The point (½, 4) is on the graph.
If we let x = –3, then y = 1/9. The point (–3, 1/9) is on the graph.
If we get enough points,
the graph will look like →

Seabrook, Texas "as grand by land as by bay, and why many choose to stay"

1. Solve by graphing: $\begin{cases} y = x + 4 \\ x + y = 1 \end{cases}$ and estimate the point of intersection.

2. Showing your work, solve the above system by substitution.

3. Name two points that are on the graph of $x^2 - y^2 = 16$.

4. $w^6 w^2 = ?$

5. $(10x^{10})^5 = ?$

6. Solve by graphing: $\begin{cases} y - x = 4 \\ y = x - 3 \end{cases}$

7. Col. C. C. Coalback fell in love with bow ties and went and bought six more of them at the local thrift store. Besides the polka dot one he already had, he got a striped one, a plain olive drab one, a checkered one, an orange one, a plaid one and one with a picture of a ship on it. He was going to wear a different tie on each day of the week. How many different ways could he arrange the ties for the week?

8. $(16x^2 yz\xi)^0 = ?$

9. You have a sphere that fits perfectly inside of a cylinder (see the diagram). What is the ratio of the volume of the sphere to the volume of the cylinder? (For fun, make a mental guess before you compute the answer.)

10. Solve by graphing: $\begin{cases} y = -2x - 4 \\ y = 3 \end{cases}$

and estimate the point of intersection.

11. Solve by graphing: $\begin{cases} y = (1/4)x^2 \\ y = 2 \end{cases}$

and estimate the TWO points of intersection.

12. Find the volume of the cardboard tube inside a roll of toilet paper. You need not actually go and measure the dimensions; just estimate them. And since you're estimating them, you may use 3 as an approximation for π.

13. $\xi^4 \xi^4 = ?$

14. $(x^2 y^5)^3 = ?$

15. Solve by graphing:
$$\begin{cases} y = -3x + 2 \\ 2y + 6x - 4 = 0 \end{cases}$$

16. The cooks at A.A.A.A. have four different dinner menu items. In the next four days, how many ways could they serve them so every dinner will be different?

17. $(5y^{-1})^2 = ?$

18. When we talked about order of operations in chapter three, the rule was: *Things inside parentheses get done before anything else. Then multiplication and division (left-to-right) and finally addition and subtraction.* Now we need to include exponents in the list. Here is the new revised list of **order of operations**:

❋ parentheses

❋ exponents

❋ multiplication and division   (done left-to-right)

❋ addition and subtraction

With those in mind, evaluate $2x^3 - 4$ when $x = 5$.

answers

1. $(-3/2, 5/2)$

2. Substituting the first equation into the second:
$$\begin{aligned} x + (x + 4) &= 1 \\ 2x + 4 &= 1 \\ 2x &= -3 \\ x &= -3/2 \end{aligned}$$

Back substituting this value into the first equation: $y = -3/2 + 4 = 2\tfrac{1}{2}$.

3. The easiest to name are $(4, 0)$ and $(-4, 0)$. These two points can be written as $(\pm 4, 0)$.

4. w^8

5. $100000x^{50}$ (See page 263 #15)

6. The lines are parallel.

The equations are inconsistent.

There is no solution.

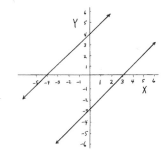

7. 7! (Or if you insist on working it out, it's 5,040 ways. We might note that 5,040 weeks is a little longer than 96 years. Coalback has a lifetime supply of bow ties.)

8. Anything to the zero power is equal to one.

9. If the radius of the sphere is r, then the radius of the cylinder is also equal to r. Since the ball is sitting snugly in the can, the height of the can is equal to the diameter of the ball which is 2r.

The volume of the sphere is $V = (4/3)\pi r^3$

and the volume of the cylinder is equal to $V = \pi r^2 h = \pi r^2(2r) = 2\pi r^3$.

To find the ratio, we divide the volume of the sphere by the volume of the cylinder: $(4/3)\pi r^3$ by $2\pi r^3$

$$\frac{(4/3)\pi r^3}{2\pi r^3} = \frac{4/3}{2} = \frac{4}{3} \div 2 = \frac{4}{3} \div \frac{2}{1} = \frac{4}{3} \times \frac{1}{2} = \frac{4}{6} = \frac{2}{3}$$

So a sphere has two-thirds of the volume of a cylinder that contains it.

10. $(-7/2, 3)$

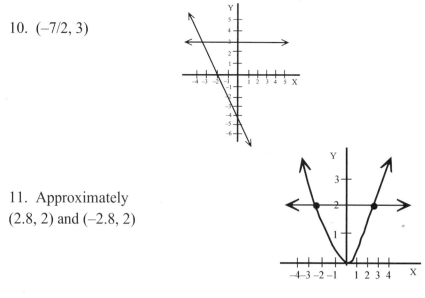

11. Approximately $(2.8, 2)$ and $(-2.8, 2)$

12. The formula for the volume of a cylinder is $V = \pi r^2 h$. Your numbers may differ from mine since you're estimating the dimensions. I went and measured. The diameter is a little bit more than 1.5" and the height is 4½".

    Then the radius is 0.75".

The volume is $V \approx 3(0.75)^2(4.5) \approx 7.6$ cubic inches.

13. ξ^8

14. $x^6 y^{15}$

15. This is the same line drawn twice.

The equations are dependent.

Any solution to one equation is a solution

to the other.

16. 4! which is 24.

17. $(5y^{-1})^2 = 25y^{-2} = 25/y^2$

18. $2x^3 - 4$

$\qquad = 2 \cdot 5^3 - 4$ when x = 5

$\qquad = 2 \cdot 125 - 4$ doing exponents first

$\qquad = 250 - 4$ doing multiplication

$\qquad = 246$ doing subtraction

Florence, South Carolina the original city limits formed a circle with a one-mile radius

1. Showing your work, solve by substitution: $\begin{cases} y = 4x - 1 \\ 5x + 16y = 122 \end{cases}$

2. Graph $y = x^3$

3. Find the volume of the earth if we assume that it's perfectly spherical and that it has a radius of 4000 miles.

4. $2a^2a^3a^4 = ?$

5. $y^7(xy^7 + 2) = ?$

6. Solve by graphing: $\begin{cases} y - 2x = -1 \\ y = 2x - 3 \end{cases}$

7. The chaplain picked out 32 different sermon topics for his next 32 hebdomadal sermons. How many different ways could he order them?

8. $(x^{-2})^{-1} = ?$

9. When Patricia gave Fred that glass of orange juice, it was in a regulation army glass that had a diameter of 4" and a height of 7". What was the volume of that glass?

10. Solve by graphing: $\begin{cases} y - x = -6 \\ x = 2 \end{cases}$ and estimate the point of intersection.

11. Explain whether or not the graph of $y = 6x^5 - 4x^3 + \pi x - 7$ passes through the origin.

12. Sergeant Snow wants to pick out two recruits to paint the exterior of the chow hall. (He wants it painted brown to remind him of his favorite beans and hot dogs.) He picks one man from the men's bunkhouse (where there are ten men) and one from the bunkhouse behind the chow hall (which has Fred, Pat, and Chris). How many ways can he pick this pair?

13. $(6x^3)^2 = ?$ (See page 263 #15)

14. $w(xy + w^6) = ?$

15. Solve by graphing: $\begin{cases} -3x + y = 4 \\ 2y = 6x + 8 \end{cases}$

16. The volume of a cone is $V = (\tfrac{1}{3})\pi r^2 h$. Fill in the blank: If you were able to fit a cone perfectly inside a cylinder it would take up ___?___ of the volume of the cylinder.

17. $16^{-1} = ?$

18. $9(x - 3 \sin x + \log \sqrt{3}\,)^0 = ?$

answers

1. Substituting the first equation into the second: $5x + 16(4x - 1) = 122$

$$5x + 64x - 16 = 122$$
$$69x - 16 = 122$$
$$69x = 138$$
$$x = 2$$

Back substituting $x = 2$ into the first equation: $y = 4(2) - 1$
$$y = 7$$

2. →

3. $V = (4/3)\pi r^3 = (4/3)\pi(4000)^3$
$= (256000000000/3)\pi$ or $(2.56\pi/3)10^{11}$ cubic miles.

4. $2a^9$

5. $xy^{14} + 2y^7$

6. Parallel lines.
The equations are
inconsistent. There is
no solution.

7. 32! That's the exact answer. If you were to approximate 32!, it would work out to about 2.6313×10^{35}.

8. $(x^{-2})^{-1} = x^2$

9. If the diameter was 4", then the radius was 2". The formula for the volume of a cylinder is $V = \pi r^2 h$, which in this case was $\pi(2)^2 7 = 28\pi$ cubic inches.

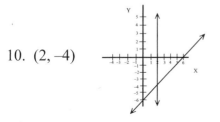

10. (2, –4)

11. It doesn't pass through the origin since (0, 0) doesn't satisfy the equation.

12. $10 \times 3 = 30$. (It's like having 10 shirts and 3 pants and asking how many ways could you assemble an outfit.)

13. $36x^6$

14. $wxy + w^7$

15. The same line drawn twice.

The equations are dependent. Any solution to one equation is a solution to the other.

$y = 3x + 4$

16. One-third. Most people on this planet don't know that a cone takes up exactly one-third the volume of a cylinder. If you think about it, there seems to be no obvious reason why the cone should be exactly one-third the volume of a cylinder. It might just as well have been 0.3987962 the volume of a cylinder. $(1/3)\pi r^2 h$ is one-third of $\pi r^2 h$.

17. $16^{-1} = 1/16$

18. Anything to the zero power is equal to one.
   $9(x - 3 \sin x + \log \sqrt{3}\,)^0 \ = \ 9 \cdot 1 \ = \ 9.$

In the **order of operations**:
   ❀ parentheses
   ❀ exponents
   ❀ multiplication and division   (done left-to-right)
   ❀ addition and subtraction

exponents come before multiplication.

That is why $9(x - 3 \sin x + \log \sqrt{3}\,)^0$ becomes $9 \cdot 1$ and not 1.

Glenmora, Louisiana on US Highway 165

1. Solve by graphing: $\begin{cases} y - 2x = 3 \\ y = 1 \end{cases}$ and estimate the point of intersection.

2. Are there any points of the graph of $y = -2x^2 + 6$ that lie in QI?

3. A cylindrical water tank at A.A.A.A. has a radius of 10 feet and a height of 40 feet. What volume of water could it hold and how much would that water weigh? (Density of water = 62.4 lbs/cu ft)

4. $(5y^4)^2 = ?$

5. $z(abc + 6xyz) = ?$

6. Solve by graphing: $\begin{cases} y = (\frac{1}{2})x - 3 \\ 2y - x = -6 \end{cases}$

7. A.A.A.A. received 12 postcards all addressed to Fred. Camp regulations permit recruits to get only one card per day. How many different ways could the camp post office arrange to deliver these cards to Fred over the next dozen days?

8. $5(33x^2y^3)^0 = ?$ (Hint: the answer is not equal to one.)

9. Is exponentiation commutative? Namely, is it always true that x raised to the y power is equal to y raised to the x power?

10. Solve by graphing: $\begin{cases} 3x + y = 2 \\ y = 2x - 2 \end{cases}$

and estimate the point of intersection.

11. Name four points that are on the graph of $x^2 + y^2 = 25$.

12. Could you pick up a lead ball with a one-foot diameter? (Lead weighs 11.34 times as much as water. The density of water is 62.4 lbs/cu ft.)

13. $z^6z^6 = ?$

14. $(a^2b^3)^4 = ?$

15. Solve by graphing: $\begin{cases} y = x/3 + 1 \\ -x + 3y = -2 \end{cases}$

16. In chemistry you're told that the order in which you pour things into a beaker sometimes matters. (For example, putting concentrated acid into a beaker first and then adding water is much more dangerous than putting the water in first.)

If you've got five compounds —A, B, C, D, and E—in how many different orders could you put them into a beaker?

17. $(3x^3)^4 = ?$

18. At Waddle Doughnuts, they always give Sergeant Snow two free doughnuts every time he visits. (After all, they figure, he is the one customer that has made their business prosper.) The two doughnuts are always different flavors. There are 20 flavors of doughnuts at Waddle's. Snow gets one doughnut when he walks in and a second one just as he's about to leave.

The management has decided to make a series of movies of Snow's visits. How many different movies might they make? (Getting a chocolate doughnut as he walks in and a glazed doughnut as he walks out will be considered a different movie than if he gets a glazed on the way in and a chocolate on the way out.)

answers

1. (–1, 1) See →
2. Yes, there's an infinite number of them. The easiest to name is (1, 4).

3. $V = \pi r^2 h = \pi 10^2(40) = 4000\pi$ cu ft. The weight would be $(62.4)(4000\pi)$ $= 249,600\pi$ lbs.

4. $25y^8$

5. $abcz + 6xyz^2$

6. The same line drawn twice. The equations are dependent. Any solution to one equation is a solution to the other.

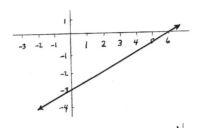

7. $12!$

8. $5(33x^2y^3)^0 = 5\cdot 1 = 5$

9. No. In fact it's almost never true.

For an easy example, we know that $2^3 \ne 3^2$.

10. $(4/5, -2/5)$ See →

11. The easiest to name are $(\pm 5, 0)$ and $(0, \pm 5)$.

12. Not many of us could. $V = (4/3)\pi r^3 = (4/3)\pi(0.5)^3 = \pi/6$ cubic feet. The weight would be $(11.34)(62.4)(\pi/6) = 117.936\pi$ lbs. Since we are trying to answer the question of whether or not we can lift that ball, we'll approximate π by 3.14 and find that the weight is ≈ 370 lbs.

13. z^{12}

14. $a^8 b^{12}$

15. Parallel lines.

The equations are inconsistent. There is no solution.

16. $5!$

17. $(3x^3)^4 = 3^4(x^3)^4 = 81x^{12}$

18. It's the same problem as if you had 20 shirts and 19 pants. On the way into the store, there were 20 possible doughnuts to hand to him, and on the way out, there were 19 possible doughnuts (since they weren't giving him the same flavor twice).

So there are $20 \times 19 = 380$ possible movies they might make.

F red and Jack had finished their Sunday afternoon jog. A couple of hours were enough for their first run together. They passed the other eleven recruits who were in the rec hall watching their third hour of reruns of "I Love Shopping" on television.

They ran past a large truck driving on the parallel concrete tracks that had been installed as a driveway to the back of Col. C. C. Coalback's mansion at the camp. The trucks usually made deliveries at night. They unloaded things like cameras, expensive furs, giant-screen televisions, cases of old wines, funny-looking plants. . . .

The parallel concrete tracks had been installed to prevent ruts from forming in the dirt since there had been so many deliveries made over the years.

This truck was carrying grass . . . the lawn kind. Maybe that's why it was operating during the daytime. The driver got out of the truck and went to the front door of the mansion and rang the bell. As the door opened the driver bellowed out his usual line, "Cubit Garden Supplies! Hey Mack, where do you want us to dump this stuff?"

He gulped as he saw that he wasn't addressing a Mack, but a robot that was dressed up as a maid. "Oh, just anywhere!" the robot answered in a mechanical Marilyn Monroe voice and vaguely pointed to where the truck was now standing.

The driver headed back to the truck and pulled the dump lever and deposited one giant piece of lawn sod into a heap across the concrete driveway strips. He backed the truck over the mess and left.

The colonel looked out the upstairs window and smiled. It was the new lawn that he had ordered to fill in the space between the concrete strips. He had ordered a piece that was 4 cubits by 9 cubits.

He wasn't sure how long a *cubit* was, but he figured that it sounded about right. He called Sergeant Snow on his cell phone and ordered him to, "Get a couple of men out here to straighten out that sod."

And you know the pair that was selected.

Fred couldn't be of much help in arranging a piece of lawn that measured 4 by 9 cubits, but for Jack LaRoad it was child's play. He lifted the rectangle of lawn and flung it into place. Of course, 4 x 9 cubits didn't exactly fit between the driveway strips as it was supposed to. Turning the strip 90° wouldn't work because the piece of lawn was too narrow.

Jack cut off the piece of lawn on the right and with some cutting and pasting, formed it into a rectangle on the top. It was a beautiful piece of work. The 4 x 9 cubits now neatly fit

between the concrete strips. Jack had to cut 18" off the right side, and after he had done a lot of cutting and fitting, that piece made the lawn between the concrete strips 9" longer.

They were about to head back to the men's bunkhouse to start working out with Jack's set of free weights when the mansion door opened and the robot maid in her breathy voice called out, "Yoo-hoo!" Jack was the first to turn around and look, but it was Fred to whom she was beckoning.

"The colonel would like to talk with you for a moment," she said.

Fred and the maid passed through a dozen rooms to get to the colonel upstairs. One room was filled with the heads of hundreds of trophy animals. Another contained a giant screen bordered by dark red curtains and a Double Doopy sound system that would put a modern movie theater to shame. There were only two chairs in the room. The rest of the "auditorium" was empty.

Another room had an indoor Olympic-sized swimming pool surrounded by a jungle of green tropical plants. Fred wondered how the Colonel, who was on his army salary, could afford such luxuries that rivaled those of the Hearst Castle.

As the maid led Fred into the upstairs office suite, two helmeted and armed guards stood at the doorway. Fred said to them, "It's okay boys!" as they passed. The guards remained expressionless until the maid and Fred passed, and then they turned to each other and laughed.

The colonel moved several piles of jewels from the top of his desk into a desk drawer and came right to the point. "I hear that you know something about math and measurements. That true?"

Fred ignored the colonel's eliding the word *is* in his question and answered, "Some people say I do."

"Okay boy. Let's cut out the small talk. I ordered a four-cubit-by-nine-cubit piece of lawn. You saw it. You worked with it. What's a cubit?"

Fred explained, "A cubit is an ancient unit of measurement of length."

"No. How long is it?" the colonel clarified his question.

"Well, for this Cubit Garden Supply company," Fred began, "it must be 18 inches." That's all the colonel wanted to know, but Fred was in his college lecture mode now.

> Fred explained how he arrived at c = 18".

We started out with a rectangle that was 4 cubits wide and 9 cubits long, so it had an area of 36 square cubits.

We cut off 18" from the right side and had added 9" to the top. The original rectangle was 4c by 9c and the new rectangle was (9c – 18) by (4c + 9) where c = the length of a cubit in inches.

The area of the new rectangle was equal to the area of the original rectangle. We hadn't lost or gained any area by rearranging the lawn's shape.

The area of the new rectangle was (9c – 18)(4c + 9) and this was equal to the area of the original rectangle (4c)(9c).

$$(9c - 18)(4c + 9) = (4c)(9c)$$

The right side of the equation was easy to work out:

$$(9c - 18)(4c + 9) = 36c^2$$

It was the left side that Fred knew would take some explaining. How do you multiply $(9c - 18)$ times $(4c + 9)$?

How to Multiply $(9c - 18)(4c + 9)$

We know how to multiply $9c$ times $(4c + 9)$. That was the distributive property. You multiply the $9c$ times the $4c$ and then you multiply the $9c$ times the $+ 9$.

It looked like this: $9c(4c + 9) = 36c^2 + 81c$.

The same would be true for the -18 times $(4c + 9)$.

That would look like: $-18(4c + 9) = -72c - 162$.

Before Fred was drafted, he used to tell his beginning algebra students that when they multiply out $(9c - 18)(4c + 9)$, they should pretend that the $(9c - 18)$ were boys and the $(4c + 9)$ were girls,

$$(9c - 18)(4c + 9)$$

boys   girls

. . . and that you arrange every possible meeting between the boys and the girls.

So $(9c - 18)(4c + 9) = 36c^2 + 81c - 72c - 162$

the first boy meets each girl ... the second boy meets each girl

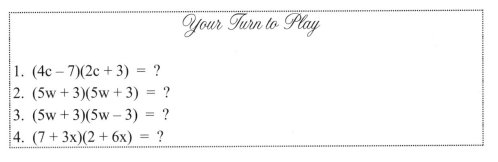

Your Turn to Play

1. $(4c - 7)(2c + 3) = ?$
2. $(5w + 3)(5w + 3) = ?$
3. $(5w + 3)(5w - 3) = ?$
4. $(7 + 3x)(2 + 6x) = ?$

5. $(5y + 4)^2 = ?$

6. $(x + 7)(x + 6) = ?$

7. $(z + 3)(z - 8) = ?$

8. $5(2x - 7) = ?$

9. $x^3(x^4 + x^7 + 4x^{11}) = ?$

Intermission

In problem 6, we multiplied $(x + 7)(x + 6)$ and got an answer of $x^2 + 13x + 42$. Several pages from now we are going to go in the opposite direction. You will be given $x^2 + 13x + 42$ and will be asked to find $(x + 7)(x + 6)$.

For now, you might notice that the 7 and 6 add to 13 and that they multiply to 42.

This "going in the opposite direction" is called factoring. In elementary school it was a little easier. When they asked you to factor 91, you said it's 7 × 13.

10. A preview: factor $x^2 + 11x + 28$

. . . . COMPLETE SOLUTIONS . . . .

1. $(4c - 7)(2c + 3) = 8c^2 + 12c - 14c - 21 = 8c^2 - 2c - 21$

2. $(5w + 3)(5w + 3) = 25w^2 + 15w + 15w + 9 = 25w^2 + 30w + 9$

3. $(5w + 3)(5w - 3) = 25w^2 - 15w + 15w - 9 = 25w^2 - 9$

4. $(7 + 3x)(2 + 6x) = 14 + 42x + 6x + 18x^2 = 14 + 48x + 18x^2$

5. $(5y + 4)(5y + 4) = 25y^2 + 20y + 20y + 16 = 25y^2 + 40y + 16$

6. $(x + 7)(x + 6) = x^2 + 13x + 42$ (I skipped a step.)

7. $(z + 3)(z - 8) = z^2 - 5z - 24$

8. $5(2x - 7) = 10x - 35$

9. $x^3(x^4 + x^7 + 4x^{11}) = x^7 + x^{10} + 4x^{14}$

10. Factor $x^2 + 11x + 28$.

Your answer will be in the form $(x + ?)(x + ?)$. The two question marks are two numbers that add to 11 and that multiply to 28.

$(x + 4)(x + 7)$

We can check to see that this is the correct factoring by multiplying it out. $(x + 4)(x + 7) = x^2 + 7x + 4x + 28 = x^2 + 11x + 28$. It works.

T hen Fred finished working out the equation to show the colonel that a cubit is 18" long by writing with his finger in the dust on the colonel's desk:

$$(9c - 18)(4c + 9) = (4c)(9c)$$

$$36c^2 + 81c - 72c - 162 = 36c^2$$

combine like terms

$$36c^2 + 9c - 162 = 36c^2$$

subtract $36c^2$ from both sides

$$9c - 162 = 0$$

$$9c = 162$$

$$c = 18$$

Fred smiled and turned to the colonel whose eyes had glazed over. His normally olive-green eyes seemed to be several shades lighter. The colonel dismissed Fred, and he was escorted by the maid back through the labyrinth of the mansion. While he is heading back to the bunkhouse, it's time for . . .

Your Turn to Play

1. $(5c - 3)(4c + 7) = ?$
2. $(8x + 6)(2x + 9) = ?$
3. $(w + 9)(w + 6) = ?$
4. $(x + 7)^2 = ?$
5. $(6x + 5)(x - 3) = ?$
6. What we've been doing in the above problems is multiplying a **binomial** times a binomial. A binomial is a **polynomial** with two **terms**. That only leaves you with two questions: What's a polynomial? and What's a term?

Here are some polynomials with one term: $3x$

$4x^2y$

These are monomials. $-189732.87x^{38}y^3z^7$

πx^2

Here are some polynomials with three terms: $3x + 4yz - 2c$

$(1/3)x^2 + 6x^4 - 3y^4$

These are trinomials. $8x + 3 + 9w$

$a + b + c/4$

The official **definition of a polynomial** is an expression formed by adding, subtracting or multiplying together numbers and letters. (Division isn't mentioned.) So things like $\dfrac{5}{x}$ or $\dfrac{x+6}{y}$ aren't polynomials since you can't form them by adding, subtracting or multiplying together numbers and letters.

Your question: How can $a + b + c/4$ be a polynomial? It looks like we used division.

7. How about multiplying a **trinomial** (3 terms) times a binomial? $(2x + 4y + 7)(5x + 8) = ?$

8. A binomial times a trinomial: $(4x^3 - y^4)(3x + 4x^8y^5 - 9) = ?$

9. A **monomial** is a polynomial with one term. Multiplying a monomial times a binomial was something that Fred did in chapter three and proved at the beginning of chapter four. What was that called?

10. Looking at the window in the paddy wagon, he proved that $a(b + c) = ab + ac$. A very similar diagram can be used to prove $(a + b)(c + d) = ac + ad + bc + bd$. First, turn back to the second page of Lesson 25 if you need to look at Fred's proof of $a(b + c) = ab + ac$. Then try to work out the diagram and proof for $(a + b)(c + d) = ac + ad + bc + bd$. You will learn a lot more and retain it much better if you attempt this yourself, rather than just read the answer.

11. If you multiply out $(x + 3)(x - 3)$, two terms drop out. You get $(x + 3)(x - 3) = x^2 - 9$. What do you multiply to get $y^2 - 25$?

.......COMPLETE SOLUTIONS.......

1. $(5c - 3)(4c + 7) = 20c^2 + 35c - 12c - 21$

    combine like terms $= 20c^2 + 23c - 21$

2. $(8x + 6)(2x + 9) = 16x^2 + 72x + 12x + 54 = 16x^2 + 84x + 54$

3. $(w + 9)(w + 6) = w^2 + 6w + 9w + 54 = w^2 + 15w + 54$

4. $(x + 7)^2 = (x + 7)(x + 7) = x^2 + 7x + 7x + 49 = x^2 + 14x + 49$

5. $(6x + 5)(x - 3) = 6x^2 - 18x + 5x - 15 = 6x^2 - 13x - 15$

6. a + b + c/4 is a polynomial because it can be formed by adding, subtracting or multiplying numbers and letters together. The term "c/4" can be written as (1/4)c and this is formed by multiplying the number 1/4 times c.

7. When we multiply out $(2x + 4y + 7)(5x + 8)$, we think of it as all the possible meetings between three boys and two girls. We first start with the 2x and multiply it times the 5x and then times the 8. Then we work with the 4y and, finally, with the 7.

$(2x + 4y + 7)(5x + 8) = 10x^2 + 16x \quad + 20xy + 32y \quad + 35x + 56.$

We look for the like terms (those that have the same exponents with the same bases), and the only ones we can spot are the 16x and the 35x. Combining these, we get

$$10x^2 + 51x + 20xy + 32y + 56.$$

8. $(4x^3 - y^4)(3x + 4x^8y^5 - 9) = 12x^4 + 16x^{11}y^5 - 36x^3 - 3xy^4 - 4x^8y^9 + 9y^4$
This is as far as we can go since there are no like terms that we can combine.

9. $a(b + c) = ab + ac$ is called the distributive property.

10. To prove $(a + b)(c + d) = ac + ad + bc + bd$.

The area of the largest rectangle is $(a + b)(c + d)$.
The areas of the small rectangles
are ac, ad, bc, and bd.

The area of the large rectangle equals
the sum of the areas of the four smaller rectangles.

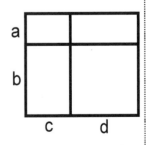

11. $(y + 5)(y - 5)$. This process is called **factoring**. It's the opposite of multiplying.

H eading back to the bunkhouse, Fred noticed that the eleven recruits had stopped watching television. They were engaged in something more constructive. On this sunny Sunday afternoon they were tearing a hole in the side of the bunkhouse. It was a nice big square hole. Fred stopped to watch. Everyone seemed busy. Pat was swinging a sledge hammer. Two of the guys were standing outside as lookouts. Fred guessed that they hadn't received permission to remodel their

a nice big hole

quarters. The only one not working on making the hole was Jack. Fred could see him through the hole in the wall. Jack was on the other side of the room working out with his weights.

Chris was on the phone with Waddle Windows (a subsidiary of Waddle Doughnuts). She was finding out the availability of various sizes of windows, and after several minutes of discussion, she ordered the largest one they offered. (Their motto is "A Wot of Window for a Wittle Price is What You Want at Waddle's.")

"You gotta make the hole a lot larger," she told Pat. "Six feet taller and six feet wider. The guy from Waddle said that that would let in four times as much light as what we've got now."

Fred decided not to go inside. He didn't want to be near what he thought might be a crime scene. He went and sat under a tree and thought about what he had heard:

It was a square. If it was made six feet larger in each dimension, its area would quadruple.

Was that enough information to find the current dimensions of the hole? Sometimes "real life" gives you enough information to solve a problem and sometimes it doesn't.

Fred started by letting x = the length of one of the sides of the square hole as it is now, before enlargement. Its current area is x^2 square feet.

After enlargement of the hole, its new dimensions will be x + 6 on each side and the new area will be $(x + 6)^2$ square feet which will be four times the area of the original hole. So $(x + 6)^2$ will be four times x^2.

$$(x + 6)^2 \ = \ 4x^2$$

We've seen an equation like this (with a squared term in it) just before the most recent *Your Turn to Play*. We multiplied everything out and subtracted off the x^2 term and solved the resulting linear equation. Will anything new pop up this time?

	$(x + 6)(x + 6) \ = \ 4x^2$
Multiply out the $(x + 6)^2$	$x^2 + 6x + 6x + 36 \ = \ 4x^2$
Combine like terms	$x^2 + 12x + 36 \ = \ 4x^2$
Subtract x^2 from both sides	$12x + 36 \ = \ 3x^2$

We're stuck. We can't get rid of the **quadratic** term (the one with the x^2 in it). When we were working with the lawn from Cubit Garden Supply, there was $36c^2$ on each side of the equation and when we subtracted, both $36c^2$ terms disappeared. As Fred sat under a tree, he was faced with a **quadratic equation** to solve.

Transpose everything to one side
$$0 \ = \ 3x^2 - 12x - 36$$

Use the symmetric law of equality to switch the sides of the equation
$$3x^2 - 12x - 36 \ = \ 0$$

He noticed that it would look easier if he divided both sides by 3
$$x^2 - 4x - 12 \ = \ 0$$

Now what? If you were to write "$x^2 - 4x - 12 \ = \ 0$" on a piece of scratch paper and shut this book and try to solve it, it might take you a small forever to solve the equation. Luckily neither you nor I have to reinvent the method for solving quadratic equations. Instead, it was Fred's problem.

And here's how he thought about it: If I have two numbers—call them a and b—and they multiply to equal zero, then what can I say about a and b? If I know that ab = 0, then certainly either a = 0 or b = 0 (or both equal zero). The product of two non-zero numbers can't be equal to zero.

Since he didn't want to carve this fact into the trunk of the tree he was sitting under, he took a stick and wrote in the ground:

$$ab = 0 \text{ implies } a = 0 \text{ or } b = 0$$

He then took $x^2 - 4x - 12$ and "unmultiplied it" and got $(x - 6)(x + 2)$. This process is called **factoring**[*]. If you take $(x - 6)(x + 2)$ and multiply it out, you'll find that it equals $x^2 - 4x - 12$. (Please do it!)

So $$x^2 - 4x - 12 = 0$$
became $$(x - 6)(x + 2) = 0$$

He had two factors that multiplied to equal zero. So, by what he had written on the ground, he knew that at least one of the factors must equal zero. Namely, either $x - 6$ must be zero or $x + 2$ must be zero.

$$x - 6 = 0 \quad \text{OR} \quad x + 2 = 0$$

———————————————

[*] We'll show you how to factor in the next lessons. As usual, we are first showing that you have a *need* to factor. Then we learn how to do it. In all of *Life of Fred* the need comes before the math is explained.

Fred had turned a quadratic equation into two linear equations and these were easy to solve:

$$x - 6 = 0 \qquad\qquad x + 2 = 0$$
$$x = 6 \qquad\qquad\quad x = -2$$

Let's check this answer to see if it's correct. The value of $x = 6$ says that the original hole was 6 ft x 6 ft (with an area of 36 sq ft). If we add 6 ft to each dimension, we get a hole that's 12 ft x 12 ft (with an area of 144 sq ft). This new area (144 sq ft) is four times the old area (of 36 sq ft). The solution "checks."

This solution doesn't make any sense. We couldn't have a square whose side equals minus two feet. We discard this solution. This will often happen in solving quadratic equations for problems that occur in the real world. The value of $x = -2$ makes the equation $x^2 - 4x - 12 = 0$ true, but the equation didn't know that we needed an answer that is positive.

Your Turn to Play

1. [Fill in one word] If we have an equation $6 = x^2 + 3$, we can say that $x^2 + 3 = 6$ by the _____ law of equality.

2. $76x^2 - 3.5x + 8979 = 0$ is called a quadratic equation since it has x^2 in it. What is $98x - 222.004 = 0$ called?

Here is how to factor 18: $18 = 2 \times 3 \times 3$

3. Factor 15.

4. Factor 24.

5. Solve $(x - 5)(x - 8) = 0$

6. Solve $(y - 4)(y + 332) = 0$

7. If I know that $ab = 5$, could either a or b be equal to zero?

8. Solve $(z + 3)(z + 7)(z - 89) = 0$

9. Solve $(x + 2.2)(x - \pi) = 0$

10. Fred's favorite watering can is in the shape of a cylinder. Its height is seven inches and its diameter is six inches. What is the volume of water that it can hold?

11. Solve

$$(x - 4)(x - ¼)(x + 10) = (\log \pi + \sin \pi/7)^0 - 1$$

. . . . **COMPLETE SOLUTIONS** . . . .

1. $\alpha = \beta$ implies $\beta = \alpha$ by the symmetric law of equality.

2. Equations that contain a variable only to the first power are called linear equations. Fred had mentioned that to the chaplain in Lesson 39.

3. $15 = 3 \times 5$

4. $24 = 2 \times 2 \times 2 \times 3$

5. $(x - 5)(x - 8) = 0$

Since $(x - 5)$ and $(x - 8)$ multiply to zero, we know that either

$$x - 5 = 0 \quad \text{OR} \quad x - 8 = 0$$
$$x = 5 \quad \text{OR} \quad x = 8$$

6. $(y - 4)(y + 332) = 0$

Either $y - 4 = 0 \quad \text{OR} \quad y + 332 = 0$
$$y = 4 \quad \text{OR} \quad y = -332$$

7. Neither *a* nor *b* could equal zero. If either was equal to zero, then their product would have to equal zero.

8. $(z + 3)(z + 7)(z - 89) = 0$

Either $z + 3 = 0 \quad \text{OR} \quad z + 7 = 0 \quad \text{OR} \quad z - 89 = 0$
$$z = -3 \quad \text{OR} \quad z = -7 \quad \text{OR} \quad z = 89$$

which is sometimes written as $z = -3, -7, 89$

9. $(x + 2.2)(x - \pi) = 0$

Either $x + 2.2 = 0$ OR $x - \pi = 0$

$x = -2.2, \pi$

Intermission

Pi, π, is a number. (It's also a Greek letter.) If you ask a mathematician to name a number between one and ten, you might get any of the following answers:
7, 4, 3½, π, $\sqrt{97}$, 7.99999999994 or log (152)

10. The volume of a cylinder is $V = \pi r^2 h$ where r is the radius of the cylinder and h is the height. Since the diameter is equal to 6 inches, the radius is equal to 3 inches. The volume is $\pi 3^2(7)$ or 63π cubic inches.

11. $(x - 4)(x - \frac{1}{4})(x + 10) = (\log \pi + \sin \pi/7)^0 - 1$ is equivalent to

$(x - 4)(x - \frac{1}{4})(x + 10) = 1 - 1 = 0$ since anything to the zero power is equal to one.

$x - 4 = 0$ OR $x - \frac{1}{4} = 0$ OR $x + 10 = 0$

$x = 4, \frac{1}{4}, -10$

A window that's 12 feet x 12 feet! That's huge. Most rooms in ordinary houses are eight feet tall. Luckily the men's bunkhouse walls were taller than eight feet. Unluckily, they were 11' tall. When the 12' x 12' window arrived from Waddle Windows, they were unable to whittle the window down to 11' so they discarded it and left a big hole in the side of their bunkhouse. Through that hole would come lots of sunlight, some birds, lots of mosquitos, and the anger of Sergeant Snow.

The only part of the math that we've left hanging is how Fred was able to take $x^2 - 4x - 12$ and factor it into $(x - 6)(x + 2)$. We'll spend several lessons explaining how to factor quadratic expressions.

There are five kinds of factoring in beginning algebra:
1. common factor (a.k.a. the distributive law in reverse)
2. easy trinomials (of the form $x^2 + bx + c$)
3. difference of squares (of the form $x^2 - y^2$)
4. grouping
5. harder trinomials (of the form $ax^2 + bx + c$ where $a \neq 1$)
   [$\neq$ means *not equal*.]

1. **Common Factor** (a.k.a. the distributive law in reverse)

We know how to use the distributive law: $a(b + c) = ab + ac$. If you want to remove the parentheses from $9c(4c + 9)$ you multiply the 9c times the 4c and then the 9c times the 9:

$$9c(4c + 9) = 36c^2 + 81c$$

Factoring out a common factor reverses this process. We start with $36c^2 + 81c$ and look for what numbers or letters divide evenly into both terms. What divides evenly into both 36 and 81? The biggest number is 9. What letter(s) divide evenly into c^2 and into c? The answer is c. So we can factor out a 9c out of each term.

$36c^2 + 81c$ factors into $9c(4c + 9)$.

When you attempt to factor anything, *it's always the common*

factor that you should look for first. ✎ important! It will make your life easier if you take out any common factors before you try any of the four other techniques.

Factor $44x^2y^5 + 66x^3y^7z$. The largest number that divides into both 44 and 66 is 22. The letter(s) that divide into both x^2y^5 and into x^3y^7z are x^2y^5.

$$44x^2y^5 + 66x^3y^7z = 22x^2y^5(2 + 3xy^2z)$$

Please work these out on paper before looking at the answers. Please.

Your Turn to Play

1. Factor $100x^2 + 50x$
2. Factor $6y^2 + 9y^4$
3. Factor $15abx - 22xyz$
4. Factor $60xy^2z^{60} + 12x^5y^2z^{16} + 20wxy^5z^{10}$

. . . . COMPLETE SOLUTIONS . . . .

1. The largest number that divides into both $100x^2$ and $50x$ is 50. The letter(s) that divide into $100x^2$ and $50x$ is x.

$$100x^2 + 50x = 50x(2x + 1)$$

If you multiply out $50x(2x + 1)$ you'll get $100x^2 + 50x$. This makes a good check that you've done the factoring correctly.

2. The largest number that divides into both 6 and 9 is 3. The largest common divisor of y^2 and y^4 is y^2. Factoring $6y^2 + 9y^4$ we get $3y^2(2 + 3y^2)$.

3. $15abx - 22xyz = x(15ab - 22yz)$

4. The largest number that divides into 60, 12, and 20 is 4.
$60xy^2z^{60} + 12x^5y^2z^{16} + 20wxy^5z^{10} = 4xy^2z^{10}(15z^{50} + 3x^4z^6 + 5wy^3)$

Easy Trinomials (of the form $x^2 + bx + c$)

For many people, this is their favorite kind of factoring. It's very similar to playing the game: Name two numbers that multiply to 12 and that add to 7.

That's what's involved in factoring $x^2 + 7x + 12$. If you can think of the two numbers that multiply to 12 and that add to 7, you can do the factoring.

The two numbers? They are +3 and +4. Now look at the factoring of $x^2 + 7x + 12$. It is $(x + 3)(x + 4)$. If you multiply out $(x + 3)(x + 4)$, you get $x^2 + 3x + 4x + 12$ which simplifies to $x^2 + 7x + 12$.

Let's try another one. If you are factoring $x^2 + 8x + 15$, you are looking for two numbers that multiply to +15 and that add to +8. In a classroom setting, we'd pause for a moment, but it's more difficult to have you stop reading while you think of the two numbers that multiply to +15 and that add to +8. Please stop reading until you've figured out the two numbers that multiply to +15 and that add to +8. If I asked you to shut your eyes that might work. If you did that, everything would start to look

like ███████████████████████████████ .

(Have I killed enough time so that you have found the answers of +3 and +5?)

So factoring $x^2 + 8x + 15$ gives an answer of $(x + 3)(x + 5)$.

A last example, before you start *Your Turn to Play*, involves negative numbers. If we're trying to factor $x^2 + x - 12$, we're looking for two numbers that multiply to -12 and that add to +1. If they multiply to -12 then one of them is positive and the other is negative.

The two numbers are +4 and -3, so $x^2 + x - 12$ factors into $(x + 4)(x - 3)$. If you wrote $(x - 3)(x + 4)$ that's okay also.

Your Turn to Play

1. Factor $x^2 + 10x + 21$

2. Factor $x^2 + 11x + 30$

3. Factor $y^2 + 9y + 20$

4. Factor $z^2 - 6z + 5$

5. Factor $3x^2 + 18x - 48$

6. Factor $x^3 - 18x^2 - 40x$

7. Solve the quadratic equation $x^2 + 11x + 24 = 0$

8. Solve $x^2 - 15x = -56$

9. The janitors arranged the chairs in a large classroom so that there were x rows with x chairs in each row. But since an extra large crowd was expected they added four more chairs to each of the rows. That gave a seating capacity of 140. How many rows of chairs are there?

10. Factor $x^2 + 7xy + 10y^2$

. . . . **COMPLETE SOLUTIONS** . . . .

1. $x^2 + 10x + 21 = (x + 3)(x + 7)$
2. $x^2 + 11x + 30 = (x + 5)(x + 6)$
3. $y^2 + 9y + 20 = (y + 4)(y + 5)$
4. $z^2 - 6z + 5 = (z - 5)(z - 1)$
5. Two pages ago you read that when you attempt to factor anything, you should always look for a common factor first.

There is a common factor in $3x^2 + 18x - 48$, which is 3.

So the first factoring of $3x^2 + 18x - 48$ is $3(x^2 + 6x - 16)$.

Now we look at the $x^2 + 6x - 16$ and try to find two numbers that multiply to -16 and that add to $+6$. They are $+8$ and -2.
The complete factoring of $3x^2 + 18x - 48$ is $3(x + 8)(x - 2)$.

6. Common factor $x^3 - 18x^2 - 40x = x(x^2 - 18x - 40)$.

Now to find two numbers that multiply to −40 and that add to −18. They are −20 and +2.

$$= x(x - 20)(x + 2).$$

7. The first step in solving a quadratic equation by factoring is to put everything on one side of the equation. This has already been done. Factoring $x^2 + 11x + 24 = 0$, we get $(x + 3)(x + 8) = 0$. Now if two numbers multiply to zero, then one of them has to be zero.

$$x + 3 = 0 \quad \text{OR} \quad x + 8 = 0.$$
$$x = -3 \quad \text{OR} \quad x = -8.$$

8. First we take $x^2 - 15x = -56$ and transpose the −56 to the left side so that the equation has everything on one side: $x^2 - 15x + 56 = 0$.

factor

$$(x - 7)(x - 8) = 0$$

set each factor equal to zero

$$x - 7 = 0 \quad \text{OR} \quad x - 8 = 0$$
$$x = 7 \quad \text{OR} \quad x = 8$$

9. The dimensions of the expanded seating are x rows with x + 4 chairs in each row. That gives us x(x + 4) chairs, which we are told is 140. So x(x + 4) = 140. We'd be heading in the wrong direction if we thought, "Great! It's already factored." In solving quadratic equations by factoring, *the first step is to put everything on one side of the equation.*

$$x(x + 4) - 140 = 0$$

Distributive property

$$x^2 + 4x - 140 = 0$$

Two numbers that multiply
to −140 and add to +4

$$(x - 10)(x + 14) = 0$$

Set each factor equal to zero

$$x - 10 = 0 \quad \text{OR} \quad x + 14 = 0$$
$$x = 10 \quad \text{OR} \quad x = -14$$

The x = −14 solution doesn't make any physical sense and we discard that solution.

So the remaining answer is x = 10. Let's **check** that solution in the original problem. If x = 10 is the correct solution then we originally had 10 rows with 10 chairs in each row. By adding four chairs to each row, we'd have 10 rows with 14 chairs in each row. That does give us 140 chairs. So the answer checks. By checking your answer you *know* that you have the correct answer.

10. $x^2 + 7xy + 10y^2 = (x + 2y)(x + 5y)$ You want two numbers that add to 7y and multiply to $10y^2$. That would be 2y and 5y.

Difference of Squares (of the form $x^2 - y^2$)

The easiest way to approach this kind of factoring is to see what happens when you multiply out some binomials:

$$(x + 7)(x - 4) = x^2 + 3x - 28$$
$$(x + 5)(x - 9) = x^2 - 4x - 45$$
$$(x + 3)(x - 8) = x^2 - 5x - 24$$
$$(x + 6)(x - 6) = x^2 - 36$$

Wait! Stop! This last one is different. The middle term (the x-term) isn't there. All we have is $x^2 - 36$. What we have is a difference (that means subtraction) of squares.

How would you factor $x^2 - 25$? It's $(x + 5)(x - 5)$. You can check to see that that works by multiplying out $(x + 5)(x - 5)$.

If it's a difference of squares, it factors instantly.

Here's one huge example: Factor $49x^2 - 100y^6$

First you set out $(\quad + \quad)(\quad - \quad)$

and then you fill in
the pieces: $(7x + 10y^3)(7x - 10y^3).$

The General Rule for Difference of Squares

$$a^2 - b^2 = (a + b)(a - b)$$

The difference of squares is so easy that the *Your Turn to Play* has only one problem in it. That's the good news . . .

Your Turn to Play

. . . and I guess the bad news is that this one problem has several parts.
1. Factor each of the following:

$y^2 - 4w^2$

$16x^2 - 25y^2$

$4 - y^2$

$x^8 - y^{10}$

$49 + z^2$

$72 - 8y^{10}$

$x^2 - 3x - 18$

. . . . **COMPLETE SOLUTIONS** . . . .

1. $y^2 - 4w^2 = (y + 2w)(y - 2w)$

$16x^2 - 25y^2 = (4x + 5y)(4x - 5y)$

$4 - y^2 = (2 + y)(2 - y)$

$x^8 - y^{10} = (x^4 + y^5)(x^4 - y^5)$

$49 + z^2$ isn't a *difference* of squares; it is the *sum* of squares. It doesn't factor.

$72 - 8y^{10} = 8(9 - y^{10}) = 8(3 + y^5)(3 - y^5)$ *Always look for the common factor first*, and in this case, if you didn't factor out the common factor first, you wouldn't have a difference of squares.

$x^2 - 3x - 18 = (x + 3)(x - 6)$ Two numbers that multiply to −18 and add to −3.

Grouping

These are easy to spot. Instead of two terms (like difference of squares) or three terms, they have more than three terms.

For example, take the 4-termed polynomial $6x^3 - 15x^2 + 4x - 10$.

The first thing you look for is a common factor, but there isn't any number or letter that divides into all four terms.

Next you group the first two terms together and the last two terms together:

$$6x^3 - 15x^2 \qquad\qquad + 4x - 10$$

and then we factor each part separately:

$$3x^2(2x - 5) \qquad\qquad +2(2x - 5)$$

Hey look! There's a $(2x - 5)$ in each of them.

Let me first move the groupings closer to each other:

$$3x^2(2x - 5) + 2(2x - 5).$$

What I'm going to do next is factor out the common factor of $(2x - 5)$, but that might be hard to see. Let me first squish the $(2x - 5)$ down into a small little unit:

$$3x^2_{(2x-5)} + 2_{(2x-5)}.$$

Now I'll factor it out:

$$_{(2x-5)}(3x^2 + 2)$$

and now I'll inflate it again:

$$(2x - 5)(3x^2 + 2).$$

Done!

In chapter two of *Life of Fred: Calculus Expanded Edition*, Fred was working on a radio contest so that he could win 500 gallons of Siberian Bear Brand Beer for his father. In order to find the limit of a function (limits are done in calculus) he had to factor $x^3 - 3x^2 + 6x - 18$. Maybe that would be a good place to begin *Your Turn to Play* .

Your Turn to Play

1. Factor $x^3 - 3x^2 + 6x - 18$ ← We'll use this one in Chapter 2 of Calculus.
2. Factor $x^3 + 4x^2 - 9x - 36$
3. Factor $y^3 + 9y - 9y^2 - 81$
4. Factor $x^2y - 5xy + 6y + 7x^2 - 35x + 42$
5. Factor $w^{68} + 2w^{34} - 35$
6. We had a square with each side equal to x inches. We added 5" to one side and took off 7" from an adjacent side. The resulting rectangle had an area of 64 square inches. How long was the side of the original square?

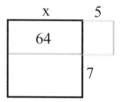

7. On my bicycle I rode at a rate of x mph for x hours. Afterwards, I got off and walked at a rate that was 5 mph slower, but I walked for six hours longer than I rode. I walked 12 miles. How fast was I riding my bicycle?

. . . . COMPLETE SOLUTIONS . . . .

1. Factor the first pair and the second pair of terms $x^2(x - 3) + 6(x - 3)$
Factor out the common factor of $(x - 3)$ $(x - 3)(x^2 + 6)$

2. Factor the first pair and the second pair of terms $x^2(x + 4) - 9(x + 4)$
Factor out the common factor of $(x + 4)$ $(x + 4)(x^2 - 9)$
We're not done. A difference of squares $(x + 4)(x + 3)(x - 3)$

3. $y^3 + 9y - 9y^2 - 81$
    $= y(y^2 + 9) - 9(y^2 + 9)$
    $= (y^2 + 9)(y - 9)$

Note: If this pairing of the first two terms and the last two terms hadn't worked, we might have rearranged the four terms, putting them in order of decreasing powers of y: $y^3 - 9y^2 + 9y - 81$.

4. Group the first three terms together $x^2y - 5xy + 6y$ + $7x^2 - 35x + 42$

and the last three terms together $y(x^2 - 5x + 6)$ + $7(x^2 - 5x + 6)$

Factor out the common factor of $(x^2 - 5x + 6)$ $(x^2 - 5x + 6)(y + 7)$

Continue by factoring $(x^2 - 5x + 6)$

Find two numbers that multiply to +6 and add to −5 $(x - 2)(x - 3)(y + 7)$

5. Experienced algebraists just look at $w^{68} + 2w^{34} - 35$ and find two numbers that multiply to –35 and add to + 2 and write $(w^{34} + 7)(w^{34} - 5)$.

 Beginning algebraists might find it easier to let $w^{34} = x$. Then $w^{68} = x^2$. Then $w^{68} + 2w^{34} - 35$ becomes $x^2 + 2x - 35$,
         which factors into $(x + 7)(x - 5)$ = $(w^{34} + 7)(w^{34} - 5)$

6. The area of the new rectangle is $(x + 5)(x - 7)$ and that is equal to 64.

To solve $(x + 5)(x - 7) = 64$

Multiply out the left side $x^2 - 2x - 35 = 64$

Transpose everything to one side $x^2 - 2x - 99 = 0$

Factor $(x - 11)(x + 9) = 0$

Set each factor equal to zero $x - 11 = 0$ OR $x + 9 = 0$
          $x = 11"$ OR $x = -9"$

The negative number is not possible as a measurement of the length of a side of a square and we discard it, leaving as our final answer $x = 11"$. We can check this answer in the original problem. If we have a square that is 11" on each side and we add five inches to one side and take seven inches off the other side, we'll have a rectangle that's 16" x 4", and this does have an area of 64 square inches.

7. *Let x = the speed I rode my bicycle.*

 Then x − 5 = the speed at which I walked.

 Then x + 6 = the number of hours I walked.

Using the d = rt formula, I covered a distance of 12 miles = $(x - 5)(x + 6)$.
          $12 = x^2 + x - 30$
          $0 = x^2 + x - 42$
          $0 = (x + 7)(x - 6)$
          $x = -7$ OR $x = 6.$

We discard the $x = -7$ solution since it doesn't make sense.

 I rode my bicycle at 6 miles per hour.

 You are an algebraist. (al-je-BRAY-ist) Did you know that?

Harder Trinomials
(of the form $ax^2 + bx + c$ where $a \neq 1$)

We've done the easy trinomials such as $x^2 + 7x + 12$, where we try to find two numbers that multiply to +12 and add to +7, and then we could write $x^2 + 7x + 12 = (x + 3)(x + 4)$.

A harder trinomial might look like $6x^2 + 29x + 35$. (The coefficient of the x^2 term isn't equal to one.)

There are two ways to tackle the harder trinomials. There's the old-fashioned approach, which has been taught to students for millions of years, and there's the new, quicker, handy-dandy approach.

In the interest of fairness we'll show both methods. Feel free, however, to ignore the old-fashioned approach which is in this type face and just skip to the modern approach.

Factoring $6x^2 + 29x + 35$ into $(2x + 5)(3x + 7)$ is a little like trying to open a combination lock by trying out various numbers. The only difference is that it is a little bit easier than opening a combination lock.

Here are the steps in factoring $6x^2 + 29x + 35$ . . .

① Name two terms that multiply to the $6x^2$ and put them in the front positions. For example $(6x \quad)(x \quad)$.

② Name two numbers that multiply to +35 and put them in the back positions. For example $(6x + 7)(x + 5)$.

③ Multiply it out and see if it "opens the lock." Namely you see if it equals $6x^2 + 29x + 35$.

Multiplying out $(6x + 7)(x + 5)$ we get $6x^2 + 30x + 7x + 35$
which simplifies to $6x^2 + 37x + 35$. Nope. That try didn't work. We have $37x$ as the middle term and not $29x$ that we want.

So you try again.

Two numbers that multiply to $6x^2$: $(2x\ \ )(3x\ \ )$

Two numbers that multiply to +35: $(2x + 35)(3x + 1)$

See if it works: $(2x + 35)(3x + 1) = 6x^2 + 2x + 105x + 35$

$$= 6x^2 + 107x + 35$$

Nope. That didn't work either. The middle term is 107x, not 29x that we want.

So you try again.

Two numbers that multiply to $6x^2$: $(2x\ \ )(3x\ \ )$

Two numbers that multiply to +35: $(2x + 5)(3x + 7)$

See if it works: $(2x + 5)(3x + 7) = 6x^2 + 14x + 15x + 35$

$$= 6x^2 + 29x + 35.$$

Bingo! We got it. The lock opened. Or as one student liked to yell out, "Manumission!" ↩ You do own a dictionary, don't you?

 If you've ever taken piano lessons, this old-fashioned approach feels like playing with both hands instead of just with one hand. It feels more than twice as tough. In factoring $6x^2 + 29x + 35$ by this old-fashioned approach, your left hand gets two numbers that multiply to $6x^2$ and your right hand gets two numbers that multiply to +35, and then you have to juggle all this stuff to see if it comes out to +29x.

The new quicker handy-dandy approach

Starting with $6x^2 + 29x + 35$, split the 29x into two numbers

?   ?

that add to 29x and that multiply to $(6x^2)(35)$, which is $210x^2$.

(Doesn't this remind you a little of factoring the "easy trinomials" like $x^2 + 7x + 12$ where you had to find two numbers that add to 7 and that multiply to 12? It's like playing the piano with just one hand.)

So we need two numbers that add to 29x and multiply to 210x². Here's where a hand calculator can smooth out some of the arithmetic. Playing with it a little bit, we find 14x and 15x do the job.

Once you've split $6x^2 + 29x + 35$

into $6x^2 + 14x + 15x + 35$

you finish the factoring
by grouping (which is a strictly mechanical procedure)

$$2x(3x + 7) + 5(3x + 7)$$
$$(3x + 7)(2x + 5)$$

Using the old-fashioned approach is about as hard as learning how to brush your teeth with your other hand. With the new, quicker, handy-dandy approach (where all you have to do is find two numbers that add to the middle term and that multiply to the product of the first and last terms) the whole process is about as difficult as figuring out which end of the toothbrush to use.

I've listed a ton of examples in this *Your Turn to Play*. Work enough of them until you get the hang of it, and then quit and go for a walk in the sunshine.

Your Turn to Play

1. $2y^2 + 11y + 5$
2. $3x^2 + 13xy + 14y^2$
3. $5w^2 - 12w + 7$
4. $3x^2 + 5x + 2$
5. $5a^2 - 7a - 6$
6. $3x^2 + 19x + 6$
7. $2w^2 + 7wz + 3z^2$
8. $4x^2 + 9x - 9$
9. $49y^2 + 84y + 36$
10. $3w^2 + 50w + 47$
11. $36x^2 + 72x - 108$
12. $16x^4 - 8x^2 + 1$

If a dozen problems is a "ton of examples," then each problem must weigh $\dfrac{2000 \text{ pounds}}{12} \doteq 167$ pounds

. . . . C O M P L E T E   S O L U T I O N S . . . .

1. Two numbers that add to 11y and that multiply to $(2y^2)(5)$.
   10y and y will do the trick.

$$2y^2 + 11y + 5$$

becomes $2y^2 + 10y + y + 5$

factor by grouping $2y(y + 5) + (y + 5)$

$$(y + 5)(2y + 1)$$

2. We split the middle term of $3x^2 + 13xy + 14y^2$ into two numbers that add to 13xy and that multiply to $42x^2y^2$ (which is the product of $3x^2$ and $14y^2$). How about 6xy and 7xy?

$$3x^2 + 13xy + 14y^2$$

$$3x^2 + 6xy + 7xy + 14y^2$$

factor by grouping $3x(x + 2y) + 7y(x + 2y)$

$$(x + 2y)(3x + 7y)$$

3. $5w^2 - 12w + 7$

$$5w^2 - 5w - 7w + 7$$

factor by grouping $5w(w - 1) - 7(w - 1)$

$$(w - 1)(5w - 7)$$

4. For $3x^2 + 5x + 2$, we look for two numbers that add to 5x and that multiply to $6x^2$. That's 2x and 3x.

$$3x^2 + 5x + 2$$

becomes $3x^2 + 2x + 3x + 2$

factor by grouping $x(3x + 2) + (3x + 2)$

$$(3x + 2)(x + 1)$$

5. For $5a^2 - 7a - 6$, we ask what two numbers add to –7a and multiply to $-30a^2$? Answer: –10a and +3a.

$$5a^2 - 7a - 6$$

$$5a^2 - 10a + 3a - 6$$

$$5a(a - 2) + 3(a - 2)$$

$$(a - 2)(5a + 3)$$

6. I don't know about you, but these are all starting to look alike to me. Here goes the complete solution to factoring $3x^2 + 19x + 6$. Two numbers

that add to 19x and that multiply to 18x² are 18x and x.

$$3x^2 + 18x + x + 6$$
$$3x(x + 6) + (x + 6)$$
$$(x + 6)(3x + 1)$$

What if I had said "x and 18x" instead of "18x and x"? Here's how the problem would work out: $3x^2 + x + 18x + 6$

$$x(3x + 1) + 6(3x + 1)$$
$$(3x + 1)(x + 6)$$

which by the commutative
law of multiplication is $(x + 6)(3x + 1)$ which is exactly what we got doing it
the other way.

7. $2w^2 + 7wz + 3z^2$

$$2w^2 + 6wz + wz + 3z^2$$
$$2w(w + 3z) + z(w + 3z)$$
$$(w + 3z)(2w + z)$$

8. $4x^2 + 9x - 9$

$$4x^2 + 12x - 3x - 9$$
$$4x(x + 3) - 3(x + 3)$$
$$(x + 3)(4x - 3)$$

9. $49y^2 + 84y + 36$

Now, finding two numbers that add to 84y and that multiply to 1764y² is something I like to use my calculator on.

I first tried y and 83y. The product was 83y², which is too small.

Then I tried 2y and 82y. The product is 164y².

Then I tried 4y and 80y. The product is 320y².

20y and 64y   The product is 1280y²

30y and 54y   The product is 1620y²

Notice that as the numbers ⬆ get closer together, the product gets larger.

When I tried 42y and 42y I got the right product of 1764y².

$$49y^2 + 42y + 42y + 36$$
$$7y(7y + 6) + 6(7y + 6)$$
$$(7y + 6)(7y + 6)$$

The problem we just finished is probably worse than any that you'll meet in Cities. If you've been doing these problems *actively* (by actually getting out a pencil and paper and attempting to do them) rather than *passively* (by just reading the solutions and nodding to yourself, "Oh yeah, I can do these things,") then the problems in the Cities section will be much easier.

10.
$$3w^2 + 50w + 47$$
$$3w^2 + 47w + 3w + 47$$
$$w(3w + 47) + (3w + 47)$$
$$(3w + 47)(w + 1)$$

11. $36x^2 + 72x - 108$

So I want two numbers that add to $72x$ and that multiply to $3888x^2$.

Stop! You always must look for a common factor first. This is a simple problem IF you noticed that $36x^2 + 72x - 108 = 36(x^2 + 2x - 3)$. Now all I have to do is factor this easy trinomial.
$$36x^2 + 72x - 108$$
$$36(x^2 + 2x - 3)$$
$$36(x + 3)(x - 1)$$

12. $16x^4 - 8x^2 + 1$

Two numbers that add to $-8x^2$ and that multiply to $16x^4$
$$16x^4 - 4x^2 - 4x^2 + 1$$
$$4x^2(4x^2 - 1) - (4x^2 - 1)$$
$$(4x^2 - 1)(4x^2 - 1)$$

Hey! There's a difference of squares.

Our final answer is $(2x + 1)(2x - 1)(2x + 1)(2x - 1)$

Philomath, Oregon *Philomath is Greek for "one who loves learning"*

1. Factor $77x^2y - 11y$

2. Factor $25x^2 + 81w^2$

3. Factor $9a^3 - 16ab^2$

4. Solve $x^2 - 9x = -20$

5. We had a square with each side equal to x inches. We added 3" to one side and took off 8" from an adjacent side. The resulting rectangle had an area of 60 square inches. How long was the side of the original square?

6. Factor $x^2 - x - 30$

7. Factor $xy - 6x + 4y - 24$

8. Solve $12y^2 - 5y - 2 = 0$

9. In a right triangle it is always true that the sum of the squares of the two smaller sides is equal to the square of the largest side. In terms of the diagram, this means that $a^2 + b^2 = c^2$.

    Now suppose we have a right triangle in which the shortest side is equal to x inches. The next shortest side is seven inches longer. And the longest side is eight inches longer than the shortest side. How long are the three sides?

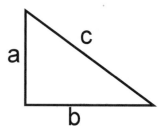

10. Factor $12x^2y^4 + 18xy^3 - 24wx^3y^2$

11. Factor $81x^2 - 16xy^2$

12. Factor $a^{50} - 4a^{25} + 4$

13. Solve $9x^2 - 25 = 0$

14. On my go-cart I rode at a rate of x mph for x hours. Afterwards I got off and walked at a rate that was 3 mph slower, but I walked for six hours longer than I rode. I walked 10 miles. How fast was I riding my go-cart?

15. Factor $x^2 + 7x + 10$

16. Factor $12ab + 21a - 20b - 35$

17. Solve $30x^2 - 19x = 5$

18. What is the quadratic equation whose solutions are x = 3 and x = 14?

answers

1. $77x^2y - 11y = 11y(7x^2 - 1)$

2. $25x^2 + 81w^2$ There's no common factor, and it's not a *difference* of squares. This can't be factored.

3. $9a^3 - 16ab^2 = a(9a^2 - 16b^2) = a(3a + 4b)(3a - 4b)$

4.
$$x^2 - 9x = -20$$

Transpose $$x^2 - 9x + 20 = 0$$

Factor $$(x - 4)(x - 5) = 0$$

$$x - 4 = 0 \quad OR \quad x - 5 = 0$$

$$x = 4, 5$$

5.   *Let x = the length of a side of the original square.*

   Then x + 3 = the length of the longer side of the rectangle.

   Then x - 8 = the length of the shorter sides of the rectangle.

The area of the rectangle is 60. $$(x + 3)(x - 8) = 60$$

$$x^2 - 8x + 3x - 24 = 60$$

Combine like terms $$x^2 - 5x - 24 = 60$$

Transpose $$x^2 - 5x - 84 = 0$$

Two numbers that multiply to −84 and add to −5 $(x - 12)(x + 7) = 0$

$$x - 12 = 0 \quad OR \quad x + 7 = 0$$

$$x = 12, -7$$

The length of a side of the original square is 12".

The other solution was discarded since you can't have a length of −7".

6. $(x + 5)(x - 6)$

7.
$$xy - 6x + 4y - 24$$

$$x(y - 6) + 4(y - 6)$$

$$(x + 4)(y - 6)$$

8. $12y^2 - 5y - 2 = 0$ Two things that add to −5y and multiply to −24y²

$$12y^2 - 8y + 3y - 2 = 0$$

$$4y(3y - 2) + 1(3y - 2) = 0$$

$$(3y - 2)(4y + 1) = 0$$

$$3y - 2 = 0 \quad OR \quad 4y + 1 = 0$$

$$y = \frac{2}{3} \quad OR \quad y = -\frac{1}{4}$$

9.   *Let x = the shortest side.*

      Then x + 7 = the next shortest side.

      Then x + 8 = the longest side.

Since the sum of the squares of the two shortest sides is equal to the square of the longest side . . .

$$x^2 + (x + 7)^2 = (x + 8)^2$$
$$x^2 + (x + 7)(x + 7) = (x + 8)(x + 8)$$
$$x^2 + x^2 + 7x + 7x + 49 = x^2 + 8x + 8x + 64$$
$$2x^2 + 14x + 49 = x^2 + 16x + 64$$
$$x^2 - 2x - 15 = 0$$
$$(x - 5)(x + 3) = 0$$
$$x - 5 = 0 \text{ OR } x + 3 = 0$$
$$x = 5 \text{ OR } x = -3$$

We discard x = −3 since we are dealing with lengths.

The shortest side (x) is equal to 5". The next shortest side (x + 7) is equal to 12". The longest side (x + 8) is equal to 13".

10. $12x^2y^4 + 18xy^3 - 24wx^3y^2$

                  $6xy^2(2xy^2 + 3y - 4wx^2)$

11. This is not a difference of squares. The only factoring to be done is a common factor $81x^2 - 16xy^2 = x(81x - 16y^2)$

12. $a^{50} - 4a^{25} + 4 = (a^{25} - 2)(a^{25} - 2)$

13. $9x^2 - 25 = 0$

                  $(3x + 5)(3x - 5) = 0$

            $3x + 5 = 0 \text{ OR } 3x - 5 = 0$

                  $x = -5/3 \text{ or } x = 5/3$

14. d = rt becomes $10 = (x - 3)(x + 6)$

                      $10 = x^2 + 3x - 18$

                      $0 = x^2 + 3x - 28$

                      $0 = (x - 4)(x + 7)$

x − 3 = rate of walking
x + 6 = hours spent walking
10 = miles walked

4 mph (We discard the solution of x = −7 mph.)

15. $(x + 2)(x + 5)$

16. $(3a - 5)(4b + 7)$

17. $30x^2 - 19x = 5 \Rightarrow 30x^2 - 19x - 5 = 0 \Rightarrow 30x^2 + 6x - 25x - 5 = 0$
$\Rightarrow 6x(5x + 1) - 5(5x + 1) = 0 \Rightarrow (5x + 1)(6x - 5) = 0 \Rightarrow x = -1/5, 5/6$

18. Working backwards . . . If the solutions are x = 3 and x = 14

the previous step would be $x - 3 = 0 \text{ OR } x - 14 = 0$

the step before that would be $(x - 3)(x - 14) = 0$

So the quadratic equation would be $x^2 - 17x + 42 = 0.$

Owensboro, Kentucky "the BBQ capital of the world"

1. Factor $30a^2 + 40a^2b$

2. Factor $4x^2 - 100y^2$

3. Factor $m^4 + m$

4. Solve $7x^2 + 14x = 105$

5. We had a square with each side equal to x inches. We added 3" to one side and took off 9" from an adjacent side. The resulting rectangle had an area of 28 square inches. How long was the side of the original square?

6. Factor $x^2 + x - 12$

7. Factor $r^3 - 2r^2 + 5r - 10$

8. Solve $21w^2 - 41w + 10 = 0$

9. Three of the following polynomials factor (and the others don't). Find those three and factor them.

$$15x^2 - 17xy + 4y^2$$
$$a^2 - 6a + 12$$
$$w^2 + 8w + 7$$
$$z^3 - 13z + z^2$$
$$b^3 + 5b^2 + 2b + 8$$

10. Factor $8r^2st^3 - 12r^3s^3t^3$

11. Factor $x^6 - y^{10}$

12. Factor $y^{18} + 4y^9 - 21$

13. Solve $y^2 + 12 = 8y$

14. On my skates I rode at a rate of x mph for x hours. Afterwards I took them off and walked at a rate that was 2 mph slower, but I walked for ten hours longer than I skated. I walked 13 miles. How fast was I skating?

15. Factor $x^2 - 2x - 48$

16. Factor $x^3 + 2x^2 - 16x - 32$

17. Solve $24z^2 - 2z = 15$

18. Three of the following polynomials factor (and the others don't). Find those three and factor them.

$$z^2 - 4z - 16$$
$$w^8 - 16$$
$$64a^4 + 25$$
$$6x^2 + 25x + 14$$
$$16y^3 - 12y^2 + 4y - 3$$

answers

1. $30a^2 + 40a^2b = 10a^2(3 + 4b)$

2. $4x^2 - 100y^2 = 4(x^2 - 25y^2) = 4(x + 5y)(x - 5y)$

3. $m^4 + m = m(m^3 + 1)$ (Later on in *Life of Fred: Advanced Algebra Expanded Edition*, we'll be able to factor this further, because we will show how to factor the sum of cubes.)

4.

$$7x^2 + 14x = 105$$

Transpose.

$$7x^2 + 14x - 105 = 0$$

It will make things a lot easier

if we divide everything by 7

$$x^2 + 2x - 15 = 0$$

An easy trinomial. Two numbers

that multiply to -15 and add to $+2$

$$(x + 5)(x - 3) = 0$$

$$x = -5 \text{ or } x = 3$$

> If you didn't see that common factor of 7, then you would have had a much harder time.
> $$7x^2 + 14x - 105$$
> a general trinomial
> We want two numbers that multiply to $-735x^2$ and add to $+14x$.
> $$7x^2 - 21x + 35x - 105$$
> etc.

5. *Let x = the length of a side of the square.*

Then x + 3 = the length of the longer side of the rectangle.

Then x - 9 = the length of the width of the rectangle.

The area of the rectangle is 28 square inches.

$$(x + 3)(x - 9) = 28$$

$$x^2 - 9x + 3x - 27 = 28$$

Combine like terms

$$x^2 - 6x - 27 = 28$$

Transpose

$$x^2 - 6x - 55 = 0$$

Two numbers that multiply to -55 and add to -6

$$(x - 11)(x + 5) = 0$$

$$x - 11 = 0 \quad \text{OR} \quad x + 5 = 0$$

The length of the original square was 11". (The other solution of $x = -5$" was discarded because we are working with lengths. The side of a square cannot be -5" long.)

6. $(x - 3)(x + 4)$

7.

$$r^3 - 2r^2 + 5r - 10$$

$$r^2(r - 2) + 5(r - 2)$$

$$(r - 2)(r^2 + 5)$$

8. Solve

$$21w^2 - 41w + 10 = 0$$

Two numbers that multiply to $210w^2$ and add to $-41w$

$$21w^2 - 35w - 6w + 10 = 0$$

$$7w(3w - 5) - 2(3w - 5) = 0$$

$$(3w - 5)(7w - 2) = 0$$

$$3w - 5 = 0 \quad \text{OR} \quad 7w - 2 = 0$$

$$w = 5/3, \ 2/7$$

9. $15x^2 - 17xy + 4y^2 = (5x - 4y)(3x - y)$

$w^2 + 8w + 7 = (w + 7)(w + 1)$

$z^3 - 13z + z^2 = z(z^2 - 13 + z)$

10. $8r^2st^3 - 12r^3s^3t^3 = 4r^2st^3(2 - 3rs^2)$ common factor

11. $x^6 - y^{10} = (x^3 + y^5)(x^3 - y^5)$ difference of squares

12. $y^{18} + 4y^9 - 21 = (y^9 - 3)(y^9 + 7)$ easy trinomial

13. $y^2 + 12 = 8y$

$y^2 - 8y + 12 = 0$

$(y - 2)(y - 6) = 0$

$y - 2 = 0 \ \ \text{OR} \ \ y - 6 = 0$

$y = 2 \text{ or } y = 6$

14. $d = rt$ becomes $13 = (x - 2)(x + 10)$

$13 = x^2 + 10x - 2x - 20$

$0 = x^2 + 8x - 33$

$0 = (x - 3)(x + 11)$

$x - 3 = 0 \ \ \text{OR} \ \ x + 11 = 0$

I was skating at 3 mph. (We discard the solution of x = –11 mph.)

15. $(x + 6)(x - 8)$

16. $x^3 + 2x^2 - 16x - 32$

$x^2(x + 2) - 16(x + 2)$

$(x + 2)(x^2 - 16)$

$(x + 2)(x + 4)(x - 4)$

17. $24z^2 - 2z = 15$

$24z^2 - 2z - 15 = 0$

$24z^2 + 18z - 20z - 15 = 0$ Two things that multiply to $-360z^2$ and add to $-2z$

$6z(4z + 3) - 5(4z + 3) = 0$

$(4z + 3)(6z - 5) = 0$

$z = -3/4 \text{ or } z = 5/6$

18. $w^8 - 16 = (w^4 + 4)(w^4 - 4) = (w^4 + 4)(w^2 + 2)(w^2 - 2)$

$6x^2 + 25x + 14 = (2x + 7)(3x + 2)$

$16y^3 - 12y^2 + 4y - 3 = (4y^2 + 1)(4y - 3)$

A Short Note to You

You have probably noticed by now that after you have done two cities, the third city can be done with much less effort.

The problems are not any easier. The difference is that you are mastering the math.

Slatyfork, West Virginia   a haven for fly fisherman and mountain bikers

---

1. Factor $14x^2y - 14x^3y^2$

2. Factor $100a^4 - w^2$

3. Factor $50a^2 - 98$

4. Solve $z^2 - 20 = -z$

5. We had a square with each side equal to x inches. We added 6" to one side and took off 2" from the other side. The resulting rectangle had an area of 48 square inches. How long was the side of the original square?

6. Factor $y^2 + 9y + 8$

7. Factor $y^3 + 3y^2 - 25y - 75$

8. Solve $40x^2 + 11x - 2 = 0$

9. Three of the following polynomials factor (and the others don't). Find those three and factor them.

$$7a^2 + 35ab - 42$$
$$r^2 - 20r + 36$$
$$2y^2 + 6y + 3$$
$$z^2 + 5z + 6 + w$$
$$18x^2 + 21x - 4$$

10. Factor $36w^2x^{40} + 18wx^2$

11. Factor $64x^2 - 9w^6$

12. Factor $2x^2 + 12x + 16$

13. Solve $x^2 + 56 = -15x$

14. On my skateboard I rode at a rate of x mph for x hours. Afterwards I

got off and walked at a rate that was 7 mph slower, but I walked for eight hours longer than I skated. I walked 34 miles. How fast was I skateboarding?

15. Factor $w^2 + 10w + 24$

16. Factor $5a^3 - 15a^2 + 5a - 15$

17. Solve $16a^2 + 62a + 21 = 0$

18. Three of the following polynomials factor (and the others don't). Find those three and factor them.

$3z^2 + 452z + 5$
$c^2 - 20c + 40$
$2y^2 + 5y + 3$
$x^2w + 5xw + 6w + 4x^2 + 20x + 24$
$24a^2 + 14a - 3$
$16b^2 - 3a^2$
$d^2 + 14d - 40$

answers

1. $14x^2y - 14x^3y^2 = 14x^2y(1 - xy)$
2. $100a^4 - w^2 = (10a^2 + w)(10a^2 - w)$
3. $50a^2 - 98 = 2(25a^2 - 49) = 2(5a + 7)(5a - 7)$ Common factor first
4.
$$z^2 - 20 = -z$$
$$z^2 + z - 20 = 0$$
$$(z - 4)(z + 5) = 0$$
$$z = 4 \text{ or } z = -5$$

5. *Let x = the length of a side of the original square.*
Then x + 6 = the length of the longer side of the resulting rectangle.
The x - 2 = the width of the rectangle.

The area of the rectangle is 48 square inches.
$$(x + 6)(x - 2) = 48$$
$$x^2 - 2x + 6x - 12 = 48$$
$$x^2 + 4x - 60 = 0$$
$$(x - 6)(x + 10) = 0$$

The length of a side of the original square was 6". (We discard the solution of $x = -10$".)

6. $(y + 1)(y + 8)$

7.
$$y^3 + 3y^2 - 25y - 75$$
$$y^2(y + 3) - 25(y + 3)$$
$$(y + 3)(y^2 - 25)$$
$$(y + 3)(y + 5)(y - 5)$$

8.
$$40x^2 + 11x - 2 = 0$$
$$40x^2 + 16x - 5x - 2 = 0$$
$$8x(5x + 2) - (5x + 2) = 0$$
$$(5x + 2)(8x - 1) = 0$$
$$5x + 2 = 0 \quad \text{OR} \quad 8x - 1 = 0$$
$$x = -2/5 \text{ or } x = 1/8$$

9.
$$7a^2 + 35ab - 42 = 7(a^2 + 5ab - 6)$$
$$r^2 - 20r + 36 = (r - 18)(r - 2)$$
$$18x^2 + 21x - 4 = (3x + 4)(6x - 1)$$

10. $36w^2x^{40} + 18wx^2 = 18wx^2(2wx^{38} + 1)$

11. $64x^2 - 9w^6 = (8x + 3w^3)(8x - 3w^3)$

12. $2x^2 + 12x + 16 = 2(x^2 + 6x + 8) = 2(x + 2)(x + 4)$ Common factor first

13.
$$x^2 + 56 = -15x$$

Transpose
$$x^2 + 15x + 56 = 0$$
$$(x + 7)(x + 8) = 0$$
$$x = -7 \text{ OR } x = -8$$

14. $d = rt$ becomes
$$34 = (x - 7)(x + 8)$$
$$34 = x^2 + 8x - 7x - 56$$
$$0 = x^2 + x - 90$$
$$0 = (x - 9)(x + 10)$$
$$x - 9 = 0 \quad \text{OR} \quad x + 10 = 0$$

I skateboarded at 9 mph. (We discard the solution of $x = -10$ mph.)

15. $(w + 4)(w + 6)$

16. $5a^3 - 15a^2 + 5a - 15 = 5(a^3 - 3a^2 + a - 3) = 5(a^2 + 1)(a - 3)$

17.
$$16a^2 + 62a + 21 = 0$$
$$16a^2 + 6a + 56a + 21 = 0 \qquad \text{Arithmetic is harder than algebra!}$$
$$2a(8a + 3) + 7(8a + 3) = 0$$
$$(8a + 3)(2a + 7) = 0$$
$$a = -3/8 \text{ OR } a = -7/2$$

18. $2y^2 + 5y + 3 = (2y + 3)(y + 1)$

$$x^2w + 5xw + 6w + 4x^2 + 20x + 24$$
$$= w(x^2 + 5x + 6) + 4(x^2 + 5x + 6)$$
$$= (x^2 + 5x + 6)(w + 4)$$
$$= (x + 2)(x + 3)(w + 4)$$

$$24a^2 + 14a - 3$$
$$= 24a^2 + 18a - 4a - 3 \qquad \text{Arithmetic is harder than algebra!}$$
$$= 6a(4a + 3) - (4a + 3)$$
$$= (4a + 3)(6a - 1)$$

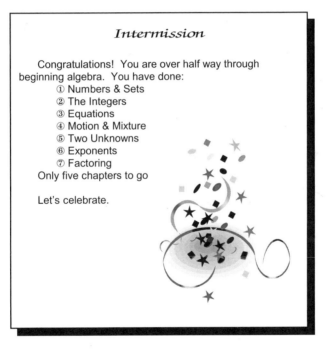

Intermission

Congratulations! You are over half way through beginning algebra. You have done:
① Numbers & Sets
② The Integers
③ Equations
④ Motion & Mixture
⑤ Two Unknowns
⑥ Exponents
⑦ Factoring
Only five chapters to go

Let's celebrate.

Chapter Eight
Lesson Sixty-four—Job Problems

Four o'clock in the afternoon and the hole in the side of the bunkhouse was something to behold. Pat looked at it and admired her work. Chris was delighted, and the nine other guys who had spent part of the afternoon creating the space for the Waddle Window (which wouldn't work) wondered at their workmanship.

a nice big hole

They stood in a semicircle around the black opening giving each other high fives and chatting about what they'd be doing this evening when they received their get-off-the-base passes at 6 p.m. Chris said to the guy next to her, "What about the old Snowman? Won't he be bent out of shape when he sees what we've done?"

What Chris had failed to notice was that there were now twelve in the semicircle—not eleven. Worse yet, the twelfth had the smell of beans and hot dogs on his breath.

"At–ten–shun!" someone shouted. Chris suddenly realized that the guy she had been talking to had sergeant's stripes on his sleeve.

Now we'll give you the chance to select which way the story will proceed. Pick any one of the columns below.

Choice #1

Sergeant Snow smiled and looked over the assembled group of recruits and began in his sweetest voice, "Ladies and gentlemen, I see that you've been hard at work attempting to

Choice #2

Sergeant Snow met their glances with an icy cold stare. His silence, more piercing than the loudest effete screams, brought terror to their hearts.

Each of the eleven pictured the most

Choice #3

Sergeant Snow snorted like an old bull and blasted out his hackneyed curses: #%#%@%# &#&^$$@&@%@ %$@@%%@%% @%$@*^%^^#^#^ #^%#^#^#^%#^#^

improve the architectural landscape of our fair A.A.A.A., and I wish to commend you on your choice of a project. Would you be so kind as to explain to me the aim of your zestful exertions?

diabolically inspired punishments that awaited. Chris looked at Sergeant Snow and thought of Clint Eastwood . . .
—Forget it. Sergeant Snow, with bean drippings on his shirt, could never play this role.

@&!^%@^%#%^^ #^$$^^$^$&$*&& ^$&$$&&$&$&$ &#$&$$&$&$##!^ !^!%%@&@^&$^ ^#&!!^!^~$^!~^!@ ^^#%^^#^^#@#@ %%#%@!!!!!!

"You, you, and you," pointed Sergeant Snow, "fix this bunkhouse up and the other eight of you report to the chow hall to help with the KP." It was always important to him that enough help was available in the kitchen so that none of his meals would ever be delayed.

The three individuals selected to rebuild the house were Pat, Chris, and Carol. None of them would get their get-off-the-base passes until their reconstruction was finished.

Wait! Carol!?!? Where'd she come from? you, my reader, ask.

The answer is the same as how Pat and Chris were mistakenly thought to be males. Carol Alfredo Thumbs (he likes to be called "the C.A.T.") is the proud possessor of both X and Y chromosomes; i.e., Carol is male. About 2% of all Carols are male. You may have heard of Carol Reed, the famous English film director. He won an Oscar for *Oliver!* which was the musical version of Charles Dickens' *Oliver Twist*.

After the rest had left, Pat looked at Chris and Carol and asked, "Will we ever get this put back together?"

Carol picked up a couple of pieces of wood and a hammer. He dropped one of the pieces. "All Thumbs" is what some of the guys called him. The C.A.T. said, "Hey man, working alone I bet I could put this whole thing back together in 24 hours."

Chris said, "I could finish the whole job in 16 hours."

"And I, in 12 hours," Pat added.

"I got it figured out," proclaimed the C.A.T. "If I can do it in 24 hours and you can do it in 16 (pointing to Chris) and you can do it in 12 hours (pointing to Pat), then if we all work together we can do it in 24 + 16 + 12 hours, which is 52 hours." Carol smiled at his math acumen.

Pat and Chris looked at each other and silently agreed that Carol didn't have keen insight when it came to math. They knew, even before they started thinking seriously about the problem, that if they all worked together it would certainly take less than 12 hours because Pat working all alone could finish the whole job in 12 hours and with the help of Chris and Carol, it would be finished even quicker than that.

Six-year-olds need more than the army allotment of eight hours of sleep per night, and Fred was no exception. He had conked out under the tree and was dreaming of roses that were the color of milkshakes. A pink rose, the color of a strawberry shake. A white rose, vanilla. A yellow rose, pineapple. He was about to run into trouble with the brown rose/chocolate combination, when the threesome came over to wake him up.

"Hey Fred, I hear you're great in math," Carol shouted a little too loudly.

Fred woke with a start and mumbled something about muddy roses. After a second or two, he was more or less awake.

Chris explained the problem to him: Pat, the whole job in 12; Chris, the whole job in 16; and Carol, in 24. How long will it take if we all work together?

Fred glanced at his watch and said, "Done at 9:20 tonight" and turned over and shut his eyes. He was really sleepy and, after all, he had answered their question.

They rolled him back over, and Carol asked, "How'd ja get that?"

Fred mumbled something about lavender roses and blueberry shakes. Someone suggested that they shake Fred to wake him up. Another thought that caffeine might work. They compromised and poured some caffeinated cola down his throat and then shook him. Fred fizzed.

He was more fully awake now, but had a little difficulty talking through the bubbles. Drifting into lecture mode, he asked, "Okay, what are we trying to find out?"

Carol answered, "X."

"No, dummy," Chris (whose manners could use a little improvement) corrected. "We're trying to find out how long it'll take us if we all work together."

"Okay, call that x," Carol retorted.

Fred held up his arms trying to indicate that they shouldn't quarrel, and then he continued, "I have four questions. They're all alike. Pat, you say you can do the whole job in 12 hours. *What part of the job can you do in one hour?*"

"That's easy," she responded. "1/12th."

Fred then turned to Chris, "And what part of the job can you do in one hour if you work alone?"

"That's cinchy. If it takes me 16 hours to do the whole job, I can do 1/16th of it in an hour," said Chris.

"And Carol, what part of the job can you do in one hour?" Fred asked.

He didn't know, but Chris volunteered, "1/24th."

Fred's last question was, "And if it takes x hours for you all, working together, to do the job, what part of the job can you all do in one hour?"

He received two different answers. Pat said, "1/x," and Carol said that they could do 1/12 + 1/16 + 1/24 of the job if they all worked together.

They were both right, and here is the equation:

$$\frac{1}{12} + \frac{1}{16} + \frac{1}{24} = \frac{1}{x}$$

Your Turn to Play

1. If it takes Sandy 6 hours to paint the outside of the bunkhouse, what part of the job would be done in one hour?

2. If Drew could paint the outside of the bunkhouse in 10 hours, what part of the job would be done in one hour?

3. (Continuing the previous two questions) what part of the whole job could Sandy and Drew do in one hour if they worked together?

4. If it took x hours for Sandy and Drew, working together, to paint the outside of the bunkhouse, what part of the job would be done in one hour?

5. Put the results of the previous two questions together and write the equation which solves: *If Sandy could paint the outside of the bunkhouse in 6 hours and Drew could paint it in 10 hours, how long would it take both of them working together to do the job?*

6. Sergeant Snow wanted to install a hot-dog-and-bean cooker in the kitchen. It was the new 6600-volt model. Since the kitchen only had electrical outlets for 110 and 220 volts, Snow ordered Sandy, Drew, and Cameron to dig a ditch between the kitchen and the main power house so that a 6600-volt cable could be laid.

　Sandy could dig that ditch in 2 hours.

　Drew could do it in 4 hours.

　Cameron could do the digging in 3 hours.

If we let x equal the number of hours that all three would take if they worked together, find the equation. (We'll learn how to solve equations containing fractions in the next lesson.)

7. Find the equation for: *Sandy could clean up the entire American Army Amusement Arena (A.A.A.A.) in 8 hours. Drew could do it in 8 hours. Cameron could do it in 12 hours. Jackie in 5 hours. Dana in 6 hours. If all five soldiers worked together it would take them x hours.*

8. In the time between breakfast and his first morning snack, Sergeant Snow uttered 3888 cuss words. When he's not angry, he issues them at the rate of 14/minute. Angry, at the rate of 55/minute. He was angry 18 minutes more than he was not angry. How long was he not angry?

.......COMPLETE SOLUTIONS.......

1. Sandy could do one-sixth of the job in one hour.

2. Drew could do one-tenth of the job in one hour.

3. $\dfrac{1}{6} + \dfrac{1}{10}$

4. $\dfrac{1}{x}$

5. $\dfrac{1}{6} + \dfrac{1}{10} = \dfrac{1}{x}$

6. $\dfrac{1}{2} + \dfrac{1}{4} + \dfrac{1}{3} = \dfrac{1}{x}$

7. $\dfrac{1}{8} + \dfrac{1}{8} + \dfrac{1}{12} + \dfrac{1}{5} + \dfrac{1}{6} = \dfrac{1}{x}$

8. *Let x = the number of minutes (between breakfast and his first morning snack) in which he wasn't angry.*

Then x + 18 = the number of minutes he was angry (since he was angry 18 minutes longer than he was not angry.)

He uttered 14x cuss words when he wasn't angry (which equals 14 cuss words per minute times x minutes).

He uttered 55(x + 18) cuss words when he was angry (which equals 55 cuss words per minute times x + 18 minutes).

$14x + 55(x + 18)$ = total cuss words uttered.

3888 = total cuss words uttered. (This was given in the problem.)

So the equation is $\qquad 14x + 55(x + 18) = 3888$

Using the distributive law: $\quad 14x + 55x + 990 = 3888$

$$69x + 990 = 3888$$
$$69x = 2898$$
$$x = 42$$

So Sergeant Snow wasn't angry for 42 minutes.

Intermission

Let's check our answer.

If he wasn't angry for 42 minutes, he was angry for 60 minutes (42 + 18).

While he wasn't angry, he uttered 42(14) cuss words (= 588).

While angry, he uttered 60(55) cuss words (= 3300).

588 + 3300 equals 3888 which is what was given in the problem. Our answer checks.

$$\frac{1}{12} + \frac{1}{16} + \frac{1}{24} = \frac{1}{x}$$

ractions! An "x" in the denominator (which is the bottom part of a fraction). The responses were predictable. Carol passed out. Chris frowned, and Pat waited for Fred to explain how to solve it.

"Just get rid of the fractions," was Fred's first comment.

"How?" asked Pat.

Fred wanted that response. He continued, "We have an equation. If you do the same thing to both sides of an equation, everything stays in balance. Remember in the old days when you first started studying equations, and you had something like $y - 4 = 9$. You could add 4 to each side and get $y = 13$. What we're going to do to get rid of all the fractions is multiply both sides by some big number."

Pat suggested, "How about 24?"

"Let's try it," Fred said.

He multiplied every term by **24**:

$$\frac{1 \cdot \mathbf{24}}{12} + \frac{1 \cdot \mathbf{24}}{16} + \frac{1 \cdot \mathbf{24}}{24} = \frac{1 \cdot \mathbf{24}}{x}$$

There was some simplification:

$$2 + \frac{24}{16} + 1 = \frac{24}{x}$$

but there were still two fractions in the equation.

"Name a bigger number," Fred urged.

"How big?" Pat wondered.

"Big enough to do the job," Fred continued.

"How about 100000000?" Pat suggested.

At this point Chris interrupted. She saw what Fred was driving at.

The number had to be *big enough so that all the denominators would divide evenly into it.* She thought for a moment more and then offered, "Let's try 48x."

So Fred multiplied every term of $\frac{1}{12} + \frac{1}{16} + \frac{1}{24} = \frac{1}{x}$ by **48x**.

$$\frac{1 \cdot 48x}{12} + \frac{1 \cdot 48x}{16} + \frac{1 \cdot 48x}{24} = \frac{1 \cdot 48x}{x}$$

All the fractions disappeared!

We are left with a
good ol' equation: $4x + 3x + 2x = 48$

$$9x = 48$$

$$x = 48/9 = 5\frac{3}{9} = 5⅓ \text{ hours or 5 hours and 20 minutes.}$$

Just an aside. Some students, instead of marking up the book with highlighters, take notes as they read. So far, in this chapter, an industrious student might have written:

1. Solve by figuring _per hour_;

2. Carol can be a boy's name;

3. Eliminate fractions from equations by multiplying thru by some big-enough number.

Your Turn to Play

1. Solve $\frac{1}{4x} - \frac{1}{8x} = \frac{1}{40}$

2. Solve $\frac{1}{x-1} - \frac{6}{35} = \frac{1}{x^2-1}$

3. The hot water faucet runs more slowly than the cold water faucet. (If you don't drain your water heater for a long time, sediment may build up. They say that you should drain your heater once for every 365 times you

---

floss your teeth.) It takes 4 minutes to fill the tub using just the cold water faucet. It takes 36 minutes using just the hot water faucet. If you turn both faucets on, how long will it take to fill the tub?

.......**COMPLETE SOLUTIONS**.......

1. $\dfrac{1}{4x} - \dfrac{1}{8x} = \dfrac{1}{40}$

All these denominators divide evenly into 40x.

(You could also use 80x or 160x or 1280x, but that would just make your work harder.)   So we multiply each term by 40x.

$$\frac{1(40x)}{4x} - \frac{1(40x)}{8x} = \frac{40x}{40}$$

All the denominators disappear

$$10 - 5 = x$$
$$5 = x$$

If you have a fractional equation, it takes one step to get rid of the fractions. This is nice. This is sweet. Life is good.

When I taught how to solve equations containing fractions, I said the magic words: All you needed to do in order to get rid of the fractions is multiply every term by an expression that each denominator would divide into evenly.

When you had $\dfrac{1}{12} + \dfrac{1}{16} + \dfrac{1}{24} = \dfrac{1}{x}$ you multiplied by **48x**.

$$\frac{1 \cdot 48x}{12} + \frac{1 \cdot 48x}{16} + \frac{1 \cdot 48x}{24} = \frac{1 \cdot 48x}{x}$$

One student looking at this once commented to me, "And that's not hard to remember. Just think about the secular celebration of Christmas."

"Okay," I responded. "But I don't get the connection."

The student smiled and explained, "You know. When you've got an equation that contains fractions, think of Santa Claus coming and dropping a little present on each rooftop."

← Note present in bag.

Another student raised his hand and said, "I don't believe in Santa Claus, but we do have pigeons here in San Francisco."

2.
$$\frac{1}{x-1} - \frac{6}{35} = \frac{1}{x^2-1}$$

The trick here is to find *the smallest expression* that x − 1, 35, and $x^2 - 1$ will all evenly divide into. If you use $(x-1)(35)(x^2-1)$, the resulting equation will have an x^3 in it.*

In order to find the smallest expression that x − 1, 35, and $x^2 - 1$ will all evenly divide into, factor the $x^2 - 1$.

$$\frac{1}{x-1} - \frac{6}{35} = \frac{1}{(x+1)(x-1)}$$

The smallest expression that all the denominators will divide evenly into is **35(x + 1)(x − 1)**. This is the present that Santa Claus (or the pigeons) drops on each rooftop.

$$\frac{1(35)(x+1)(x-1)}{x-1} - \frac{6(35)(x+1)(x-1)}{35} = \frac{1(35)(x+1)(x-1)}{(x+1)(x-1)}$$

---

* If you would care to try to solve $6x^3 - 41x^2 + 29x + 6 = 0$, you might quickly see what the difficulty is. Even most second-year algebra students can't solve it.

Then in one step all the denominators disappear.

$$35(x + 1) - 6(x + 1)(x - 1) = 35$$

Multiply two binomials
(See the Your Turn to Play of Lesson 53 if you want a refresher on how to do this.)

$$35(x + 1) - 6(x^2 - 1) = 35$$

Distributive property

$$35x + 35 - 6x^2 + 6 = 35$$

Transpose everything to one side

$$0 = 6x^2 - 35x - 6$$

Factor
(A harder trinomial)
Two expressions that multiply to $-36x^2$
and that add to $-35x$

$$0 = 6x^2 - 36x + x - 6$$

Factor by grouping

$$0 = 6x(x - 6) + x - 6$$

$$0 = (x - 6)(6x + 1)$$

If two numbers multiply to zero, then
one of them must be zero.

$$x - 6 = 0 \quad OR \quad 6x + 1 = 0$$

$$x = 6 \quad OR \quad x = -\frac{1}{6}$$

There is a rule in algebra that should be mentioned at this point . . .

> Whenever you multiply an equation by an expression containing x, you must check any answers you get back in the original equation.

I, your reader, have a question.

Yes. What is your question?

Why? That seems like extra work.

It isn't extra work. It is necessary work. When you multiply an equation by something like **35(x + 1)(x − 1)**, extra answers are sometimes introduced. You need to check each answer you get so that you can weed out any extra answers. Fancy algebra books call these extra answers **extraneous roots**. (ex-TRAIN-ee-us)

Okay. Since you made up this rule, Mr. Author, you can check those answers. I'll just watch.

Hey! Wait a minute! I didn't invent that rule.

Next, you're going to tell me that back in the first chapter of the first book of this series, Life of Fred: Apples, you didn't make up the rule 5 + 2 = 7. Here's the picture from that book:

5 + 2 is 7

Now you are really getting silly. I didn't invent 5 + 2 = 7 either. My first grade teacher invented that.*

You are wasting time. I'm waiting. Do the arithmetic and check x = 6 and x = – 1/6 in the original equation. I'm not going anywhere until you get that work done.

Okay.

Checking x = 6 in $\dfrac{1}{x-1} - \dfrac{6}{35} = \dfrac{1}{x^2-1}$

$$\dfrac{1}{6-1} - \dfrac{6}{35} \stackrel{?}{=} \dfrac{1}{36-1}$$

$$\dfrac{1}{5} - \dfrac{6}{35} \stackrel{?}{=} \dfrac{1}{35}$$

$$\dfrac{7}{35} - \dfrac{6}{35} \stackrel{?}{=} \dfrac{1}{35} \quad \text{yes} \quad \text{x = 6 is a valid answer.}$$

---

* Do not believe everything you read in a book.

Checking $x = -\dfrac{1}{6}$ in $\dfrac{1}{x-1} - \dfrac{6}{35} = \dfrac{1}{x^2-1}$

$$\frac{1}{(-1/6)-1} - \frac{6}{35} \stackrel{?}{=} \frac{1}{(1/36)-1}$$

Yucky arithmetic! May I please be excused?

Are you kidding? You wrote this problem. Now finish it!

$$\frac{1}{-7/6} - \frac{6}{35} \stackrel{?}{=} \frac{1}{-35/36}$$

Recalling division of fractions from arithmetic* . . .

$$-\frac{6}{7} - \frac{6}{35} \stackrel{?}{=} -\frac{36}{35}$$

Before I finish this arithmetic . . .

We need to take a break for a second.
Negative signs within fractions is something new.

Fractions like $-\dfrac{+1}{-8}$ need to be talked about.

There are three signs in a fraction:
        The one on the top (the numerator)
        The one on the bottom (the denominator) and
        The one in front.

Recall that $+\dfrac{-20}{-5}$ was the same as $+\dfrac{+20}{+5}$

since a negative divided by a negative gave the same answer as
a positive divided by a positive.

---

* $\dfrac{a}{b} \div \dfrac{c}{d}$ becomes $\dfrac{a}{b} \times \dfrac{d}{c}$ and $1 \div \dfrac{-7}{6}$ becomes $1 \times \dfrac{6}{-7}$

The general rule is that you can change any two of the three signs of a fraction.

So $+ \dfrac{-18}{+9} = + \dfrac{+18}{-9}$ (changing the sign of the numerator and denominator)

So $- \dfrac{-7}{+4} = + \dfrac{+7}{+4}$ (changing the sign in front and the sign of the numerator)

Now I'll finish the arithmetic.

I was at $- \dfrac{6}{7} - \dfrac{6}{35} \stackrel{?}{=} - \dfrac{36}{35}$

To add fractions I need to make the denominators (the bottoms) all alike.

$$- \dfrac{6(5)}{7(5)} - \dfrac{6}{35} \stackrel{?}{=} - \dfrac{36}{35}$$

$$- \dfrac{30}{35} - \dfrac{6}{35} \stackrel{?}{=} - \dfrac{36}{35} \quad \text{yes} \quad x = - \dfrac{1}{6} \text{ is also a valid answer}$$

3. Now, of course, you wouldn't normally just turn both faucets on to fill the tub since you'd probably **FREEZE TO DEATH** with so little hot water running into the tub. But let's suppose you're filling the tub so that you can wash your goldfish.

If the cold tap can fill the tub in 4 minutes, in one minute it can fill $\dfrac{1}{4}$ of the tub. If the hot tap can fill the tub in 36 minutes, in one minute it can fill $\dfrac{1}{36}$ of the tub.

If both taps can fill the tub in x minutes, in one minute they can fill $\dfrac{1}{x}$ of the tub. $\dfrac{1}{4} + \dfrac{1}{36} = \dfrac{1}{x}$

And now Santa comes with his little present to drop on every rooftop. All the denominators divide evenly into 36x.

$$\dfrac{1(36x)}{4} + \dfrac{1(36x)}{36} = \dfrac{(36x)}{x}$$
$$9x + x = 36$$
$$10x = 36$$
$$x = 3.6 \text{ minutes}$$

Checking x = 3.6 in $\dfrac{1}{4} + \dfrac{1}{36} = \dfrac{1}{x}$ I don't have enough room to do all the fraction work.

$0.25 + 0.027777777777777777777 \stackrel{?}{=} 0.277777777777777777777$ yes

P at, Chris, and Carol chimed in unison, "Thanks, Fred!" They picked up their hammers and headed back to repair the big hole in the side of the bunkhouse.

Fred rolled back over to catch another hour's nap. Fred wanted to be fully awake when he and Jack LaRoad got their passes at 6 p.m. for their Sunday night trip into town.

Dreaming (again!), Fred imagined that he had muscles as big as those of Jack LaRoad. He dreamed that he could do a thousand pushups in 20 minutes. Then he thought of one of the poems of Christina Rossetti (perhaps the most important woman poet in England before the 20th century).

It was one of Fred's favorite Rossetti poems, "Maggie A Lady," which begins:

> You must not call me Maggie, you must not call me Dear,
> For I'm Lady of the Manor now stately to see;
> And if there comes a babe, as there may some happy year,
> 'Twill be little lord or lady at my knee.

He could recite the whole poem in four minutes in his little voice, giving special dramatic emphasis when he came to the line:

> His mother said "fie," and his sisters cried "shame,"

and then Fred (in his dream) wondered how long it would take him to do both a thousand pushups and recite "Maggie A Lady."

The answer was easy. If it took 20 minutes to do the pushups and four minutes to recite the poem, he could do both of them in 20 minutes since he could do them both at the same time.

It's important, occasionally, to have the brain engaged so that you don't "go on automatic" and start to write something like: *If Fred can* do all of the pushups in 20 minutes, he can do 1/20 of the job in one minute. . . .

But there are things that you can do almost on automatic pilot. Working with **rational expressions** (which means a polynomial divided by a polynomial, such as $\frac{x^2 - 5x + 6}{x - 3}$) is as easy as falling off a log *if you know how to do arithmetic.*

For example, simplifying rational expressions follows the same steps as simplifying fractions in arithmetic.

Arithmetic and Algebra		
In English	Arithmetic example	Algebra example
Simplify 1. Factor numerator. 2. Factor denominator. 3. Cancel like factors.	$\dfrac{6}{8}$ $= \dfrac{3 \cdot 2}{4 \cdot 2}$ $= \dfrac{3 \cdot 2}{4 \cdot 2}$ $= \dfrac{3}{4}$	$\dfrac{x^2 + 5x + 6}{x^2 + 6x + 8}$ $= \dfrac{(x + 3)(x + 2)}{(x + 4)(x + 2)}$ $= \dfrac{(x + 3)(x + 2)}{(x + 4)(x + 2)}$ $= \dfrac{(x + 3)}{(x + 4)}$

Your Turn to Play

1. Simplify $\dfrac{y^2 + 9y + 20}{y + 5}$

2. Simplify $\dfrac{x^2 - 6x + 5}{x^2 - 25}$

3. Simplify $\dfrac{x^3 + 4x^2 - 9x - 36}{x - 3}$

4. Simplify $\dfrac{4z^2 - 25}{2z - 5}$

5. $\dfrac{3a^2 + 19a + 6}{3a + 1}$

6. $\dfrac{5x^2 + 10y^2}{30x^2}$

7. Fred has a book of poetry entitled *Some Collected Poems of Christina Rossetti and Charlie Baudelaire.* Fred went through the book and marked the poems that he really liked and wanted to memorize. For example, he really liked Rossetti's "Maggie A Lady" and he really liked Baudelaire's "The Cat." He liked 10% of CR's poems and 2% of CB's poems for a total of 16 poems out of the 520 in the book. How many Rossetti poems were in the book?

Intermission

Ever get "lost" in a math class? There is a reason why that can happen.

Math builds upon itself.

Lesson 58 dealt with factoring differences of squares: $x^2 - y^2 = (x + y)(x - y)$.

Now in Lesson 66 we have used factoring differences of squares to simplify fractions.

Simplifying fractions →

Factoring →

If you were to walk into a calculus class, you would see unintelligible symbolism:

$$\int \frac{dx}{x^3 + x}$$

Of course, a first grader looking at $x^2 - y^2$ would have the same reaction. In math, the stuff ahead always looks very mysterious . . . but it really isn't so tough.

. . . . COMPLETE SOLUTIONS . . . .

1. $\dfrac{y^2 + 9y + 20}{y + 5} = \dfrac{(y + 4)(y + 5)}{y + 5} = y + 4$

2. $\dfrac{x^2 - 6x + 5}{x^2 - 25} = \dfrac{(x - 5)(x - 1)}{(x + 5)(x - 5)} = \dfrac{x - 1}{x + 5}$

3. $\dfrac{x^3 + 4x^2 - 9x - 36}{x - 3} = \dfrac{x^2(x + 4) - 9(x + 4)}{x - 3}$

$= \dfrac{(x + 4)(x + 3)(x - 3)}{x - 3} = (x + 4)(x + 3)$

4. $\dfrac{4z^2 - 25}{2z - 5} = \dfrac{(2z + 5)(2z - 5)}{2z - 5} = 2z + 5$

5. $\dfrac{3a^2 + 19a + 6}{3a + 1} = \dfrac{(3a + 1)(a + 6)}{3a + 1} = a + 6$

Please note: In the previous problem $\dfrac{3a^2 + 19a + 6}{3a + 1}$ you can't just cancel the 3s. The seven famous words for simplifying fractions are:

Factor numerator, Factor denominator, Cancel like factors,

It is only factors that can be canceled.

Factors are things that multiply together to make the whole expression.

For example, 7 and x + 4 and $z^2 + 8$ are all factors of $7(x + 4)(z^2 + 8)$.

In contrast, **terms** are things that add together to make the whole expression. For example, 5 and w^{88} and 3xyz are terms of $5 + w^{88} + 3xyz$.

6. $\dfrac{5x^2 + 10y^2}{30x^2} = \dfrac{5(x^2 + 2y^2)}{5(6)x^2} = \dfrac{x^2 + 2y^2}{6x^2}$

and you can't cancel the x^2 since x^2 is *not* a factor of the numerator.

7. *Let x = the number of Christina Rossetti poems in the book.* This is what we are trying to find.

Then 520 − x = the number of Baudelaire poems. There are a total of 520 poems and x of them are Rossetti's.

Then 0.10x of the Rossetti poems Fred really liked.

Then 0.02(520 − x) of the Baudelaire poems he liked.

16 = the total number of poems he liked. (This was given.)

The equation	$0.1x + 0.02(520 - x) = 16$
Distributive property	$0.1x + 10.4 - 0.02x = 16$
Subtract 10.4 from both sides	$0.1x - 0.02x = 5.6$
Combine like terms	$0.08x = 5.6$
Divide both sides by 0.08	$x = 70$

There are 70 Christina Rossetti poems in the book.

T here are five things that you learned to do with arithmetic fractions:

    i) simplify them

    ii) add them

    iii) subtract them

    iv) multiply them

    v) divide them.

The only one that has any trace of excitement is the addition of rational expressions. Subtraction takes two lines to explain. Multiplication takes one line to explain. Division takes one line to explain.

One time when I put $\dfrac{3}{x+4} + \dfrac{x-7}{x+3}$ on the blackboard, one of my students looked at it and said that it was just a matter of Interior Decorating. I looked at him and knew I was in for a treat when I asked him what he meant by Interior Decorating.

He asked, "May I?" as he went to the blackboard and took the chalk. He wrote at the top of the board: Addition is like Interior Decorating.

The student continued, "When I see an addition problem like $\dfrac{3}{x+4} + \dfrac{x-7}{x+3}$ I think of it as a couple of windows, you know, the old-fashioned wooden windows like my grandma has that you can move up and down." I winced when he said "my grandma" since the windows in my house were just like that.

He went on, drawing a picture of an old-fashioned window:

 using part of the window frame as the fraction bar.

"Now," he explained, "it is time for some Interior Decorating. As everyone knows, if you add a curtain to the top half of the window

good taste dictates that one must add a similar curtain to the bottom half in order to maintain an aesthetic balance in window treatment."

not aesthetic

fantastically aesthetic

It suddenly became clear what my Interior Decorating student was driving at. "That's brilliant!" I cried. "If I ever write a book I'll tell everyone about your concept of Interior Decorating."

"And please don't forget to mention my name," he asked.

"I won't forget, Jean," I told him.

So when Jean saw $\dfrac{3}{x+4} + \dfrac{x-7}{x+3}$ he mentally put them into

two windows:

$$\boxed{\begin{array}{c} 3 \\ \hline x+4 \end{array}} \quad + \quad \boxed{\begin{array}{c} x-7 \\ \hline x+3 \end{array}}$$

and his Interior Decorating technique of adding curtains to the top and bottom of each window gave him

$$\frac{3(x+3)}{(x+4)(x+3)} \;+\; \frac{(x-7)(x+4)}{(x+3)(x+4)}$$

Now with the same denominators, we may add the fractions to obtain

$$\frac{3(x+3) + (x-7)(x+4)}{(x+4)(x+3)}$$

and the rest is simplifying fractions, which we already know how to do.

Your Turn to Play

1. Combine (and simplify if possible)
$$\frac{x+4}{x+3} + \frac{x+8}{x+5}$$

2. Combine (and simplify if possible)
$$\frac{y-3}{y+4} + \frac{2y+1}{y+6}$$

3. If Sandy could clean up the bunkhouse in 3 hours and Drew could clean it up in 4 hours and Cameron could clean it up in 5 hours, how long would it take them if they all worked together?

Intermission

Conversion factors are used frequently in biology, chemistry, and physics.

You know, for example, that 1 day = 24 hours.

Therefore, the fraction $\frac{1\ day}{24\ hours}$ is equal to one (since the numerator and denominator are equal to each other.

$\frac{1\ day}{24\ hours}$ is called a conversion factor.

If we have 168 hours and I want to convert it to days, I multiply by that conversion factor.

$$\frac{168\ hours}{1} \times \frac{1\ day}{24\ hours} = \frac{168\ days}{24} = 7\ days$$

Notice that the "hours" cancel

---

4. One foot equals 30.48 centimeters. Approximately how many feet are 40 centimeters? Give your answer to the nearest hundredth of a foot.

5. One pound of water occupies 27.68 cubic inches. What is the volume of 8 pounds of water? Give your answer to the nearest hundredth of a cubic inch.

6. $\left(\dfrac{44.48 + 6\log 54.8 + 8!}{\sin(\pi/6)} \right)^0 = ?$

. . . . COMPLETE SOLUTIONS . . . .

1. $\dfrac{x+4}{x+3} + \dfrac{x+8}{x+5}$

$= \dfrac{(x+4)\mathbf{(x+5)}}{(x+3)\mathbf{(x+5)}} + \dfrac{(x+8)\mathbf{(x+3)}}{(x+5)\mathbf{(x+3)}}$

$= \dfrac{(x+4)(x+5) + (x+8)(x+3)}{(x+3)(x+5)}$

$= \dfrac{x^2 + 5x + 4x + 20 + x^2 + 3x + 8x + 24}{(x+3)(x+5)}$

$= \dfrac{2x^2 + 20x + 44}{(x+3)(x+5)}$

Now it's time to use the seven famous words for simplifying fractions: Factor numerator, Factor denominator, Cancel like factors.

$= \dfrac{2(x^2 + 10x + 22)}{(x+3)(x+5)}$

That is as far as it goes. I can't think of two numbers that multiply to 22 and add to 10.

2. $\dfrac{y-3}{y+4} + \dfrac{2y+1}{y+6} = \dfrac{(y-3)\mathbf{(y+6)}}{(y+4)\mathbf{(y+6)}} + \dfrac{(2y+1)\mathbf{(y+4)}}{(y+6)\mathbf{(y+4)}}$

$= \dfrac{y^2 + 6y - 3y - 18 + 2y^2 + 8y + y + 4}{(y+4)(y+6)} = \dfrac{3y^2 + 12y - 14}{(y+4)(y+6)}$ The numerator doesn't factor.

3. $\dfrac{1}{5} + \dfrac{1}{4} + \dfrac{1}{3} = \dfrac{1}{x}$

This is not an addition of fractions problem. This is a fractional equation. The equals sign tells you that.

In Lesson 65 . . .

Multiply by 60x (which all the denominators divide into)

$\dfrac{1\mathbf{(60x)}}{5} + \dfrac{1\mathbf{(60x)}}{4} + \dfrac{1\mathbf{(60x)}}{3} = \dfrac{1\mathbf{(60x)}}{x}$

All the denominators disappear $12x + 15x + 20x = 60$

$$47x = 60$$

$$x = \frac{60}{47} \text{ hours}$$

If we multiply by a conversion factor of $\dfrac{60 \text{ minutes}}{1 \text{ hour}}$

we get $\dfrac{60 \text{ hours}}{47} \times \dfrac{60 \text{ minutes}}{1 \text{ hour}} = \dfrac{3600 \text{ minutes}}{47} \doteq 77 \text{ minutes}$

$\doteq$ means rounded to

4. $\dfrac{40 \text{ centimeters}}{1} \times \dfrac{1 \text{ foot}}{30.48 \text{ centimeters}} \doteq 1.31 \text{ feet.}$

5. $\dfrac{8 \text{ pounds of water}}{1} \times \dfrac{27.68 \text{ cubic inches}}{1 \text{ pound of water}} \doteq 221.44 \text{ cubic inches}$

If you had written the conversion factor as $\dfrac{1 \text{ pound of water}}{27.68 \text{ cubic inches}}$

it wouldn't have worked since the "pounds of water" wouldn't have canceled.

6. Anything (except zero) raised to the zero power is equal to one.

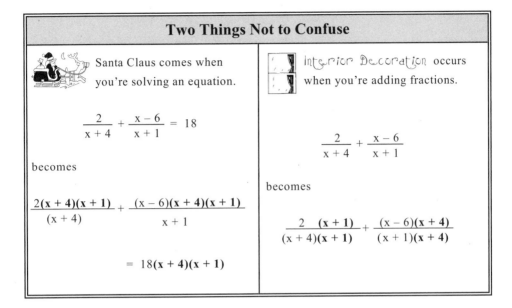

Two Things Not to Confuse	
Santa Claus comes when you're solving an equation.	Interior Decoration occurs when you're adding fractions.

$$\frac{2}{x+4} + \frac{x-6}{x+1} = 18$$

becomes

$$\frac{2(x+4)(x+1)}{(x+4)} + \frac{(x-6)(x+4)(x+1)}{x+1}$$

$$= 18(x+4)(x+1)$$

$$\frac{2}{x+4} + \frac{x-6}{x+1}$$

becomes

$$\frac{2\ (x+1)}{(x+4)(x+1)} + \frac{(x-6)(x+4)}{(x+1)(x+4)}$$

The memory aids of Santa Claus, pigeons, and interior decorating are never mentioned in most ordinary algebra books. That's a shame since inventing such aids makes the process of learning algebra much more pleasurable, and you tend to remember things a lot longer. It's been more than thirty-five years since Jean went to the blackboard in my classroom and passed along his insight to me and my class, but I shall remember it.

In mnemonics (neh-MON-iks) the first *m* is silent—the science of learning how to memorize effectively—they've found that the *sillier* the association you can make, the better it sticks in your mind.

One student, in taking notes on adding fractions, changed Jean's mnemonic slightly: She said it was easier for her to memorize things that had a touch of violence in them.

"After all," she explained, "when you go to the movies, will you remember the violent scenes or will you remember how they decorated the living room?"

The important thing is that you get in the habit of choosing memory aids *even when your teacher or your book don't suggest it.* It will make your life a lot easier.

(In case you're wondering, I prefer rubber duckies.)

Subtraction of algebra fractions only takes two lines to explain.

Make $\dfrac{a}{b} - \dfrac{c}{d}$ into $\dfrac{a}{b} + \dfrac{-c}{d}$ *

Arithmetic and Algebra		
In English	**Arithmetic example**	**Algebra example**
Add 1. Figure out what the least common denominator is. 2. Multiply top and bottom of each fraction to make that least common denominator. 3. Add.	$\dfrac{2}{5} + \dfrac{1}{4}$ $= \dfrac{2\cdot 4}{5\cdot 4} + \dfrac{1\cdot 5}{4\cdot 5}$ $= \dfrac{8}{20} + \dfrac{5}{20}$ $= \dfrac{13}{20}$	$\dfrac{x+2}{x+5} + \dfrac{x+1}{x+4}$ $= \dfrac{(x+2)(x+4)}{(x+5)(x+4)} + \dfrac{(x+1)(x+5)}{(x+4)(x+5)}$ $= \dfrac{(x+2)(x+4) + (x+1)(x+5)}{(x+5)(x+4)}$ $= \dfrac{2x^2 + 12x + 13}{(x+5)(x+4)}$

---

* Oops. I guess that proves that I can't count.

Arithmetic and Algebra		
Subtract	$\dfrac{37}{449} - \dfrac{8}{11}$	$\dfrac{x^3 - 3x + 7}{x - 5} - \dfrac{x^2 + 9}{x - 2}$
1. Change two signs of the fraction and turn it into an addition problem.	$= \dfrac{37}{449} + \dfrac{-8}{11}$	$= \dfrac{x^3 - 3x + 7}{x - 5} + \dfrac{-(x^2 + 9)}{x - 2}$ Note: we stick parentheses around the $x^2 + 9$ so the whole numerator receives the minus sign.

Your Turn to Play

1. Combine (and simplify if possible) $\dfrac{w + 6}{w + 3} - \dfrac{w - 4}{w - 5}$

2. $\dfrac{10x - 13}{(x - 4)(x + 5)} - \dfrac{7}{x + 5}$

Intermission

See how high our stack of blocks has become!

✓ Subtracting fractions requires you be able to do addition.
✓ Adding fractions requires you be able to simplify.
✓ Simplifying fractions requires factoring.
✓ Factoring requires knowing multiplication and addition of numbers.

None of these steps is particularly hard. The hardest was learning the addition and multiplication tables.
Quick! What's 7 × 8?

3. Before lunch Jack LaRoad did some pushups at the rate of 60/minute. After lunch he did some more at the rate of 40/minute. He exercised for a total of one hour and completed a total of 2840 pushups. How long was his before-lunch workout?

4. The nice big hole in the side of the bunkhouse was a square that was 11 feet on each side. What is its perimeter?

5. If you have a rectangle for which length and width are both natural numbers, must the perimeter be a natural number?

6. Reading the poems of Christina Rossetti gave Fred *r* spine-tingling thrills per hour. Reading Charles-Pierre Baudelaire gave him *b* spine-tingling thrills per hour. One day he read 3 hours of Rossetti and 2 hours of Baudelaire and received 22 thrills. Another day he read 6 hours of Rossetti and 1 hour of Baudelaire and received 38 thrills. What are the thrill-rates (*r* and *b*) for Fred's reading of those two famous poets?

. . . . COMPLETE SOLUTIONS . . . .

1. $\dfrac{w+6}{w+3} - \dfrac{w-4}{w-5}$

$= \dfrac{w+6}{w+3} + \dfrac{-(w-4)}{w-5}$ We put parentheses around the $w-4$ so that the whole numerator receives the minus sign.

$= \dfrac{w+6}{w+3} + \dfrac{-w+4}{w-5}$

$= \dfrac{(w+6)(w-5)}{(w+3)(w-5)} + \dfrac{(-w+4)(w+3)}{(w-5)(w+3)} = \dfrac{w^2-5w+6w-30-w^2-3w+4w+12}{(w-5)(w+3)}$

$= \dfrac{2(w-9)}{(w+3)(w-5)}$

2. $\dfrac{10x-13}{(x-4)(x+5)} - \dfrac{7}{x+5}$

$= \dfrac{10x-13}{(x-4)(x+5)} + \dfrac{-7}{x+5}$

$= \dfrac{10x-13}{(x-4)(x+5)} + \dfrac{-7(x-4)}{(x+5)(x-4)}$ I only had to do Interior Decorating on the second fraction.

$= \dfrac{10x-13-7x+28}{(x-4)(x+5)}$

$= \dfrac{3x+15}{(x-4)(x+5)} = \dfrac{3(x+5)}{(x-4)(x+5)} = \dfrac{3}{x-4}$

3. Using "six pretty boxes" . . .

	pushups	rate	time
before lunch			t
after lunch			

He exercised for a total of 60 minutes.* **See the box on first page of Lesson 26.**

	pushups	rate	time
before lunch			t
after lunch			60 – t

The rates were given in the problem.

	pushups	rate	time
before lunch		60	t
after lunch		40	60 – t

The number of pushups equals the rate times the time. **If, for example, you did 5 pushups per minute for 7 minutes, you would have done 35 pushups. This is very similar to d = rt.**

	pushups	rate	time
before lunch	60t	60	t
after lunch	40(60 – t)	40	60 – t

He did a total of 2840 pushups.

$$60t + 40(60 - t) = 2840$$

Distributive property $\qquad 60t + 2400 - 40t = 2840$

Combine like terms $\qquad\qquad 20t + 2400 = 2840$

Subtract 2400 from both sides $\qquad 20t = 440$

Divide both sides by 20 $\qquad\qquad t = 22$

Jack's before-lunch workout took 22 minutes.

---

* Since the rate was in terms of pushups per *minute*, the time must also be expressed in minutes.

Instead of using "six pretty boxes," the problem could have been done using "Let x equal" statements.

Let t = the number of minutes Jack exercised before lunch. (This is the quantity we are trying to find.)

Then 60 – t = the number of minutes Jack exercised after lunch. (The total amount of time he exercised was 60 minutes. Notice that $t + (60 - t) = 60$.)

Then 60t = the number of pushups done before lunch.

Then 40(60 – t) = the number of pushups done after lunch.

Since the total number of pushups he did was 2840 . . .

$$60t + 40(60 - t) = 2840$$

4. $11 + 11 + 11 + 11 = 44$ feet

5. The natural numbers are $\{1, 2, 3, 4, 5, 6, \ldots\}$. The perimeter of a rectangle which has a length **l** and width **w** is $p = 2l + 2w$. Or, if you prefer, $p = l + l + w + w$. Adding up four natural numbers will always give you a natural number. (You couldn't add up four natural numbers and get 84.3 or get 103¾ or get –99.)

6. $\begin{cases} 3r + 2b = 22 \\ 6r + b = 38 \end{cases}$ � $\begin{cases} 6r + 4b = 44 \\ -6r - b = -38 \end{cases}$ � $3b = 6$ � $b = 2$

and back substituting into the first equation: $3r + 2(2) = 22$

$$r = 6$$

When Fred taught subtraction, multiplication, and division of rational expressions at KITTENS University, it only took him three minutes. Each topic took only one line to explain.

Subtraction: Make $\dfrac{a}{b} - \dfrac{c}{d}$ into $\dfrac{a}{b} + \dfrac{-c}{d}$

Multiplication: Make $\dfrac{a}{b} \times \dfrac{c}{d}$ into $\dfrac{ac}{bd}$

Division: Make $\dfrac{a}{b} \div \dfrac{c}{d}$ into $\dfrac{a}{b} \times \dfrac{d}{c}$

For rest of the lecture hour, he just played. Instead of using baby blocks he brought in four giant hotdogs and stacked them up.

"The top hotdog is division," he explained. "You turn every division problem into a multiplication problem. Every multiplication problem becomes a simplification problem."

He then said the seven famous words for simplifying fractions:
Factor top, Factor bottom, Cancel like factors,

"And the bottom hotdog was knowing your multiplication tables."

At this point, most of his class was thinking about food rather than math. Fred cut the giant hotdogs into one-pound servings and passed them out to his students. He also passed around some mustard and ketchup. And since they had ketchup, he also passed around a basket of onion rings.

Fred was the most popular teacher at KITTENS.

While the students were having their snack, Fred put up the chart for multiplication and division.

Arithmetic and Algebra		
In English	**Arithmetic example**	**Algebra example**
Multiply	$\dfrac{2}{3} \times \dfrac{4}{7}$	$\dfrac{x+2}{x+3} \cdot \dfrac{x+4}{x+7}$
1. Top times top and bottom times bottom.	$= \dfrac{8}{21}$	$= \dfrac{(x+2)(x+4)}{(x+3)(x+7)}$
In English	**Arithmetic example**	**Algebra example**
Divide	$\dfrac{22}{7} \div \dfrac{5}{2}$	$\dfrac{x+22}{x+7} \div \dfrac{x+5}{x+2}$
1. Invert the divisor and multiply.	$= \dfrac{22}{7} \times \dfrac{2}{5}$	$= \dfrac{x+22}{x+7} \cdot \dfrac{x+2}{x+5}$

"Sometimes," he explained, "a division problem comes disguised as a fraction." A fraction means division. So you might see fractions inside of fractions. (These are called **complex fractions**.)

For example, $\dfrac{\dfrac{1}{x-5}}{\dfrac{7}{x^2-25}}$

That can be turned into $\dfrac{1}{x-5} \div \dfrac{7}{x^2-25}$ and you know what to do from there.

Your Turn to Play

1. $\dfrac{x^2 - 9}{2x + 5} \div \dfrac{x + 3}{6x^2 + 7x - 20}$

2. $8x \div \dfrac{4x^2 - 12x - 112}{x^2 + 5x + 4}$

3. Fill in one number:
A binomial has _____ *terms.*

4. $\dfrac{\dfrac{30x^2 + 27x + 6}{9x^2 - 49}}{\dfrac{30x + 15}{30x - 70}}$

5. $\dfrac{2y - 3}{y - 3} - \dfrac{y + 3}{y + 4}$

6. Which of the following is/are equal to $\dfrac{-5}{7}$?

$-\dfrac{5}{7}$ $-\dfrac{-5}{-7}$ $\dfrac{5}{-7}$

7. Solve $\dfrac{1}{x - 2} + \dfrac{3}{x + 6} = \dfrac{1}{2}$

8. Leslie wanted 1500 polka dots on each curtain he sewed for his Interior Decorating. Each curtain required 5 yards of cloth. Unfortunately, at the fabric store he could only find two kinds of cloth. One had 200 dots/yard and the other had 450 dots/yard. How many yards of each should he purchase?

9. What is the name of the set of all the poems that Charlie Baudelaire wrote in Finnish? (Hint: He was born in Paris in 1821 and probably didn't ever learn a single word of Finnish.)

200 dots/yard   450 dots/yard

    This is also the same set as the set of all 3,498,992,404,555-word poems written by Christina Rossetti.

. . . . COMPLETE SOLUTIONS . . . .

1. $\dfrac{x^2 - 9}{2x + 5} \div \dfrac{x + 3}{6x^2 + 7x - 20} = \dfrac{x^2 - 9}{2x + 5} \times \dfrac{6x^2 + 7x - 20}{x + 3}$

$= \dfrac{(x + 3)(x - 3)(2x + 5)(3x - 4)}{(2x + 5)(x + 3)} = (x - 3)(3x - 4)$

2. $8x \div \dfrac{4x^2 - 12x - 112}{x^2 + 5x + 4} = \dfrac{8x}{1} \div \dfrac{4x^2 - 12x - 112}{x^2 + 5x + 4}$

$= \dfrac{8x}{1} \times \dfrac{x^2 + 5x + 4}{4x^2 - 12x - 112}$ ← Always look for a common factor first!

$= \dfrac{8x}{1} \times \dfrac{(x + 4)(x + 1)}{4(x^2 - 3x - 28)}$

$= \dfrac{8x}{1} \times \dfrac{(x + 4)(x + 1)}{4(x - 7)(x + 4)} = \dfrac{2x(x + 1)}{(x - 7)}$

3. A monomial has one term. A binomial has two terms. A trinomial has three terms. A polynomial has one or more terms.

4. $\dfrac{30x^2 + 27x + 6}{9x^2 - 49} \div \dfrac{30x + 15}{30x - 70}$

$= \dfrac{30x^2 + 27x + 6}{9x^2 - 49} \times \dfrac{30x - 70}{30x + 15}$

$= \dfrac{3(10x^2 + 9x + 2)}{9x^2 - 49} \times \dfrac{10(3x - 7)}{15(2x + 1)}$ Always look for a common factor first.

$= \dfrac{3(5x + 2)(2x + 1)(10)(3x - 7)}{(3x + 7)(3x - 7)(15)(2x + 1)}$

$= \dfrac{2(5x + 2)}{3x + 7}$

5.

$\dfrac{2y - 3}{y - 3} - \dfrac{y + 3}{y + 4}$

Subtraction problem into addition

$= \dfrac{2y - 3}{y - 3} + \dfrac{-(y + 3)}{y + 4}$

Interior Decorating

$= \dfrac{(2y - 3)(y + 4)}{(y - 3)(y + 4)} + \dfrac{-(y + 3)(y - 3)}{(y + 4)(y - 3)}$

$= \dfrac{2y^2 + 8y - 3y - 12 - y^2 + 3y - 3y + 9}{(y - 3)(y + 4)}$

Combine like terms $$= \frac{y^2 + 5y - 3}{(y-3)(y+4)}$$

The numerator does not factor, so we are done.

6. They all are equal to $\dfrac{-5}{7}$

The general rule is that you can change any two of the three signs of a fraction.

7. $$\frac{1}{x-2} + \frac{3}{x+6} = \frac{1}{2}$$

This is a fractional equation.

$$\frac{1(x-2)(x+6)(2)}{x-2} + \frac{3(x-2)(x+6)(2)}{x+6} = \frac{(x-2)(x+6)(2)}{2}$$

All the fractions disappear. $(x+6)(2) + 3(x-2)(2) = (x-2)(x+6)$

Multiply everything out $2x + 12 + 6x - 12 = x^2 + 6x - 2x - 12$

Transpose $0 = x^2 - 4x - 12$

Factor $0 = (x-6)(x+2)$

$$x - 6 = 0 \ \ \text{OR} \ \ x + 2 = 0$$

$$x = 6 \text{ or } x = -2$$

8. *Let y = the number of yards of 200 dots/yard cloth he used.*

Then 5 – y = the number of yards of 450 dots/yard cloth he used.

Then 200y = the number of dots that the 200 dots/yard cloth will give him.

Then 450(5 – y) = the number of dots that the 450 dots/yard cloth will give him.

He needs a total of 1500 dots for the curtain.

$$200y + 450(5 - y) = 1500$$
$$200y + 2250 - 450y = 1500$$
$$-250y = -750$$
$$y = 3 \text{ yards of the 200 dots/yard cloth}$$

5 – y which is 2 yards of the 450 dots/yard cloth

9. The set could be written as { } and is called the empty set or the null set.

Before we start the first city, here is a little trick that may come in helpful.

Suppose we want to simplify $\dfrac{x^2 - 7x + 12}{4x - x^2}$

Factor top,

Factor bottom, $= \dfrac{(x - 3)(x - 4)}{x(4 - x)}$

The last step is to Cancel like factors.

The $(x - 4)$ and the $(4 - x)$ look soooo close to each other.
They are negatives of each other.
So the trick is to factor out $-x$
instead of x in the denominator $\dfrac{(x - 3)(x - 4)}{-x(x - 4)}$

$x - 4$ $4 - x$

Now we can cancel $= \dfrac{x - 3}{-x}$

You could play with this by changing the
signs of the numerator and denominator $= \dfrac{-(x - 3)}{x}$

and then multiply out the numerator $= \dfrac{-x + 3}{x}$

And since $a + b = b + a$
(which is called the **commutative law of addition)**
you could write the answer as $= \dfrac{3 - x}{x}$ Any of these are okay.

In chapter 6 we first mentioned the commutative law of addition. Now it's mentioned here a second time. What mnemonic (memory aid) did you think of to associate *commutative* with a + b = b + a?

Commutative, commutative, commutative. . . .   It sounds like people commuting to work. I picture two cars—one commuting east and the other commuting west. That's how I remember the commutative law of addition.

Winnemucca, Nevada "there's more than meets the eye"

1. Solve $\dfrac{1}{x-1} + \dfrac{1}{2} = \dfrac{2}{x^2-1}$

2. Combine and simplify

$$\dfrac{4w^2}{4w^2z + 6wz^2} - \dfrac{9z^2}{4w^2z + 6wz^2}$$

3. Solve $\dfrac{6z}{z-2} = 8$

4. It takes Betty 20 minutes to squeeze all the grease out of a bucket of KFF fries. It takes Alexander 30 minutes to do the same job. How long would it take them if they both worked together?

5. Combine and simplify

$$\dfrac{3}{y} - \dfrac{3}{y+1}$$

6. Solve $\dfrac{w^2-10}{6} = w^2 - 15$

7. Combine and simplify

$$\dfrac{\dfrac{x^2-5x+4}{3x^2-10x-8}}{\dfrac{3x^2-x-2}{3x+2}}$$

8. Simplify $\dfrac{x+1}{x^2-1}$

9. Solve $\dfrac{5-x}{x} + \dfrac{3}{4} = \dfrac{7}{x}$

10. Combine (and simplify if possible)

$$\dfrac{3x}{x+2} + \dfrac{x}{x-2} - \dfrac{4x}{x^2-4}$$

11. Pat can put up a sheet of plywood over the hole in the bunkhouse in two minutes. Chris could do that in three minutes. Working together, how long would it take them to put up a sheet of plywood?

12. Combine (and simplify if possible)

$$\frac{y+2}{y+1} \div \frac{y^2-4}{y^2+y}$$

13. Combine (and simplify if possible)

$$\frac{6}{x} - \frac{-7}{x^2}$$

14. Solve $\dfrac{z^2}{z^2-20} = 5$

15. Combine (and simplify if possible)

$$\frac{6}{w-9} - \frac{4}{9-w}$$

answers

1. $\dfrac{1}{x-1} + \dfrac{1}{2} = \dfrac{2}{x^2-1}$ is a fractional equation. We want to multiply through by an expression that $x-1$, 2, and x^2-1 all evenly divide into.

First, factor all the denominators

$$\frac{1}{x-1} + \frac{1}{2} = \frac{2}{(x+1)(x-1)}$$

$$\frac{1(2)(x+1)(x-1)}{x-1} + \frac{1(2)(x+1)(x-1)}{2} = \frac{2(2)(x+1)(x-1)}{(x+1)(x-1)}$$

Cancel like factors $2(x+1) + (x+1)(x-1) = 4$

Multiply out $2x+2+x^2-1 = 4$

Transpose everything to one side $x^2+2x-3 = 0$

Factor $(x+3)(x-1) = 0$

$$x+3=0 \;\; OR \;\; x-1=0$$

$$x = -3, 1$$

> Whenever you multiply an equation by an expression containing x, you must check any answers you get back in the original equation.

Checking $x = -3$ . . . $\dfrac{1}{x-1} + \dfrac{1}{2} = \dfrac{2}{x^2-1}$

becomes $-\dfrac{1}{4} + \dfrac{1}{2} \overset{?}{=} \dfrac{2}{8}$ true

Checking x = 1 . . . $\dfrac{1}{x-1} + \dfrac{1}{2} = \dfrac{2}{x^2-1}$

becomes $\dfrac{1}{0}$ Stop right here! In case you have never

heard of the "No zero denominators" rule, we'll discuss that in detail near the end of this book. x = −3 is the only answer.

2. $\dfrac{4w^2}{4w^2z+6wz^2} - \dfrac{9z^2}{4w^2z+6wz^2} = \dfrac{4w^2}{4w^2z+6wz^2} + \dfrac{-9z^2}{4w^2z+6wz^2}$

$$= \dfrac{4w^2 - 9z^2}{4w^2z + 6wz^2}$$

$$= \dfrac{(2w+3z)(2w-3z)}{2wz(2w+3z)}$$

$$= \dfrac{2w-3z}{2wz}$$

3. $\dfrac{6z}{z-2} = 8$

$\dfrac{6z}{z-2} = \dfrac{8}{1}$ This gives me a rooftop for each of the Santa gifts.

$\dfrac{6z(\mathbf{z-2})}{z-2} = \dfrac{8(\mathbf{z-2})}{1}$

$6z = 8z - 16$

$16 = 2z$

$8 = z$

Checking z = 8 in the original problem: $\dfrac{6(8)}{6} \overset{?}{=} 8$ yes

4. We have never before used "six pretty boxes" for a "job" problem. We have always started with *Let x = the minutes it takes them to do the job if they are working together.* Let's see if you like this approach better.

	together	Betty	Alexander
the whole job	x		
the part of the job that can be done in one minute			

	together	Betty	Alexander
the whole job	x	20	30
the part of the job that can be done in one minute			

	together	Betty	Alexander
the whole job	x	20	30
the part of the job that can be done in one minute	$\dfrac{1}{x}$	$\dfrac{1}{20}$	$\dfrac{1}{30}$

The part of the job that Betty can do in one minute plus the part of the job that Alexander can do in one minute will equal the part of the job that both of them working together can do.

$$\frac{1}{20} + \frac{1}{30} = \frac{1}{x}$$

Did you ever notice that the amount of room that English takes is more than the math takes?

Did you ever notice that you have to capitalize *English*, but not *math* ?

Multiple-choice question: Who made the rules for capitalization?   A) the English department

B) the math department

It's a fractional equation.

The smallest expression that 20, 30, and x all evenly divide into is 60x.

$$\frac{1(60x)}{20} + \frac{1(60x)}{30} = \frac{1(60x)}{x}$$

$$3x + 2x = 60$$

$$5x = 60$$

$$x = 12$$

Together, they can squeeze all the grease out of a bucket of KFF fries in 12 minutes.

5.

$$\frac{3}{y} - \frac{3}{y+1}$$

Subtraction problem into addition

$$= \frac{3}{y} + \frac{-3}{y+1}$$

Interior Decorating

$$= \frac{3(y+1)}{y(y+1)} + \frac{-3y}{(y+1)y}$$

$$= \frac{3y+3-3y}{y(y+1)}$$

$$= \frac{3}{y(y+1)}$$

6.

$$\frac{w^2-10}{6} = w^2-15$$

$$\frac{w^2-10}{6} = \frac{w^2-15}{1}$$ This gives me a rooftop for each of the Santa gifts.

$$\frac{(w^2-10)6}{6} = \frac{(w^2-15)6}{1}$$ Put parentheses so the entire numerators receive the "gifts."

$$w^2-10 = 6w^2-90$$

$$80 = 5w^2$$ A pure quadratic equation

$$16 = w^2$$

$$\pm 4 = w$$

$$w = \pm 4$$ The symmetric law of equality (Lesson 55, Your Turn to Play, question 1)

7.

$$\frac{x^2-5x+4}{3x^2-10x-8} \div \frac{3x^2-x-2}{3x+2}$$

$$= \frac{x^2-5x+4}{3x^2-10x-8} \times \frac{3x+2}{3x^2-x-2}$$

$$= \frac{(x-4)(x-1)(3x+2)}{(3x+2)(x-4)(3x+2)(x-1)}$$

$$= \frac{1}{3x+2}$$

8. $$\frac{x+1}{x^2-1} = \frac{x+1}{(x+1)(x-1)} = \frac{(x+1)}{(x+1)(x-1)} = \frac{1}{x-1}$$

9.
$$\frac{5-x}{x} + \frac{3}{4} = \frac{7}{x}$$

$$\frac{(5-x)4x}{x} + \frac{3(4x)}{4} = \frac{7(4x)}{x}$$

$$20 - 4x + 3x = 28$$

$$-8 = x$$

Checking $x = -8$ in the original problem $\frac{13}{-8} + \frac{3}{4} \overset{?}{=} -\frac{7}{8}$ yes

10. $\dfrac{3x}{x+2} + \dfrac{x}{x-2} - \dfrac{4x}{x^2-4}$

$= \dfrac{3x}{x+2} + \dfrac{x}{x-2} + \dfrac{-4x}{(x+2)(x-2)}$ Subtraction into addition

                                                       Factor the denominators

$= \dfrac{3x(x-2)}{(x+2)(x-2)} + \dfrac{x(x+2)}{(x-2)(x+2)} + \dfrac{-4x}{(x+2)(x-2)}$]Interior Decorating

$= \dfrac{3x^2 - 6x + x^2 + 2x - 4x}{(x+2)(x-2)}$

$= \dfrac{4x^2 - 8x}{(x+2)(x-2)}$

$= \dfrac{4x(x-2)}{(x+2)(x-2)}$

$= \dfrac{4x}{x+2}$

11. Using "six pretty boxes" . . .

	together	Pat	Chris
the whole job	t	2	3
the part of the job that can be done in one minute	$\frac{1}{t}$	$\frac{1}{2}$	$\frac{1}{3}$

The part of the job that Pat can do in one minute plus the part of the job that Chris can do in one minute is equal to the part of the job that both of them working together can do in one minute.

$$\frac{1}{2} + \frac{1}{3} = \frac{1}{t}$$

$$\frac{6t}{2} + \frac{6t}{3} = \frac{6t}{t}$$

$$3t + 2t = 6$$

$$5t = 6$$

t = one and one-fifth minutes (or 1 minute and 12 seconds)

12. $\dfrac{y+2}{y+1} \div \dfrac{y^2-4}{y^2+y}$

$= \dfrac{y+2}{y+1} \times \dfrac{y^2+y}{y^2-4}$

$= \dfrac{(y+2)y(y+1)}{(y+1)(y+2)(y-2)}$

$= \dfrac{y}{y-2}$

13. $\dfrac{6}{x} - \dfrac{-7}{x^2}$

$= \dfrac{6}{x} + \dfrac{+7}{x^2}$

$= \dfrac{6\mathbf{x}}{\mathbf{xx}} + \dfrac{+7}{x^2}$

$= \dfrac{6x+7}{x^2}$

14. There are three different ways to solve $\dfrac{z^2}{z^2-20} = \dfrac{5}{1}$

Santa Claus approach

$$\dfrac{z^2(\mathbf{z^2-20})}{z^2-20} = \dfrac{5(\mathbf{z^2-20})}{1}$$

$$z^2 = 25$$
$$z = \pm 5$$

Cross-multiplying

$$z^2 = 5(z^2-20)$$
$$z^2 = 5z^2 - 100$$
$$100 = 4z^2$$
$$25 = z^2$$
$$\pm 5 = z$$

Algebra

$$\dfrac{z^2}{z^2-20} = 5$$

Multiply both sides by $z^2 - 20$

$$z^2 = 5(z^2-20)$$
$$100 = 4z^2$$
$$25 = z^2$$
$$\pm 5 = z$$

15. $\dfrac{6}{w-9} - \dfrac{4}{9-w} = \dfrac{6}{w-9} - \dfrac{4}{-(-9+w)} = \dfrac{6}{w-9} + \dfrac{4}{w-9}$

$= \dfrac{10}{w-9}$ Or you might have done $\dfrac{6}{w-9} - \dfrac{4}{9-w} = \dfrac{6}{w-9} + \dfrac{-4}{9-w} = \dfrac{6}{w-9} + \dfrac{4}{w-9}$ etc.

Livingston, Montana Calamity Jane was a resident

1. Solve $\dfrac{z+7}{8} = \dfrac{z+8}{9}$

2. Combine (and simplify if possible)

$$\dfrac{-2xy + 2y^2}{x^2 - y^2} + 1$$

3. Dr. Thurow N. Speck can give physical exams to every soldier in camp (all 4,589 of them) in four hours. The old camp doctor who retired last year was slower in giving physicals. Old doc Gravely Dug took five hours to examine all of the soldiers in camp. If they worked together, how long would it take them to examine the whole camp?

4. Combine (and simplify if possible)

$$\dfrac{x-3}{12x+12} \div \dfrac{4x-12}{9}$$

5. Combine (and simplify if possible)

$$\dfrac{4}{x^2y} - \dfrac{-5}{xy^2}$$

6. Solve $\dfrac{y^2}{y+2} = \dfrac{8}{3}$

7. Combine and simplify

$$\dfrac{\dfrac{5-x}{4x^2+20x+21}}{\dfrac{x^2-25}{2x^2+13x+15}}$$

8. Combine (and simplify if possible) $\dfrac{x}{x+2} - \dfrac{4}{x+3}$

9. Combine (and simplify if possible)

$$\dfrac{4y}{y^2-36} - \dfrac{4}{y+6}$$

10. Solve $\dfrac{x}{x+1} + \dfrac{3}{4} = \dfrac{17}{x+10}$

(Hint: if you get to $7x^2 + 5x - 38 = 0$ you're on the right track.)

11. The plumbing doesn't work very well at A.A.A.A. Sergeant Snow's bathtub is probably the best example of the mess things are in. If you turn on the faucets full force, it takes 20 minutes to fill his tub. After he's done taking a bath, he pulls the plug and it takes 30 minutes to drain the tub.

    (Over the years the grease from his ablutions has clogged most of the drain piping.)

    One time he turned on the faucets to fill the tub but forgot to put in the plug. How long did it take to fill the tub?

(Hint: $\dfrac{1}{20} - \dfrac{1}{30} = \dots$)

12. Combine (and simplify if possible)

$$\frac{y^2 + 7y + 12}{y^2 + 9y + 20} \div \frac{y^2 + 9y + 18}{y^2 + y - 30}$$

13. Combine (and simplify if possible) $\dfrac{4}{7} + \dfrac{3}{x+5}$

14. Solve $\dfrac{2x - 14}{5} = \dfrac{12 - x}{10}$

15. The colonel's robot maid was chatting with Carol, and they both realized that they had a big craving for some chocolate chip cookies. They sneaked* up to the colonel's private office suite and found in his desk drawer an old chocolate bar and a six-pound bag of what they thought was flour. "We're in luck!" they cried and headed off to the mess hall kitchen.

    Together they could bake up the whole batch of cookies in four hours. The robot maid, working alone, could have done it in six hours. How long would it have taken Carol to bake all the cookies alone?

---

* Both *sneaked* and *snuck* are in my *Random House Webster's College Dictionary.*

answers

1.
$$\frac{z+7}{8} = \frac{z+8}{9}$$

Multiply each term by 72
$$\frac{(z+7)\mathbf{72}}{8} = \frac{(z+8)\mathbf{72}}{9}$$
$$9(z+7) = 8(z+8)$$
$$9z+63 = 8z+64$$
$$z = 1$$

2.
$$\frac{-2xy+2y^2}{x^2-y^2}+1$$

$$= \frac{-2xy+2y^2}{x^2-y^2} + \frac{1}{1}$$

Interior Decorating
$$= \frac{-2xy+2y^2}{x^2-y^2} + \frac{(x^2-y^2)}{(x^2-y^2)}$$

Add the fractions
$$= \frac{x^2-2xy+y^2}{x^2-y^2}$$

Factor top and factor bottom
$$= \frac{(x-y)(x-y)}{(x+y)(x-y)}$$

Cancel like factors
$$= \frac{x-y}{x+y}$$

3.   *Let x = the number of hours if they work together.*

Then $\frac{1}{x}$ = the amount of the job they do in one hour working together.

Dr. Speck can do $\frac{1}{4}$ of the job in one hour.

Dr. Dug can do $\frac{1}{5}$ of the job in one hour.

The amount of the job that Dr. Speck can do in one hour plus the amount of the job that Dr. Dug can do in one hour equals the amount of the job that both of them can do if they work together.

$$\frac{1}{4} + \frac{1}{5} = \frac{1}{x}$$

$$\frac{\mathbf{20x}}{4} + \frac{\mathbf{20x}}{5} = \frac{\mathbf{20x}}{x}$$

$$5x + 4x = 20$$

$$9x = 20$$

$$x = \frac{20}{9} \text{ hours or } 2\frac{2}{9} \text{ hours}$$

4.

$$\frac{x-3}{12x+12} \div \frac{4x-12}{9}$$

Invert the divisor

$$= \frac{x-3}{12x+12} \times \frac{9}{4x-12}$$

Top times top and bottom times bottom

$$= \frac{(x-3)(9)}{(12x+12)(4x-12)}$$

Factor top and factor bottom

$$= \frac{(x-3)(9)}{12(x+1)(4)(x-3)}$$

Cancel like factors

$$= \frac{3}{16(x+1)}$$

5. $\dfrac{4}{x^2y} - \dfrac{-5}{xy^2}$

$$= \frac{4}{x^2y} + \frac{5}{xy^2}$$

$$= \frac{4\mathbf{y}}{x^2y\mathbf{y}} + \frac{5\mathbf{x}}{xy^2\mathbf{x}}$$

$$= \frac{4y+5x}{x^2y^2}$$

6.

$$\frac{y^2}{y+2} = \frac{8}{3}$$

This is a proportion. I'm going to cross-multiply.

$$3y^2 = 8(y+2)$$

$$3y^2 = 8y+16$$

Solving this equation by factoring
Transpose everything to one side

$$3y^2 - 8y - 16 = 0$$

$$(3y+4)(y-4) = 0$$

$$3y+4=0 \quad \text{OR} \quad y-4=0$$

$$y = -\frac{4}{3} \quad \text{OR} \quad y=4$$

7. $\dfrac{5-x}{4x^2+20x+21} \div \dfrac{x^2-25}{2x^2+13x+15} = \dfrac{5-x}{4x^2+20x+21} \times \dfrac{2x^2+13x+15}{x^2-25}$

$$= \frac{(5-x)(2x+3)(x+5)}{(2x+7)(2x+3)(x+5)(x-5)} = \frac{5-x}{(2x+7)(x-5)} = \frac{-(x-5)}{(2x+7)(x-5)}$$

$$= \frac{-1}{2x+7}$$

8.
$$\frac{x}{x+2} - \frac{4}{x+3}$$

Subtraction changed to addition
$$= \frac{x}{x+2} + \frac{-4}{x+3}$$

Interior Decorating
$$= \frac{x(x+3)}{(x+2)(x+3)} + \frac{-4(x+2)}{(x+3)(x+2)}$$

$$= \frac{x(x+3) - 4(x+2)}{(x+2)(x+3)}$$

$$= \frac{x^2 + 3x - 4x - 8}{(x+2)(x+3)}$$

$$= \frac{x^2 - x - 8}{(x+2)(x+3)}$$

9.
$$\frac{4y}{y^2 - 36} - \frac{4}{y+6}$$

$$= \frac{4y}{y^2 - 36} + \frac{-4}{y+6}$$

$$= \frac{4y}{(y+6)(y-6)} + \frac{-4}{y+6}$$

$$= \frac{4y}{(y+6)(y-6)} + \frac{-4(y-6)}{(y+6)(y-6)}$$

$$= \frac{4y - 4y + 24}{(y+6)(y-6)}$$

$$= \frac{24}{(y+6)(y-6)}$$

10.
$$\frac{x}{x+1} + \frac{3}{4} = \frac{17}{x+10}$$

$$\frac{x(4)(x+1)(x+10)}{x+1} + \frac{3(4)(x+1)(x+10)}{4} = \frac{17(4)(x+1)(x+10)}{x+10}$$

$$4x(x+10) + 3(x+1)(x+10) = 68(x+1)$$
$$4x^2 + 40x + 3(x^2 + 10x + x + 10) = 68x + 68$$
$$4x^2 + 40x + 3x^2 + 30x + 3x + 30 = 68x + 68$$
$$7x^2 + 5x - 38 = 0$$
$$(7x + 19)(x - 2) = 0$$
$$7x + 19 = 0 \quad \text{OR} \quad x - 2 = 0$$
$$x = -\frac{19}{7} \quad \text{OR} \quad x = 2$$

11. Let's use "six pretty boxes" . . .

	both at the same time	filling the tub	emptying the tub
time to fill/empty	t	20	30
part of the job done per minute	$\dfrac{1}{t}$	$\dfrac{1}{20}$	$\dfrac{1}{30}$

The fraction of the tub being filled each minute minus the fraction of the tub being emptied each minute equals the part of the job being done each minute when the faucets are on and the drain plug is not inserted.

$$\frac{1}{20} - \frac{1}{30} = \frac{1}{t}$$

$$\frac{1(60t)}{20} - \frac{1(60t)}{30} = \frac{60t}{t}$$

$$3t - 2t = 60$$

$$t = 60 \text{ minutes}$$

12. $\dfrac{y^2 + 7y + 12}{y^2 + 9y + 20} \div \dfrac{y^2 + 9y + 18}{y^2 + y - 30} = \dfrac{y^2 + 7y + 12}{y^2 + 9y + 20} \times \dfrac{y^2 + y - 30}{y^2 + 9y + 18}$

$= \dfrac{(y+3)(y+4)(y+6)(y-5)}{(y+4)(y+5)(y+3)(y+6)} = \dfrac{y-5}{y+5}$

13. $\dfrac{4}{7} + \dfrac{3}{x+5} = \dfrac{4(x+5)}{7(x+5)} + \dfrac{3(7)}{(x+5)7} = \dfrac{4x+41}{7(x+5)}$

14. $\dfrac{2x-14}{5} = \dfrac{12-x}{10}$ ⇨ $\dfrac{(2x-14)10}{5} = \dfrac{(12-x)10}{10}$

⇨ $(2x-14)2 = 12 - x$ ⇨ $4x - 28 = 12 - x$ ⇨ $5x = 40$ ⇨ $x = 8$

15. *Let t = the number of hours for Carol to make all the cookies.*

Then $\dfrac{1}{t}$ = the part of the job Carol could do in one hour.

Then $\dfrac{1}{4}$ = the part of the job they could do in one hour working together.

Then $\dfrac{1}{6}$ = the part of the job the robot maid could do in one hour.

The part of the job that Carol could do in one hour plus the part of the job that the robot maid could do in one hour equal the part of the job that the pair of them could do in one hour.

$\dfrac{1}{t} + \dfrac{1}{6} = \dfrac{1}{4}$ ⇨ $\dfrac{12t}{t} + \dfrac{12t}{6} = \dfrac{12t}{4}$ ⇨ $12 + 2t = 3t$ ⇨ $12 = t$

Darlington, Wisconsin where the Ringling Brothers' Circus began

1. Simplify $\dfrac{y^2 - 6y + 9}{y^2 - 5y + 6}$

2. Combine (and simplify if possible) $\dfrac{7}{2xy^2} - \dfrac{5}{4xy}$

3. Solve $\dfrac{5}{z-3} = \dfrac{2}{z}$

4. Pat can stucco over the hole in the bunkhouse in five hours. Chris can do the same job in seven hours. How long would it take to do the stuccoing if they work together?

5. Combine (and simplify if possible) $\dfrac{5x}{x+3} + \dfrac{15x - 45}{x^2 - 9}$

6. Solve $\dfrac{x+3}{8} = \dfrac{x-2}{3}$

7. It takes five hours and 20 minutes to build a fence. How much of the job could be done in one hour? (Hint: Express everything in hours.)

8. Combine (and simplify if possible)

$$\dfrac{y^2 + 3y}{y^2 - 3y - 4} \div \dfrac{y^2 + 2y - 3}{y^2 - 5y + 4}$$

9. Simplify $\dfrac{x^2 - xy + 3x - 3y}{x^2 - 2xy + y^2}$

10. Simplify $\dfrac{x^2 - 49}{x^2 - 4x - 21}$

11. Solve $\dfrac{8}{x} = \dfrac{3}{x+1} + 3$

12. Jack doesn't eat pizza very often because he cannot stand the thought of all that grease in his system, but when he does, he can finish off a 16" pizza in three minutes. Sergeant Snow can polish off a 16" pizza alone in one minute flat. (He's well practiced in wolfing down his food.) How long would it take the two of them together to finish off a 16" pizza?

After you've computed your answer in minutes, change this answer to seconds.

13. Combine (and simplify if possible)

$$\dfrac{y^2 - 4}{10y - 60} \div \dfrac{(y+2)^2}{y^2 - 4y - 12}$$

14. Combine (and simplify if possible)

$$\frac{6}{x + 5y} - \frac{5x + 7y}{x^2 + 7xy + 10y^2}$$

15. Solve $\frac{z}{2} - \frac{z}{5} = 6$

16. Combine (and simplify if possible)

$$\frac{6}{w - 7} + \frac{5w}{7 - w}$$

answers

1.
$$\frac{y^2 - 6y + 9}{y^2 - 5y + 6}$$
$$= \frac{(y - 3)(y - 3)}{(y - 3)(y - 2)}$$
$$= \frac{y - 3}{y - 2}$$

2.
$$\frac{7}{2xy^2} - \frac{5}{4xy}$$
$$= \frac{7}{2xy^2} + \frac{-5}{4xy}$$
$$= \frac{7(2)}{2xy^2(2)} + \frac{-5y}{4xyy}$$
$$= \frac{14 - 5y}{4xy^2}$$

3.
$$\frac{5}{z - 3} = \frac{2}{z}$$
$$\frac{5(z - 3)z}{z - 3} = \frac{2(z - 3)z}{z}$$
$$5z = 2(z - 3)$$
$$5z = 2z - 6$$
$$3z = -6$$
$$z = -2$$

When you multiply an equation by an expression containing a variable, you must check your answer in the original problem to make sure that it isn't an extraneous root.

Putting $z = -2$ into $\frac{5}{z - 3} = \frac{2}{z}$ ⇨ $\frac{5}{-5} \overset{?}{=} \frac{2}{-2}$ true

4. "Six pretty boxes" . . .

	together	Pat	Chris
the whole job	t	5	7
the part of the job that can be done in one hour	$\dfrac{1}{t}$	$\dfrac{1}{5}$	$\dfrac{1}{7}$

The part of the job that Pat can do in one hour plus the part of the job that Chris can do in one hour equals the part of the job that they can do when working together.

$$\frac{1}{5} + \frac{1}{7} = \frac{1}{t}$$

$$\frac{35t}{5} + \frac{35t}{7} = \frac{35t}{t}$$

$$7t + 5t = 35$$

$$12t = 35$$

$t = \dfrac{35}{12}$ hours, which is $2\dfrac{11}{12}$ hours, which is 2 hours and 55 minutes

5.
$$\frac{5x}{x+3} + \frac{15x-45}{x^2-9}$$

$$= \frac{5x}{x+3} + \frac{15x-45}{(x+3)(x-3)}$$

$$= \frac{5x(x-3)}{(x+3)(x-3)} + \frac{15x-45}{(x+3)(x-3)}$$

$$= \frac{5x^2 - 15x + 15x - 45}{(x+3)(x-3)}$$

$$= \frac{5x^2 - 45}{(x+3)(x-3)}$$

Always look for a common factor first!

$$= \frac{5(x^2-9)}{(x+3)(x-3)}$$

$$= \frac{5(x+3)(x-3)}{(x+3)(x-3)}$$

$$= 5$$

6.
$$\frac{x+3}{8} = \frac{x-2}{3}$$

Cross-multiply
$$3(x+3) = 8(x-2)$$
$$3x + 9 = 8x - 16$$
$$25 = 5x$$
$$5 = x$$

7. Five hours and 20 minutes is equal to $5\frac{1}{3}$ hours.

If the fence could be built in 9 hours, then in one hour $\frac{1}{9}$ of the fence could be built.

If the fence could be built in 7 hours, then in one hour $\frac{1}{7}$ of the fence could be built.

If the fence could be built in $\frac{8}{3}$ hours, then in one hour $\frac{3}{8}$ of the fence could be built.

In each case, we take the **reciprocal** of the number.

The reciprocal of 82 is $\frac{1}{82}$

To find the reciprocal of $5\frac{1}{3}$ first turn it into an improper fraction.

Three times five plus one ... $5\frac{1}{3} = \frac{16}{3}$

The reciprocal of $5\frac{1}{3}$ is $\frac{3}{16}$ Three-sixteens of the fence could be built in one hour.

8.
$$\frac{y^2 + 3y}{y^2 - 3y - 4} \div \frac{y^2 + 2y - 3}{y^2 - 5y + 4}$$

$$= \frac{y^2 + 3y}{y^2 - 3y - 4} \times \frac{y^2 - 5y + 4}{y^2 + 2y - 3}$$

$$= \frac{(y^2 + 3y)(y^2 - 5y + 4)}{(y^2 - 3y - 4)(y^2 + 2y - 3)}$$

$$= \frac{y(y + 3)(y - 1)(y - 4)}{(y - 4)(y + 1)(y - 1)(y + 3)}$$

$$= \frac{y}{y + 1}$$

Now it's important not to go **Cancel Crazy** and cancel the y's. What you can cancel are *like factors*. A factor multiplies the whole expression, not just part of it. In the fraction $\frac{2w + 7}{4}$ the "2" does not multiply the entire 2w + 7. It just multiplies the w. You can't cancel the 2 with the 4 in the denominator.

The ultimate in **Cancel Crazy** came one year when a student wrote that $\dfrac{x+y}{w+z}$ was equal to $\dfrac{xy}{wz}$. He had canceled the plus signs!

9. $\dfrac{x^2 - xy + 3x - 3y}{x^2 - 2xy + y^2}$

$$= \dfrac{x(x-y) + 3(x-y)}{x^2 - 2xy + y^2}$$

$$= \dfrac{(x-y)(x+3)}{(x-y)(x-y)}$$

$$= \dfrac{x+3}{x-y}$$

And, of course, you can't cancel the x's. That would be **Cancel Crazy**.

10. $\dfrac{x^2 - 49}{x^2 - 4x - 21}$

$$= \dfrac{(x+7)(x-7)}{(x-7)(x+3)}$$

$$= \dfrac{x+7}{x+3}$$

11. $\dfrac{8}{x} = \dfrac{3}{x+1} + 3$

$$\dfrac{8(x)(x+1)}{x} = \dfrac{3(x)(x+1)}{x+1} + 3(x)(x+1)$$

$$8(x+1) = 3x + 3x(x+1)$$

$$8x + 8 = 3x + 3x^2 + 3x$$

$$0 = 3x^2 - 2x - 8$$

$$0 = (3x+4)(x-2)$$

$$3x + 4 = 0 \ \ OR \ \ x - 2 = 0$$

$$x = -\dfrac{4}{3} \ \ OR \ \ x = 2$$

We must check the answers in the original problem since we multiplied the equation by an expression, **(x)(x + 1)**, that contained a variable.

Checking $x = -\dfrac{4}{3}$ $\dfrac{8}{(-4/3)} \overset{?}{=} \dfrac{3}{(-4/3)+1} + 3$

$$-6 \overset{?}{=} -9 + 3 \qquad \text{yes}$$

Checking x = 2 $\dfrac{8}{2} \overset{?}{=} \dfrac{3}{2+1} + 3$

$4 \overset{?}{=} 1 + 3$ yes

The final answer is $x = -\dfrac{4}{3}$ or x = 2.

12.   *Let t = the number of minutes it would take for both of them to finish a pizza.*
       Then $\dfrac{1}{t}$ = the amount of the pizza they could finish in one minute.

Jack can eat a pizza in 3 minutes. He can eat $\dfrac{1}{3}$ of a pizza in a minute.

Snow can eat a pizza in 1 minute. He can eat 1 pizza in a minute.

$$\dfrac{1}{3} + 1 = \dfrac{1}{t}$$

$$\dfrac{3t}{3} + 3t = \dfrac{3t}{t}$$

$$t + 3t = 3$$

$$4t = 3$$

$$t = \dfrac{3}{4} \text{ of a minute.}$$

To convert minutes into seconds, multiply by the conversion factor $\dfrac{60 \text{ seconds}}{1 \text{ minute}}$

$$\dfrac{\text{¾ minute}}{1} \times \dfrac{60 \text{ seconds}}{1 \text{ minute}}$$

$$= \dfrac{\text{¾ minute}}{1} \times \dfrac{60 \text{ seconds}}{1 \text{ minute}}$$

$$= \text{¾} \times 60 \text{ seconds}$$

$$= 45 \text{ seconds}$$

Two notes about **conversion factors** …

1. The numerator and denominator of any conversion factor are equal.

2. The conversion factor is chosen so that the old units (in this case, minutes) will cancel and be replaced by the new units (seconds).

13.
$$\frac{y^2 - 4}{10y - 60} \div \frac{(y + 2)^2}{y^2 - 4y - 12}$$

$$= \frac{y^2 - 4}{10y - 60} \times \frac{y^2 - 4y - 12}{(y + 2)^2}$$

$$= \frac{(y + 2)(y - 2)(y - 6)(y + 2)}{10(y - 6)(y + 2)^2}$$

$$= \frac{y - 2}{10}$$

14.
$$\frac{6}{x + 5y} - \frac{5x + 7y}{x^2 + 7xy + 10y^2}$$

$$= \frac{6}{x + 5y} + \frac{-(5x + 7y)}{x^2 + 7xy + 10y^2}$$

$$= \frac{6}{x + 5y} + \frac{-5x - 7y}{(x + 2y)(x + 5y)}$$

$$= \frac{6(x + 2y)}{(x + 5y)(x + 2y)} + \frac{-5x - 7y}{(x + 2y)(x + 5y)}$$

$$= \frac{6x + 12y - 5x - 7y}{(x + 2y)(x + 5y)}$$

$$= \frac{x + 5y}{(x + 2y)(x + 5y)}$$

$$= \frac{1}{x + 2y}$$

15. $\frac{z}{2} - \frac{z}{5} = 6$

$$\frac{10z}{2} - \frac{10z}{5} = 6(10)$$

$$5z - 2z = 60$$

$$3z = 60$$

$$z = 20$$

16.
$$\frac{6}{w - 7} + \frac{5w}{7 - w}$$

$$= \frac{6}{w - 7} + \frac{-5w}{w - 7}$$

$$= \frac{6 - 5w}{w - 7}$$

⋙⋘⋙⋘⋙⋘⋙⋘⋙⋘⋙⋘⋙⋘⋙⋘

Peach milkshakes. There's nothing like them. Made from peaches and cream, their delicate coolness in the hot Texas afternoon sun is a foretaste of heaven. Fred was just starting to wake up from his afternoon nap. He had been dreaming of milkshakes and roses and finally thought of *Ophelia*—one of the most important modern roses—and its just off-white color. That was a good match. (There is a rose called *Peaches 'n' Cream*, but it's pink and wouldn't have been a good match for a peach milkshake.)

Fred awoke and stretched. It was 5:30. That nap felt good and the taste of a peach milkshake lingered in his mouth. He fantasized that if he ever got back to KITTENS, besides teaching, he might open a *Roses & Milkshakes* restaurant on campus. With every milkshake (except chocolate) you'd get a rose of the same color. It would appeal to all the senses: touch, taste, smell, sight. The cold tall glass would appeal to the sense of touch. Hearing? Maybe he could have one of the music students compose a song entitled *Der Rosen Milkenshaken* loosely based on *Der Rosenkavalier*. (*The Knight of the Rose* was Richard Strauss's first successful comic opera.)

If *Roses & Milkshakes* was successful, he could expand nationwide by putting his shakes into square milk cartons and selling them in grocery

stores. They would have to be 10 inches tall since that's the right height for the shelves. (*Height* rhymes with *light* and does not rhyme with *width*.)

They would hold 90 cubic inches of milkshake (in one of the 17 rose flavors) and would have a plastic rose of the appropriate color taped to the carton.

What size would the square base be? Call one of the edges x. The other edge of the base would also have to be x since the base is square. The volume of a box (after you take geometry you can call it a

rectangular parallelepiped) is length times width times height. In the case of our milk carton, its volume = $10x^2$.

Since Fred wanted to have the volume equal to 90 cubic inches he knew that $10x^2 = 90$

Divide both sides by 10 $x^2 = 9$

It's a quadratic equation (since it contains a squared term), so he could have solved it by the factoring method of chapter seven, which I place in a box so that you can ignore it if you remember how to solve quadratic equations by factoring.

Fred noticed that $x^2 = 9$ is a **pure quadratic** (which means that it didn't have a linear term, but just the squared term and a number).

Solving pure quadratics is duck soup.

> Solving by factoring $x^2 = 9$
>
> a) transpose everything to one side of the equation $x^2 - 9 = 0$
>
> b) factor $(x + 3)(x - 3) = 0$
>
> c) set each factor equal to zero $x + 3 = 0$ OR $x - 3 = 0$
>
> $x = -3$ OR $x = 3$
>
> (which can be written $x = \pm 3$)

It takes one step to solve $x^2 = 9$. The answer is $x = \pm 3$. So the sides of the milkshake carton would be three inches. (-3 inches doesn't make sense as a measurement.) He decided he'd wait till after he opened *Roses & Milkshakes* and it became a success before he ordered all the empty cartons.

Adding something like $\dfrac{x + 4}{x^2 - 4} + \dfrac{x - 5}{x - 2}$ is real work compared with solving pure quadratic equations. You could solve a million pure quadratics and hardly break a sweat. Just watch . . .

$x^2 = 25$	and in one step the solution is	$x = \pm 5$
$x^2 = 36$	and in one step the solution is	$x = \pm 6$
$x^2 = 100$	and in one step the solution is	$x = \pm 10$

$x^2 = 4$	and in one step the solution is	$x = \pm 2$
$x^2 = 144$	and in one step the solution is	$x = \pm 12$

$x^2 = 1,000,000$	and in one step the solution is	$x = \pm 1,000$
$x^2 = 49$	and in one step the solution is	$x = \pm 7$
$x^2 = 1$	and in one step the solution is	$x = \pm 1$

and if we add a bit of variety . . .

$y^2 = 64$	and in one step the solution is	$y = \pm 8$
$y^2 = 121$	and in one step the solution is	$y = \pm 11$

or how about . . .

$16 = x^2$	and in one step the solution is	$x = \pm 4$
$81 = x^2$	and in one step the solution is	$x = \pm 9$

You could do these pure quadratics in your sleep.

Just a couple more for fun . . .

$x^2 = 625$	and in one step the solution is	$x = \pm 25$
$x^2 = 15129$	and in one step the solution is	$x = \pm 123$

Wait! Stop! you, my reader, exclaim. **I don't know my multiplication tables up that high. How am I supposed to find the number that when squared equals 15129? In elementary school, we just memorized up to 12 × 12.**

No problem. If you've got one of those cheap hand calculators that have buttons +, −, ×, ÷ and $\sqrt{}$ on it, you're in luck. Of course, if your sister has done some payback for the time you used one of her lipsticks to do a little graffiti, you may want to get a new one (a new calculator, not a new sister).

You want to solve $x^2 = 15129$. Punch ⌶ 5 ⌶ 2 ⫾ into your hand calculator and then hit the $\sqrt{}$ button and out pops ⌶2⫾.

The **square root** key ($\sqrt{}$) finds the number that when squared gives you the number you started with.

If you input 49 and hit the square root key you get 7. All you have to remember when solving $x^2 = 49$ is to put the "±" in the answer.

The hand calculator only gives you the non-negative answer when you ask the question, "What is $\sqrt{49}$?" That's because the **radical sign** $\sqrt{}$ means the **principal square root** (which is the non-negative number).

The problem is with the English. If you ask for "a square root of 49," someone could correctly answer with "–7" or they could correctly answer, "7." Both –7 and 7 are square roots of 49.

In algebra we clear up the confusion. The radical sign $\sqrt{}$ has only one meaning, namely the principal square root.

So to solve $x^2 = 49$ we take the square root of both sides of the equation and get $x = \pm\sqrt{49}$, which is the same as $x = \pm 7$.

English is really much more complicated than algebra. It's difficult to see why someone would choose to become an English major rather than a math major. In algebra, when we write $\sqrt{25}$ we know that they mean +5, but in English, when we say "a square root of 25," it could mean either +5 or –5. In English, the rules are unbelievably complicated. When you're trying to spell the word `receive` you recite to yourself the rule you were taught in elementary school: *i before e, except after c*.

A rule with an exception.

But that's not the real rule. It was in middle school that you learned the "actual truth": *i before e, except after c or except when pronounced as a as in neighbor or weigh*.

A rule with an exception to the exception.

But that's not the real rule. Only in really good high schools will they tell you the "actual truth": *i before e, except after c or except when pronounced as a as in neighbor or weigh or except for the exceptions sentence: The weird ones of foreign leisure could neither seize nor forfeit the height*.

Of course, it's only in a math book that you can learn that there are exceptions to the exceptions to the exceptions to the exceptions—which in algebra is $(\text{exceptions})^4$—and they are: `ancient, codeine,`

conscience, counterfeit, deficient, deify, deity,
efficient, Eileen, Einstein, either, feisty,
heifer, heigh-ho, heir, heist, leitmotiv,
peignoir, prescient, protein, science, seismic,
sheik, sleight, society, sovereign, stein, surfeit
and their.

Suddenly, even adding $\frac{x+4}{x^2-4} + \frac{x-5}{x-2}$, when you do some interior decorating then add the numerators, seems trivial compared to English.

Your Turn to Play

1. $\sqrt{81}$

2. $\sqrt{144}$

3. Use your calculator and find the approximate value of $\sqrt{3.14159}$.

4. Use your calculator and find the approximate value of $\sqrt{\pi}$.

5. Can the principal square root of a natural number ever be equal to zero?

6. Some calculators have a key marked x^2. If you enter a number, say 7, and then hit the x^2 key, the display will read 49.

Now suppose you are applying to become a mathematics instructor at KITTENS University. In the interview they hand you a calculator with duct tape over the display so that you can't read it. They tell you that the display has some number on it. Your challenge is to press the keys so that if there were a minus sign on the number, the minus sign would disappear. For example, if the number under the tape were –3.8, after you did your work, the display would read 3.8. If the number you started with was positive, say 985.007, then after you were done, it would remain 985.007.

How would you do it?

7. Solve $x^2 = 36$ by factoring.

8. Solve $x^2 = 36$ as a pure quadratic.

9. Using your calculator, find the approximate solution to $x^2 = 7$.

10. Simplify $\frac{x^2 + 2x - 35}{30 - 6x}$

11. Does {Nevada, @, ⏭} = {@, Nevada, ⏭}?

12. You have a glass of pineapple milkshake. The glass is in the shape of

a cylinder. It's 8 inches tall and holds 157 cubic inches of milkshake. What is the approximate radius of the glass?

13. If $x = \sqrt{7}$, what does x^2 equal?

14. Can you find a number ξ (that's the Greek letter xi from Lesson 39) so that $(\sqrt{\xi})(\sqrt{\xi}) = \xi$ *isn't* true?

. . . . COMPLETE SOLUTIONS . . . .

1. 9

2. 12

3. It is approximately equal to 1.7724531.

4. We just did that problem. See problem 3.

5. No. The natural numbers are $\{1, 2, 3, 4, 5, \ldots\}$. If n is a natural number then $\sqrt{n}$ will always be positive.

6. Of course, you are not permitted to remove the duct tape and look at the number. The easiest way I can think of to remove the minus sign is to first press the x^2 key and then the $\sqrt{}$ key.

   Pressing that sequence of keys will not affect a positive number. Doing it to a negative number will make it positive. (And zero will remain zero under this procedure.)

7.
$$x^2 = 36$$
$$x^2 - 36 = 0$$
$$(x + 6)(x - 6) = 0$$
$$x + 6 = 0 \quad \text{OR} \quad x - 6 = 0$$
$$x = -6 \quad \text{OR} \quad x = 6 \quad \text{(which can be written } x = -6, 6 \text{ or it can be written } x = \pm 6.)$$

8. $x^2 = 36$
    $x = \pm 6$

9. $x = \pm\sqrt{7}$ which is approximately ±2.6457513.

If I use my big calculator

$$x \approx \pm 2.64575131106459059905016157536393.$$

382

10. $\dfrac{x^2 + 2x - 35}{30 - 6x}$

$\qquad = \dfrac{(x - 5)(x + 7)}{-6(-5 + x)}$

$\qquad = \dfrac{x + 7}{-6}$

11. If two sets have the same elements, they are equal. The order in which they are listed doesn't matter.

12. $V = \pi r^2 h$ becomes $\quad 157 = \pi r^2(8)$.

$$\dfrac{157}{8\pi} = r^2$$

$$\pm\sqrt{\dfrac{157}{8\pi}} = r$$

$V = \pi r^2 h$

    We discard the minus sign since the radius of a glass is always positive. My calculator gives an approximate answer of $r = 2.5$ inches.

13. x^2 would equal $(\sqrt{7})(\sqrt{7})$ which is 7.

14. It's true for every positive value of ξ. It's true for $\xi = 0$. But when ξ is negative, $\sqrt{\xi}$ doesn't make sense.

Lost in thought, Fred headed to the men's bunkhouse to see if Jack wanted to go with him on their Sunday evening leave. They were the only two that were getting passes at 6 p.m. since they were the only two who hadn't participated in tearing the giant hole in the side of the bunkhouse.

It was starting to cool off a bit (it was down to 93°) and the lawn looked inviting. Fred took off his shoes and walked across the grass.

Fred's footprints

the hot asphalt walkway that Fred avoided

He liked the feel of the blades of grass between his toes, and besides, it was a shorter trip than walking east and then north on the hot asphalt walkway.

"How much shorter?" he thought to himself. He took out the map of the army base that he'd been given by the Quartermaster. It wasn't a very big map. He turned it over and there was a directory of the places marked on the front of the map.

Directory

A Big Tree
B Minefield
    size: 3 hectometers
        by 2 hectometers
C Men's Bunkhouse
D Chow Hall
E Women and
Children's Bunkhouse

Fred studied the directory and decided he was in the minefield. Not being very well versed in the art of war he figured that a minefield was a field in which they were learning how to dig mines for coal or gold or something. His suspicions were confirmed when he noticed all the big holes in the field. They were obviously places where the soldiers had started digging mines.

He was happy to see that they had given the dimensions of the field. Fred drew a little triangle and labeled the two legs of the triangle. From the Directory, he knew that a was equal to 3 and b was equal to 2. (A hectometer is a hundred meters. A meter is a little longer than a yard, about 39", so a hectometer is about the length of a football field.)

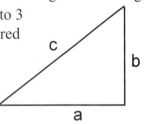

Fred had just drawn the lower right hand half of the field and not the whole rectangle.

If he had walked along the hot asphalt walkway he would have walked a + b which was 3 + 2 hectometers. Instead he had walked a distance equal to c.

How is c related to a and b?

We have a **right triangle**, which is a triangle with a right angle (also known as a 90° angle). In geometry we prove a lot of different things about right triangles. My favorite **theorem** (*theorem* = statement that is proved) is that a right triangle can't have two right angles in it. However, the most important theorem in all of geometry (the **Pythagorean theorem**) is that in any right triangle: $a^2 + b^2 = c^2$.

In *Life of Fred: Calculus Expanded Edition,* we learn that Fred had read a geometry book on his own when he was very young and knew the Pythagorean theorem.

On his piece of paper he wrote:

$$3^2 + 2^2 = c^2$$
$$9 + 4 = c^2$$
$$13 = c^2$$

Using the symmetric property of equality . . .

(x = y is the same as y = x) $c^2 = 13$

It is a pure quadratic, so all that needs to be done is to take the square root of each side and remember to put a ± in front of the number.

$$c = \pm\sqrt{13}$$

The negative sign doesn't make sense as a distance that Fred walked, so we discard it. Fred walked $\sqrt{13}$ hectometers, which is approximately 3.6 hectometers instead of the 5 hectometers on the hot asphalt walkway.

Your Turn to Play

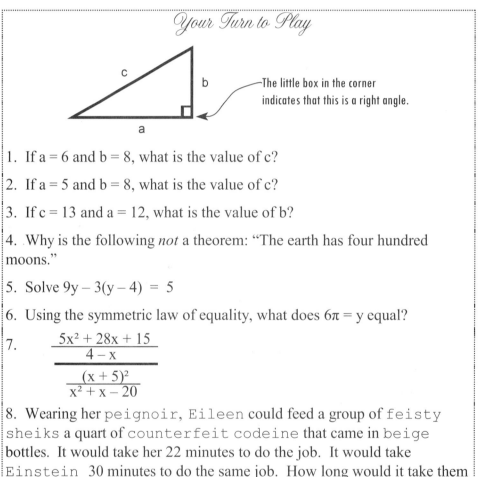

The little box in the corner indicates that this is a right angle.

1. If a = 6 and b = 8, what is the value of c?

2. If a = 5 and b = 8, what is the value of c?

3. If c = 13 and a = 12, what is the value of b?

4. Why is the following *not* a theorem: "The earth has four hundred moons."

5. Solve $9y - 3(y - 4) = 5$

6. Using the symmetric law of equality, what does $6\pi = y$ equal?

7. $$\frac{\dfrac{5x^2 + 28x + 15}{4 - x}}{\dfrac{(x + 5)^2}{x^2 + x - 20}}$$

8. Wearing her `peignoir`, `Eileen` could feed a group of `feisty sheiks` a quart of `counterfeit codeine` that came in `beige` bottles. It would take her 22 minutes to do the job. It would take `Einstein` 30 minutes to do the same job. How long would it take them if they worked together? Round your answer to the nearest hundredth of a minute.

9. What number do you add to –33 to obtain 40?

10. Is it possible to fill in the blank with a whole number that will make

this true: $-15 <$ ___?___ < 0

11. If the two short sides of a right triangle (which are called the legs) are 5 feet and 6 feet, how long is the length of the longest side (which is called the hypotenuse)?

. . . . **COMPLETE SOLUTIONS** . . . .

1. $c^2 = 6^2 + 8^2$

$c^2 = 100$

$c = \pm 10.$

We discard the -10 since we are dealing with lengths, and lengths are never negative. $c = 10$ is the final answer.

2. $c^2 = 5^2 + 8^2$

$c^2 = 89$

$c = \pm\sqrt{89}$ and we discard the negative value. $c = \sqrt{89}$

3. $13^2 = 12^2 + b^2$

$169 = 144 + b^2$

$25 = b^2$

$\pm 5 = b$ and we discard the negative value. $b = 5$

4. Theorems are statements that have been proved. In order to prove a statement, it is really nice if you start with a *true* statement. The whole idea of *proving something* is to establish that it is true. Since the earth doesn't have four hundred moons, that statement isn't true and therefore couldn't be proven and therefore couldn't be a theorem.

Intermission

The heart of geometry is learning how to prove theorems. You learn how to establish that certain statements are true.

The "stuff" of geometry—triangles, squares, parallel lines, area formulas, angles, volume formulas—is just a framework upon which we do the proofs.

In elementary school you learned that the area of a triangle is

½ (base)(height). In algebra we write it in symbols: A =(½)bh. In geometry we will prove that formula.

In geometry we will also prove the Pythagorean theorem. Many people consider the Pythagorean theorem to be the most important theorem of geometry.

After   ♥ beginning algebra

        ♥ advanced algebra

         ♥ geometry

          ♥ trig

           ♥ calculus

you will be ready for upper division (junior/senior in college) math. In those courses, most of what we'll be doing is proving things.

It's engineers—not mathematicians—who concentrate on calculating things.

5.
$$9y - 3(y - 4) = 5$$
$$9y - 3y + 12 = 5$$
$$6y = -7$$
$$y = -\frac{7}{6}$$

6. $y = 6\pi$

7.
$$\frac{5x^2 + 28x + 15}{4 - x} \div \frac{(x + 5)^2}{x^2 + x - 20}$$

$$= \frac{5x^2 + 28x + 15}{4 - x} \times \frac{x^2 + x - 20}{(x + 5)^2}$$

$$= \frac{(5x + 3)(x + 5)}{-(x - 4)} \times \frac{(x + 5)(x - 4)}{(x + 5)^2}$$

$$= \frac{(5x + 3)(x + 5)(x + 5)(x - 4)}{-(x - 4)(x + 5)^2}$$

$$= -(5x + 3) = -5x - 3$$

8. *Let t = the number of minutes it takes for them to do the job together.*

Eileen can do $\frac{1}{22}$ of the job in one minute.

Einstein can do $\frac{1}{30}$ of the job in one minute.

Together they can do $\frac{1}{t}$ of the job in one minute.

The part of the job that Eileen can do in one minute plus the part of the job that Einstein can do in one minute is equal to the part of the job that they can do in one minute working together.

$$\frac{1}{22} + \frac{1}{30} = \frac{1}{t}$$

$$\frac{330t}{22} + \frac{330t}{30} = \frac{330t}{t}$$

I used **330t** instead of **660t** because 22, 30, and t all divide evenly into 330t, and it is smaller than 660t. It made the arithmetic easier.

$$15t + 11t = 330$$

$$26t = 330$$

$$t = \frac{330}{26} \doteq 12.69 \text{ minutes}$$

9. $-33 + 73 = 40$

10. Numbers that will make $-15 < x < 0$ true are numbers such as: $-14, -5.3, -\pi, -\frac{3}{8}$, or -1. The question asks for *whole numbers* that make it true. The whole numbers are $\{0, 1, 2, 3, 4, 5 \dots \}$. There are no whole numbers that make $-15 < x < 0$ true.

11. $c^2 = 5^2 + 6^2$
        $c^2 = 25 + 36$
        $c^2 = 61$
        $c = \pm\sqrt{61}$

A length can never be negative.
The length of the hypotenuse is $\sqrt{61}$ feet.

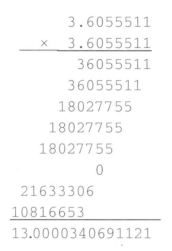

red walked $\sqrt{13}$ hectometers as he crossed the minefield.

What might have seemed strange was the cow that was grazing in the field.

Wait! Stop! I, your reader, have a different question. The cow grazing in the field indicates that there were no live mines for Fred to step on. But, tell me, how do you take the square root of 13? That's crazy! Let me outta here! $\sqrt{13}$ is nuts.

That's a very good question.

Let's take it slow and easy. First of all, we can find out *approximately* how big $\sqrt{13}$ is by using a hand calculator. Punch in 13 and hit the $\sqrt{}$ button and you get that $\sqrt{13}$ is approximately equal to 3.6055511, but that's only approximately correct. If I multiply 3.6055511 by itself, I'll get something that's close to 13. For fun, let me do the work:

$$
\begin{array}{r}
3.6055511 \\
\times \quad 3.6055511 \\
\hline
36055511 \\
36055511 \\
18027755 \\
18027755 \\
18027755 \\
0 \\
21633306 \\
10816653 \\
\hline
13.0000340691121
\end{array}
$$

Oops, a tad too big.

Of course, we could try 3.6055510, but you know what would happen:

```
          3.605551
    ×     3.605551
    ·        3605551
         18027755
        18027755
       18027755
                0
      21633306
     10816653
    12.999998013601
```

which is a bit too small for $\sqrt{13}$. Would a bigger calculator help? No, not really. Like π, the number $\sqrt{13}$ has an infinitely long, non-repeating decimal expansion. And also like π, it doesn't equal any rational number like $\frac{1388}{385}$ (although $\frac{1388}{385}$ is actually pretty close to $\sqrt{13}$).

So how far did Fred walk? That's easy. He walked $\sqrt{13}$ hectometers. We know that $\sqrt{13}$ is approximately equal to 3.6 and so instead of walking 5 hectometers on the asphalt walkway, walking on the lawn saved him approximately 1.4 hectometers.

But I, your reader, want to know how much, exactly, did he save?

That's easy. He saved exactly $5 - \sqrt{13}$ hectometers.

But how much exactly is $\sqrt{13}$?

That is an even easier question. $\sqrt{13}$ is exactly $\sqrt{13}$. It's a number just like π is a number. Just like 7/9 is a number. Just like –3 is a number. Just like 5.4 is a number.

$\sqrt{13}$ might look strange to you, but that's because it's new to you. But it is an actual, honest-to-goodness real number just like any of the other numbers you know. And it is even easier to "point to" than the negative numbers. If you want to "see" $\sqrt{13}$, just look at the longest side of a right triangle whose shorter sides are 3 and 2.

Now all of these numbers we've named (like π, 7/9, –3, $\sqrt{13}$, and 5.4) and all the other ones like the integers and the rational numbers, are called the **real numbers**. If you drew a number line, you could locate any of these numbers on that line.

the real number line

Every point on the **real number line** corresponds to some real number. Before you started this book, you might have thought that every number on the number line was rational. (Rational number = any number that can be written as an integer divided by a natural number.)

In Chapter 2 you met π, which was the first real number that you ever encountered that was not rational. Real numbers that are not rational are called **irrational numbers**. Now, two pages ago you encountered $\sqrt{13}$. It's also irrational.

Would you like to meet some more irrational numbers? All these are irrational: $\sqrt{2}, \sqrt{3}, \sqrt{5}, \sqrt{6}, \sqrt{7}, \sqrt{8}, \sqrt{10}, \sqrt{11}, \sqrt{12}, \sqrt{13}$. There are tons of them. If you take the square root of a whole number that is not a **perfect square** (perfect squares are 0, 1, 4, 9, 16, 25, 36, 49, . . .) you'll get an irrational number as an answer.

Your Turn to Play

1. Name a perfect square that is greater than 100 and less than 125.

2. Which is true:

$$\frac{1388}{385} < \sqrt{13} \quad \text{OR} \quad \sqrt{13} < \frac{1388}{385}$$

3. If the big box is the whole numbers, what would go in the box with the question mark?

whole numbers

natural numbers	?

4. Solve $x^2 = 19$.

5. If the big box is the real numbers, what would go in the box with the question mark?

real numbers

rational numbers	?

6. A cubic yard is approximately 764.6 liters. To the nearest liter, how much is π cubic yards?

7. If Sergeant Snow drank 1.4 liters of ale each day, approximately how many years would it take him to drink 5 cubic yards of ale?

8. Express the ratio 5:16 as a fraction.

9. The cow in the minefield that Fred walked through had 28 pounds of muscle, 7 pounds of fat, and 5 pounds of bone. (It was a small cow.) Express this as a continued ratio.

10. $\dfrac{ac + 2ad + bc + 2bd}{5a + 30b} \div \dfrac{(a + b)^2}{a^2 + 7ab + 6b^2}$

11. Six of Jack's pushups plus 3 of Fred's burn 39 calories. Five of Jack's plus 6 of Fred's burn 43 calories. How many calories does each kind of pushup burn?

. . . . COMPLETE SOLUTIONS . . . .

1. 121 (= 11^2)

2. If we square $\dfrac{1388}{385}$ we get approximately 12.997.

   If we square $\sqrt{13}$ we get 13. Therefore, $\dfrac{1388}{385}$ is smaller.

3. The number zero. The whole numbers = {0, 1, 2, 3 . . .}.
The natural numbers = {1, 2, 3 . . .}.

4. $\pm\sqrt{19}$

5. The irrational numbers. Every real number is either rational or irrational.

6. We use the conversion factor $\dfrac{764.6 \text{ liters}}{1 \text{ cubic yard}}$

$$\dfrac{\pi \text{ cubic yards}}{1} \times \dfrac{764.6 \text{ liters}}{1 \text{ cubic yard}} \approx (3.14159)(764.6) \text{ liters} \doteq 2402 \text{ liters}$$

$\approx$ means *approximately equal to* and $\doteq$ means *equal to after rounding.*

7. $\dfrac{5 \text{ cubic yards}}{1} \times \dfrac{764.6 \text{ liters}}{1 \text{ cubic yard}} \times \dfrac{1 \text{ day}}{1.4 \text{ liters}} \times \dfrac{1 \text{ years}}{365 \text{ days}} \approx 7\frac{1}{2} \text{ years}$

We used the first conversion factor $\left(\dfrac{764.6 \text{ liters}}{1 \text{ cubic yard}} \right)$ to convert cubic yards into liters.

The second conversion factor converted liters into days.

The third, days into years.

We are only approximating. That is why we used 365 days instead of 365¼.

8. $\dfrac{5}{16}$

9. 28:7:5

10. $$\dfrac{ac + 2ad + bc + 2bd}{5a + 30b} \div \dfrac{(a + b)^2}{a^2 + 7ab + 6b^2}$$

$$= \dfrac{ac + 2ad + bc + 2bd}{5a + 30b} \times \dfrac{a^2 + 7ab + 6b^2}{(a + b)^2}$$

$$= \dfrac{(a + b)(c + 2d)(a + 6b)(a + b)}{5(a + 6b)(a + b)^2}$$

$$= \dfrac{c + 2d}{5}$$

11. *Let x = number of calories that one of Jack's pushups burns.*

Let y = number of calories that one of Fred's pushups burns.

$$\begin{cases} 6x + 3y = 39 \\ 5x + 6y = 43 \end{cases} \Rightarrow \begin{cases} -12x - 6y = -78 \\ 5x + 6y = 43 \end{cases} \Rightarrow -7x = -35$$

$x = 5$ Substitute back into the first equation $6(5) + 3y = 39 \Rightarrow y = 3$

One of Jack's pushups burns 5 calories. One of Fred's, 3 calories.

F red arrived at the men's bunkhouse. Pat, Chris, and Carol were busy repairing the hole in the side of the building. Fred headed inside and could hear Jack counting, "3427, 3428, 3429, 3430, . . ." as he was doing his pushups.

"Hop on my shoulders!" Jack called out. Fred climbed on the upper part of Jack's back and enjoyed the ride up and down as Jack continued, "3431, 3432. . . . Jack was so strong that the extra pounds didn't slow him down a bit.

Fred loved the ride. His father had never horsed around with him. No piggyback rides. No mock wrestling. On Jack's shoulders, Fred was filling in an essential boyhood experience that he had previously missed.

Fred also took over counting: 34^3_3, 3_43^4, 34^3^5, 34^3_6. . . .

Jack had to stop at 3500 pushups since it was getting close to 6 p.m. and that's when their get-off-the-base passes became effective. Fred hopped off and Jack got up. Jack headed into the bathroom to wash his face and comb his hair.

It's almost time for *Your Turn to Play*, but before you start there's two little items you should note:

♪#1: $6\sqrt{x} + 5\sqrt{x} = 11\sqrt{x}$. You could have guessed that. It's just like adding like terms: $6x^2y + 5x^2y = 11x^2y$. Recall, if the terms aren't alike you can't add them. You can't add $4x^2y$ and $7xy$. The same is true with radicals. You can't add $4\sqrt{x}$ and $7\sqrt{w}$.

♪#2: $\sqrt{a}\sqrt{b} = \sqrt{ab}$. So, for example, $\sqrt{8}\sqrt{11} = \sqrt{88}$. It's that simple.

Your Turn to Play

1. $\sqrt{3}\ \sqrt{xyz}$

2. $\sqrt{3} + \sqrt{xyz}$

3. Solve $x^2 = 7$

4. What is the length of the **hypotenuse** (the longest side) of a right triangle whose shorter sides (the **legs**) are each equal to one?

5. $\sqrt{6xz} + 30\sqrt{6xz} + 5\sqrt{6xy}$

6. Is it true that $\sqrt{18} = \sqrt{9}\ \sqrt{2}$?

7. Simplify $\sqrt{28}$ (Hint: Look at the answer to question 6.)

8. For English majors and other non-math majors: State in a complete English sentence the procedure for simplifying square roots. Speling and puktuation will count.

9. Simplify $\sqrt{27x^3y^2}$

10. Solve $x^2 = -4$

. . . . **C O M P L E T E   S O L U T I O N S** . . . .

1. 1. $\sqrt{3}\ \sqrt{xyz} = \sqrt{3xyz}$

2. $\sqrt{3} + \sqrt{xyz}$ can't be added. Unless the stuff under the radical sign (which is called the **radicand**) is the same for each term, you can't combine them.

3. $x^2 = 7$. Take the square root of both sides and remember to put the $\pm$ in front of the number and you get $x = \pm\sqrt{7}$.

4. By the Pythagorean theorem $(a^2 + b^2 = c^2)$ we have $1^2 + 1^2 = c^2$. So

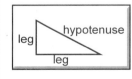

$2 = c^2$ and solving this pure quadratic by taking the square root of both sides and remembering to put the $\pm$ in front of the number, $\pm\sqrt{2} = c$. Since we're looking for the length of the hypotenuse, we couldn't have an answer of $-\sqrt{2}$ and we should eliminate that answer.

The final answer is $c = \sqrt{2}$.

5. The first two radicands are identical so we can combine the first two terms . . .

$\sqrt{6xz} + 30\sqrt{6xz} + 5\sqrt{6xy} = 31\sqrt{6xz} + 5\sqrt{6xy}$

That's all that can be done. As you might have noticed, working with square roots has much less computation than adding rational expressions as we did in the previous chapter. That's one reason why many students wave the happy square root flag.

6. $\sqrt{18} = \sqrt{9}\ \sqrt{2}$ is true by ♪#2: $\sqrt{a}\ \sqrt{b} = \sqrt{ab}$.

Wait a minute. If $\sqrt{18} = \sqrt{9}\ \sqrt{2}$, that would mean that $\sqrt{18}$ is equal to $3\sqrt{2}$. That gives us a way of simplifying square roots. In dealing with square roots, we want the radicand to be as small as possible.

Some examples:
$$\sqrt{20} = \sqrt{4}\ \sqrt{5} = 2\sqrt{5}$$
$$\sqrt{200} = \sqrt{100}\ \sqrt{2} = 10\sqrt{2}$$

7. $\sqrt{28} = \sqrt{4}\ \sqrt{7} = 2\sqrt{7}$

8. *To simplify a square root, express, if possible, the original square root as the product of two square roots where the radicand of the first factor is a perfect square, and then make the natural simplification of the first factor.*

See how incredibly dense it becomes in English? Something like $\sqrt{a^2b} = \sqrt{a^2}\ \sqrt{b} = a\sqrt{b}$ is much more direct.

9. $\sqrt{27x^3y^2} = \sqrt{9x^2y^2}\ \sqrt{3x} = 3xy\sqrt{3x}$

10. $x^2 = -4$. This is a pure quadratic. If we blindly followed the procedure for solving pure quadratics, we arrive at $x = \pm\sqrt{-4}$. We've never had a negative number as a radicand before. This is new. It also feels like nonsense. The original problem asks for a number which when squared gives an answer of -4.

✔ If x^2 were equal to -4, could x be a *positive number*? No, of course not. Any positive number times itself gives a positive answer.

✔ If x^2 were equal to -4, could x be a *negative number*? No, of course not. Any negative number times itself gives a positive answer.

✔ If x^2 were equal to –4, could x be equal to *zero*? No, of course not. Zero times itself is equal to zero.

So the answer to $x^2 = -4$ can't be positive, it can't be negative, and it can't be equal to zero. It can't be any number on the real number line. Beginning algebra students are taught that $x^2 = -4$ has no solution.

One of the real joys of mathematics is how it keeps breaking out of any cage that you try to put it in. It's like our understanding of *Love* or *Justice* or *God.* Our conceptions of those important words are different at the age of 18 than at the age of seven. My English teacher in high school told us that we would have to wait till we were 35 years old before we could really read *Moby Dick.* Until then, it's just a great fishing story with a whale and a crazy captain.

So does that mean that $x^2 = -4$ has a solution? you ask. The answer is yes. In fact, it has two solutions. And neither solution is equal to zero. Neither solution is to the left of zero on the real number line, and neither solution is to the right of zero on the real number line.

In *Life of Fred: Advanced Algebra Expanded Edition,* which is your next algebra course, we name those solutions. We study them further in trig and use them in parts of calculus.

Well where are they!? I, your reader, can't wait till advanced algebra. Tell me.

Okay. One solution to $x^2 = -4$ is *above* the real number line and the other solution is *below* the real number line.

This is nuts! Next, I guess, you're going to tell me that there are numbers to the right of the arrow on the real number line.

$$\longleftarrow \underset{-3\ -2\ -1\ 0\ 1\ 2\ 3\ 4\ ...}{} \longrightarrow \quad \text{nos. out here}$$

How did you guess? We'll name numbers in the first chapter of *Life of Fred: Calculus Expanded Edition* that are bigger than any real number you could ever name—infinite numbers.

Why do some men and women become mathematicians? It's because of adventures like this. It's not because they enjoy balancing their checkbooks.

ack came out of the bathroom with his face washed and his hair combed. He could make the straightest parts in his hair that Fred had ever seen. It reminded Fred of the real number line.

It was 5:55 p.m. As they headed out of the bunkhouse, they passed a television that had been left on. Across the 25" (diagonal measure) screen was splashed the closing credits of H.A.S.H. (Hospitals Aren't So Hot, which dealt with medical services during the Crimean War). Fred turned it off in order to lower the noise level. He mumbled "15" x 20"."

That mystified Jack. He couldn't figure out how Fred got from a 25" diagonal measurement to the dimensions of the screen.

Fred explained. There was one other fact that Fred knew about television screens, namely, that the "aspect ratio" of many screens is 3:4.

On the blackboard in Fred's mind he had drawn:

Using the Pythagorean theorem

$$(3x)^2 + (4x)^2 = 25^2$$
$$9x^2 + 16x^2 = 625$$
$$25x^2 = 625$$
$$x^2 = 25$$
$$x = \pm 5$$

He threw out the −5 answer, since it doesn't make sense to have a negative distance. So, since the dimensions of the screen are 3x × 4x, the screen is 15" × 20".

"I've got something new for you Fred," Jack began. "You remember when I found a meaning for x^{-3} when I showed that

$$\frac{x^4}{x^7} = \frac{xxxx}{xxxxxxx} = \frac{\text{xxxx}}{\text{xxxxxxx}} = \frac{1}{x^3} \quad \text{and that}$$

$\frac{x^4}{x^7}$ also had to equal x^{-3} by $\frac{x^m}{x^n} = x^{m-n}$. So $x^{-3} = \frac{1}{x^3}$ "

Fred nodded.

.   "Well," Jack continued, "I've got something now that'll knock your ears off." (Jack had never noticed that Fred wasn't real big in the "ear department.")

"Try this," Jack said as he drew in the wet paint that Carol Thumbs had accidently spilt on the floor. "What's $\sqrt{a}$ really mean?" He paused and waited for Fred to answer.

Fred dutifully responded, "$\sqrt{a}$ means the number that when multiplied by itself equals a."

Jack continued writing in the wet paint:

$$\sqrt{a}\ \sqrt{a} = a$$

"And guess what $a^{\frac{1}{2}}$ is. By the exponent law $a^m\, a^n = a^{m+n}$ we have:

$$a^{\frac{1}{2}}a^{\frac{1}{2}} = a$$

"So," Jack concluded, "we could define $a^{\frac{1}{2}}$ to mean $\sqrt{a}$."

"I like that," Fred smiled. In the five years that Fred had taught at KITTENS,[*] there had been several times that students "caught fire" and began creating math rather than just passively receiving it.

Jack's mind was steaming with ideas. He reached for the bunkhouse door knob and tore it off by accident. This would happen occasionally when he wasn't paying attention. It was one of the very few disadvantages of being healthy and strong.

As they stepped outside, Jack drew in the air with his finger:

$$b^{\frac{1}{3}}$$

and exclaimed, "And $b^{\frac{1}{3}}$ has gotta be the number so that $b^{\frac{1}{3}}b^{\frac{1}{3}}b^{\frac{1}{3}} = b$. We'll call it the **cube root** of b, and he wrote in the air:

$$\sqrt[3]{b}$$

(The little "3" sitting in the neck of the radical sign is called the **index**.)

Fred walked next to Jack as he led the way toward the Big Tree, going along the asphalt walkway instead of across the field. Jack knew the difference between a coal mine and a claymore mine.

We'll put Jack's finger waving on a handsome scroll since it's too

---

★ The story of how Fred began teaching at KITTENS University at a "fairly young" age is chronicled in *Life of Fred: Calculus Expanded Edition.*

hard to read what he was writing in the air.

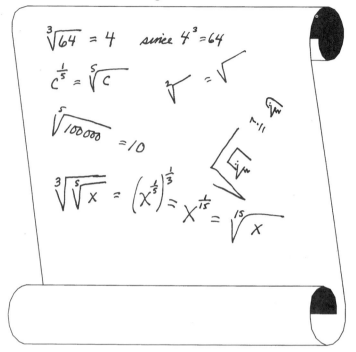

$$\sqrt[3]{64} = 4 \quad \text{since } 4^3 = 64$$

$$c^{\frac{1}{5}} = \sqrt[5]{c}$$

$$\sqrt[5]{100000} = 10$$

$$\sqrt[3]{\sqrt[5]{x}} = \left(x^{\frac{1}{5}}\right)^{\frac{1}{3}} = x^{\frac{1}{15}} = \sqrt[15]{x}$$

If you have a scientific calculator that has buttons with sin, cos, ln, and log on it, look to see if it has a y^x key. If so, you can find the cube root of 64:

type in   б4
hit the y^x key
type in   .ЗЗЗЗЗЗЗЗЗ   (which is the best we can do for 1/3)
hit   =   and out should come something real close to 4.

If you don't have such a calculator, it's no big loss. You already knew that $\sqrt[3]{64}$ is equal to 4.

Your Turn to Play

1. $144^{\frac{1}{2}}$
2. What is the index of $\sqrt[5]{3.14159}$?
3. $\sqrt[6]{y} \ \sqrt[7]{y}$

4. $\sqrt[6]{\sqrt[7]{y}}$

5. $16^{\frac{1}{4}}$

6. If the aspect ratio of your television screen is 3:4 and the diagonal measurement of the screen is 21 inches, what are the dimensions of the screen?

7. What is the cube root of 125?

8. On your calculator, enter some very large number. Then divide it by 6. Then divide it by 6 again. Each time the number on the display gets smaller. If you were to do that a zillion times, what would your answer approach?

9. On your calculator, enter some positive number that is very close to zero. Then hit the $\sqrt{}$ key. Then hit the $\sqrt{}$ key again. Each time the number on the display gets larger.

   If you were to hit the $\sqrt{}$ key a zillion times, what number would your answer approach?

. . . . COMPLETE SOLUTIONS . . . .

1. $144^{\frac{1}{2}} = \sqrt{144} = 12$

   $(x^a)^b = x^{ab}$ see page 269

2. 5

3. $\sqrt[6]{y} \, \sqrt[7]{y} = y^{1/6} \, y^{1/7} = y^{1/6 + 1/7} = y^{13/42} = (y^{13})^{1/42} = \sqrt[42]{y^{13}}$

4. $\sqrt[6]{\sqrt[7]{y}} = (y^{1/7})^{1/6} = y^{1/42} = \sqrt[42]{y}$

   $\sqrt[b]{x^a} = x^{a/b}$

5. $16^{\frac{1}{4}} = \sqrt[4]{16} = 2$ This is true since $2^4 = 16$.

6.
$$(3x)^2 + (4x)^2 = 21^2$$
$$9x^2 + 16x^2 = 441$$
$$25x^2 = 441$$
$$x^2 = \frac{441}{25}$$
$$x = \frac{21}{5} = 4.2$$

   $(3x) = 12.6$ and $(4x) = 16.8$

   12.6" × 16.8"

7. Since $5^3 = 125$, the cube root of 125 is 5.

8. We will start with some big number such as 986298440124.

    Dividing 986298440124 by 6 *n* times is $\dfrac{986298440124}{6^n}$

    The limit as n approaches infinity of $\dfrac{986298440124}{6^n}$ is zero.

    In calculus we will write this as $\lim_{n \to \infty} \dfrac{986298440124}{6^n} = 0.$

    The idea of limit is the only really new idea in all of calculus. Everything else is based on that idea.

    The only really new idea in algebra is x—having a letter stand in place of an unknown number. The rest can be considered arithmetic.

9. If you started with 0.00000039, for example, it would get larger every time you hit the $\sqrt{}$ key.

    The millionth root of 0.00000039 would be very close to 1.

    In calculus, we will be able to write $\lim_{n \to \infty} \sqrt[n]{x} = 1.$ And this will be true for any positive number x.

    Since you are not yet in calculus, we won't mention that in this book.[*]

---

[*] Do you remember what irony means?

T hey arrived at the big tree where Fred had taken his Sunday afternoon nap. A big blackish-brown circle was next to the tree. It had a radius of about five feet, and all the vegetation in that circle had died.

Jack started to panic. "Toxic waste!" he thought to himself. "Or maybe chemical warfare! This looks serious." He knew from his *Induction Handbook* that any suspicious area, especially if defoliation has occurred that is more than 70 square feet must be reported to the authorities.

Fred said that he didn't have to worry. No defoliation had happened since defoliation means loss of leaves.

"But the grass here is burned to smithereens!" Jack exclaimed.

"No, that's not true," Fred answered. "*Smithereens* means in small pieces or bits. Since grass comes in small pieces or bits, that's not the correct word. But it is clear that the grass is severely burned within this circle."

Fred didn't want to talk about that circle of destruction. He wanted to keep moving.

Jack insisted. He asked Fred, "What's the radius of a circle in terms of its area?"

Fred replied, "$\sqrt{\dfrac{A}{\pi}}$ = r, but really Jack, we don't need to worry about this at all." Fred was getting embarrassed for some reason.

"It's okay little guy," Jack said as he took control of the situation. "I'll get it figured out."

Jack knew that the radius of the circle was equal to 5 feet so he had the **radical equation** $\sqrt{\dfrac{A}{\pi}}$ = 5, which he wanted to solve for A.

(A radical equation is one in which the unknown is inside a radical sign. In advanced algebra, we'll solve exponential equations where the unknown is in the exponent, such as $2^x = 6$, but right now we've got our hands full with Jack solving this radical equation.)

He squared both sides and got $\frac{A}{\pi} = 25$. (One step and he was done with the square root sign. It kind of reminds you of how we got rid of all the denominators in a fractional equation by multiplying through by the least common denominator—a.k.a. having Santa Claus visit each term and leave a present.)

He multiplied both sides by π: $A = 25\pi$ square feet.

The critical thing according to the *Induction Handbook* was whether this denuded area was greater than 70 square feet. Since π is a little larger than 3, it was obvious that 25π would be larger than 75 square feet.

Jack whipped out his cell phone and before Fred could stop him, he notified the CBN authorities. (CBN = chemical, biological, nuclear.)

Fred pleaded with Jack, "It's nothing serious. Call them back and tell them it's okay. It really is."

But it was too late for that. The CBN squad arrived in a giant truck with sirens screaming and lights flashing.

"I'm here to the rescue!" the leader of the crew of 18 declared. "You, you, and you secure the perimeter. The rest of you divide into groups of five and discuss the issues."

Special group discussion coordinators (with special badges declaring them as "facilitators") gathered their groups. Fred could hear comments like, "Do you think it's nuclear? It looks like it might glow if it were dark enough," and "It's definitely DNA," and "Do you think we should call the president?"

This was getting to be too much for Fred. He summoned up the courage to overcome his embarrassment, walked up to the leader, and tugged on his cape. He began, "Excuse me, sir. I know what caused that blackish-brown circle."

"That's toxic waste, son," the leader warned Fred. "Unless you're an expert, I'd advise you to stand back while our men take care of it. One of them is going in now to get a sample."

But Fred really did know what had caused that blackish-brown circle. It had happened when he got sick to his stomach after Pat, Chris, and Carol had forced all that cola into him and then had shaken him. In kindergarten he had seen kids throw up all the time, but there was never such a fuss made over it.

Maybe it's time for *Your Turn to Play* while the toxic waste crew discusses whether or not to evacuate all of Texas.

Your Turn to Play

1. $49^{1/2}$

2. $27^{1/3}$

3. Solve $\sqrt{x+7} = 9$

4. If you were trying to solve the radical equation $\sqrt{w} + 5 = 20$ and you simply squared both sides, would you get rid of the square root?

5. Using what you learned in the answer to question 4, solve
$$\sqrt{w} + 5 = 20.$$

6. Solve $\sqrt{2y-3} + 3 = y$

7. In the olden days when they did long division using pencil and paper, expressions like $\dfrac{3}{\sqrt{2}}$ were agony,

since they would have a big divisor: $1.4142135\overline{)3.0000000000}$

To avoid such difficulties, fractions were considered to be in simplest form if there were no radicals in the denominator. To eliminate radicals from the denominator is called **rationalizing the denominator**. If we multiply the numerator and denominator of $\frac{3}{\sqrt{2}}$ by $\sqrt{2}$ it will eliminate the square root from the denominator. Do it.

8. Here's a significantly harder problem. Suppose you have $\dfrac{5}{3 - \sqrt{11}}$ and you want to rationalize the denominator.

Again, you'll multiply the numerator and denominator by the same expression (which keeps the value of the fraction the same), but what do you use? $\sqrt{11}$ won't work.

If you tried that you'd get $\dfrac{5\sqrt{11}}{3\sqrt{11} - 11}$ and that still has a square root in the denominator.

Play with it. See if *you* can figure it out rather than being told. Spend five minutes. It'll make your brain stronger.

. . . . COMPLETE SOLUTIONS . . . .

1. $49^{1/2} = \sqrt{49} = 7$ (Not -7)

2. $27^{1/3} =$ the cube root of 27, which is 3.

3. To solve a radical equation, we square both sides. $\sqrt{x + 7} = 9$ becomes $x + 7 = 81$. Subtracting 7 from each side we have $x = 74$. To check the answer we substitute $x = 74$ in the original equation.
$$\sqrt{74 + 7} \stackrel{?}{=} 9$$
$$\sqrt{81} \stackrel{?}{=} 9 \qquad \text{Yes. The answer checks.}$$

4. Nope. If you square both sides of $\sqrt{w} + 5 = 20$ you get
$$(\sqrt{w} + 5)^2 = 20^2$$
$$(\sqrt{w} + 5)(\sqrt{w} + 5) = 400$$

and multiplying out these two binomials (remember the boys and girls with every possible meeting)
$$(\sqrt{w})^2 + 5\sqrt{w} + 5\sqrt{w} + 25 = 400$$

and since, by definition of $\sqrt{w}$, we know that $\sqrt{w}\,\sqrt{w} = w$,
$$w + 10\sqrt{w} + 25 = 400$$

and we haven't gotten rid of the square root. What's different? Why didn't it work this time? The answer is that in the previous examples, we had the square root *all alone* on one side of the equation. In this case we had $\sqrt{w} + 5$ and that messed things up.

So the new, expanded rule for solving radical equations has two steps instead of just one:

> To solve radical equations:
>   1. Isolate the square root on one side of the equation.
>   2. Square both sides.

5. We start with

$$\sqrt{w} + 5 = 20$$

Transpose the 5 to isolate the $\sqrt{w}$

$$\sqrt{w} = 15$$

Square both sides

$$w = 225$$

To check our answer we substitute w = 225 into the original equation

$$\sqrt{225} + 5 \overset{?}{=} 20$$

$$15 + 5 \overset{?}{=} 20 \quad \text{Yes. The answer checks.}$$

One last word on radical equations. Whenever you square both sides of an equation, it is the same as multiplying by an expression containing a variable. Here was the box from Lesson 70:

> Whenever you multiply an equation by an expression containing x, you must check any answers you get back in the original equation.

We need to get rid of any **extraneous** answers. These are "extra" answers that are accidentally introduced which don't work in the original equation.

To get rid of the extraneous answers, you have to check each of the answers you get back in the original equation. *This checking is not optional.*

To tell the truth, solving radical equations is four steps long. So the new, expanded (and now final) rule for solving radical equations is:

> To solve radical equations:
>   1. Isolate the square root on one side of the equation.
>   2. Square both sides.
>   3. Solve for x.
>   4. Check each answer in the original equation.

6. $\sqrt{2y-3} + 3 = y$

Isolate the square root $\sqrt{2y-3} = y - 3$

Square both sides $2y - 3 = (y-3)(y-3)$

                                                                      $2y - 3 = y^2 - 3y - 3y + 9$

Combine like terms $2y - 3 = y^2 - 6y + 9$

To solve a quadratic equation by factoring we put everything

on one side of the equation $0 = y^2 - 8y + 12$

Factor $0 = (y-6)(y-2)$

$$y - 6 = 0 \quad \text{OR} \quad y - 2 = 0$$

$$y = 6 \quad \text{OR} \quad y = 2$$

Now we must check each answer since we may have
introduced extraneous answers when we squared both
sides of the equation.

We start with y = 6 $\sqrt{2 \cdot 6 - 3} + 3 \overset{?}{=} 6$

$\sqrt{12 - 3} + 3 \overset{?}{=} 6$

$3 + 3 \overset{?}{=} 6$ Yes. y = 6 checks.

Now y = 2 $\sqrt{2 \cdot 2 - 3} + 3 \overset{?}{=} 2$

$\sqrt{1} + 3 \overset{?}{=} 2$ No. The y = 2 answer is discarded.

7. $\dfrac{3}{\sqrt{2}} \cdot \dfrac{\sqrt{2}}{\sqrt{2}} = \dfrac{3\sqrt{2}}{2}$ done!

8. If you tried multiplying $\dfrac{5}{3 - \sqrt{11}}$ by $\dfrac{3 - \sqrt{11}}{3 - \sqrt{11}}$ (which is a logical
choice) it wouldn't work.
The denominator will be $9 - 6\sqrt{11} + 11$.

   Instead, the trick is to multiply by **the conjugate** of $3 - \sqrt{11}$,
which is $3 + \sqrt{11}$.   (a + $\sqrt{b}$ and a − $\sqrt{b}$ are conjugates.)

$$\frac{5}{3 - \sqrt{11}} \cdot \frac{3 + \sqrt{11}}{3 + \sqrt{11}} = \frac{15 + 5\sqrt{11}}{9 - 11} \text{ which is } \frac{15 + 5\sqrt{11}}{-2}$$

A small note to the reader. Each of these cities contains only one and a half dozen problems.

Eighteen problems and you are done with your day's work. That is not a lot of problems to ask of you. Many other textbooks contain almost twice that number of problems.

Please do not skip any of these three cities. By the third day, the problems should look increasingly easy. That is our goal.

Scottsbluff, Nebraska located on the "Fossil Freeway"

1. Solve $2z^2 = 98$

2. Name a natural number that is greater than 21 that is not a perfect square.

3. Show (prove, demonstrate) that the square of a rational number must be rational. "Demonstrate" is different than "illustrate." Illustrating that the square of a rational number is rational can be done by noting that $(3/4)^2$, which is 9/16, is rational.

Proving something is true is a different kind of math than most of what you have seen in arithmetic and beginning algebra. This question will not be easy since it will involve English. You will see a lot of proofs when you get to geometry. The heart of geometry is doing proofs. This problem is a little preview of coming attractions.

4. The area, A, of any circle with radius r is given by $A = \pi r^2$. If you have a pizza with an area of 80 square inches, what is its radius?

5. $(4 + 7\sqrt{w})^2$

6. Solve $\sqrt{x + 7} = 2x - 1$

7. $1000^{1/3}$

8. Simplify $\dfrac{5x + 1}{6\sqrt{x}}$

9. Simplify $\dfrac{1}{\sqrt{7wxyz}}$

10. Using set-builder notation list the set of all natural numbers that are perfect squares.

11. Solve $x^2 - 17 = 0$

12. Suppose you had the pure quadratic $x^2 = c$ (where c is some number) and you were told that it has no solutions. What can you say about c?

13. Simplify $\sqrt{50a^4 b^3}$

14. Solve $\sqrt{x} + 2 = x$

15. $81^{1/4}$

16. Simplify $\dfrac{\sqrt{y} + \sqrt{z}}{\sqrt{yz}}$

17. If you have a TV screen that has a 30" diagonal measurement, what are the dimensions of the screen? Assume the aspect ratio of your television screen is 3:4.

18. $(7 + \sqrt{2})(8 - \sqrt{6})$

answers

1. $2z^2 = 98$
     $z^2 = 49$
      $z = \pm 7$

2. You might have written 22 or 23 or 24 or 26 or. . . but not numbers like 25 or 36 or 49.

3. Suppose we have a rational number. Call it r.

   ✻ By definition, a rational number r must be equal to a/b where a is an integer and b is a natural number.

   ✻ Then r^2 must equal a^2/b^2.

   ✻ If a is an integer then a^2 is an integer. If b is a natural number then b^2 is a natural number.

   ✻ Therefore r^2 can be written as a fraction where the top is an integer and the bottom is a natural number and hence r^2 is rational.

4. Put $A = 80$ into $A = \pi r^2$ $80 = \pi r^2$

$$\frac{80}{\pi} = r^2$$

No minus sign since we are $\sqrt{\dfrac{80}{\pi}} = r$
looking for a length

5. $(4 + 7\sqrt{w})^2 = (4 + 7\sqrt{w})(4 + 7\sqrt{w})$
$$= 16 + 28\sqrt{w} + 28\sqrt{w} + 49w = 16 + 56\sqrt{w} + 49w$$

6. $\sqrt{x + 7} = 2x - 1$

Square both sides $x + 7 = 4x^2 - 4x + 1$

$$0 = 4x^2 - 5x - 6$$
$$0 = (4x + 3)(x - 2)$$
$$4x + 3 = 0 \quad \text{OR} \quad x - 2 = 0$$
$$x = -\frac{3}{4} \quad \text{OR} \quad x = 2$$

Checking the answers is mandatory after we square both sides of an equation.

Checking $x = -\frac{3}{4}$

in the original equation $\sqrt{-0.75 + 7} \overset{?}{=} 2(-0.75) - 1$
$$\sqrt{6.25} \overset{?}{=} -2.5$$

Square roots can't equal negative numbers. (Principal square roots were defined in Lesson 73.) $x = -\frac{3}{4}$ is an extraneous root.

Checking $x = 2$ $\sqrt{2 + 7} \overset{?}{=} 2(2) - 1$
$$\sqrt{9} \overset{?}{=} 3 \quad \text{yes}$$

The final answer is $x = 2$.

7. That's the cube root of 1000, which can be written as $\sqrt[3]{1000} = 10$

8. $\dfrac{5x + 1}{6\sqrt{x}} = \dfrac{5x + 1}{6\sqrt{x}} \cdot \dfrac{\sqrt{x}}{\sqrt{x}} = \dfrac{(5x + 1)\sqrt{x}}{6x}$ OR $\dfrac{5x\sqrt{x} + \sqrt{x}}{6x}$

9. $\dfrac{1}{\sqrt{7wxyz}} = \dfrac{1}{\sqrt{7wxyz}} \cdot \dfrac{\sqrt{7wxyz}}{\sqrt{7wxyz}} = \dfrac{\sqrt{7wxyz}}{7wxyz}$

10. $\{x \mid \sqrt{x} \text{ is a natural number}\}$ or
        $\{x \mid x = n^2 \text{ where } n \text{ is a natural number}\}$ or
            $\{n^2 \mid n \text{ is a natural number}\}$

   There may be others ways to list the set of all natural numbers that are perfect squares. These are the three ways I could think of.

11. $x^2 - 17 = 0$
$$x^2 = 17$$
$$x = \pm\sqrt{17}$$

Some problems in life are really easy to solve.

12. c must be negative.

13. $\sqrt{50a^4b^3} = \sqrt{25a^4b^2}\sqrt{2b} = 5a^2b\sqrt{2b}$

14.
$$\sqrt{x} + 2 = x$$

Isolate the square root
$$\sqrt{x} = x - 2$$

Square both sides
$$x = x^2 - 4x + 4$$
$$0 = x^2 - 5x + 4$$
$$0 = (x - 4)(x - 1)$$
$$x - 4 = 0 \quad OR \quad x - 1 = 0$$
$$x = 4, 1$$

We must check these roots in the original equation since we squared both sides of the equation.

Checking x = 4 $\quad\quad \sqrt{4} + 2 \overset{?}{=} 4 \quad\quad$ yes

Checking x = 1 $\quad\quad \sqrt{1} + 2 \overset{?}{=} 1 \quad\quad$ no

The final answer is x = 4.

15. This is the fourth root of 81, which can be written as $\sqrt[4]{81} = 3$.

16. $\dfrac{\sqrt{y} + \sqrt{z}}{\sqrt{yz}} = \dfrac{(\sqrt{y} + \sqrt{z})\sqrt{yz}}{\sqrt{yz}\,\sqrt{yz}} = \dfrac{(\sqrt{y} + \sqrt{z})\sqrt{yz}}{yz}$.

17. *Let 3x = the height of the screen.*
    Then 4x = the width of the screen.

By the Pythagorean theorem $\quad (3x)^2 + (4x)^2 = 30^2$
$$9x^2 + 16x^2 = 900$$
$$25x^2 = 900$$
$$x^2 = 36$$

Not −6 since we are working with length $\quad\quad x = 6$

The dimensions (*3x* × *4x*) are 18" × 24".

18. $(7 + \sqrt{2})(8 - \sqrt{6})$
$$= 56 - 7\sqrt{6} + 8\sqrt{2} - \sqrt{12}$$
$$= 56 - 7\sqrt{6} + 8\sqrt{2} - 2\sqrt{3} \quad\quad \text{since } \sqrt{12} = \sqrt{4}\sqrt{3} = 2\sqrt{3}$$

Chamberlain, South Dakota home of the South Dakota Hall of Fame

1. Simplify $\sqrt{20x^3}$

Questions 2, 3, and 4 involve thinking. Heavens! It may take you a whole minute to play with each of these before you figure out the answer. When you were learning arithmetic, you would often answer the questions as fast as you could write.

$$3 + 5 = ?$$
$$6 + 8 = ?$$
$$9 + 7 = ?$$

When you decide to major in mathematics in college, you may have to play with some of the questions for ten minutes before you find the solution.

Thought is more difficult than physical work. Your 3-pound brain is only 2% of your body weight. $\frac{3}{150} = \frac{1}{50} = \frac{2}{100} = 2\%$ but it uses 20% of the oxygen. $20\% = \frac{20}{100} = \frac{1}{5}$

Thinking is much less popular than mindless work. Look at the people in the world around you. Most of what they do is the same stuff they did yesterday and the day before and the day before.

Look at the jobs in the world around you. How much brain power is required to say, "Would you like fries with your order?" or "For here or to go?"

There are jobs out there in the world that require people who have exercised their brains a lot. It's often difficult to find those people because there are so few of them. If you are one of the few who have worked hard developing your mental strength, your financial future is much brighter. Much brighter.

If it's a job that takes only a week to learn, it probably won't pay $50/hour.

2. Suppose you had the pure quadratic $x^2 = c$ (where c is some number) and you were told that there is exactly one solution to this equation. What can you say about the number c?

3. The symbol $\geq$ means "greater than or equal to." So, for example, $6 \geq 4$, $3 \geq 0$, $5 \geq 5$, $-8 \geq -10$.

What's the simplest name for the set $\{x \mid x^2 \geq 0$ where x is a real number$\}$?

4. What is the simplest name for the set $\{x \mid \sqrt{x^2} \geq 0$ where x is a real number$\}$?

5. $\sqrt{y^3}$ can be written as $(y^3)^{1/2}$ or—since $(x^a)^b = x^{ab}$—as $y^{3/2}$. Given all this, what does $8^{5/3}$ equal?

6. The perimeter of a triangle is the "distance around" the outside. If you have a triangle whose sides are a, b, and c, then the perimeter is a + b + c.

The triangle in the diagram is your piece of property, which is bordered by three square pieces of property. The black square has an area of 200 square miles. The gray square has an area of 50 square miles.
The white square has an area of 72 square miles.
What is the perimeter of your property?

7. Solve $4x = \sqrt{2x + 6} + 3$

8. Simplify $\sqrt{63} + \sqrt{7} + \sqrt{64}$

9. Simplify $\dfrac{6}{\sqrt{a} + \sqrt{b}}$

10. In a right triangle where the hypotenuse has a length of 13 feet and one of the legs has a length of 12 feet, how long is the other leg?

11. Can a real number ever be a perfect square?

12. $18\sqrt{3xy} - \sqrt{3xy} + 3\sqrt{xy}$

13. If you know that $\sqrt{v} = w$,
    which of the following is true: $w = v^2$ or $v = w^2$?

14. $\sqrt{14xy}\,\sqrt{2}$

15. If you have a calculator and you punch in some big number and then keep hitting the square root key many times, what number is the display's answer tending toward?

16. There are some numbers that get larger when you take their square roots. In symbols: $\sqrt{x} > x$. Can you think of one of these numbers? They are not any of the natural numbers.

You will *lose the benefit* of this problem by just looking at the answer. The purpose of this problem is not to "get the answer" but to strengthen your mind.

17. Simplify $\dfrac{x}{\sqrt{x} - \sqrt{y}}$

18. Solve $4 = \sqrt{9 - x}$

answers

1. $\sqrt{20x^3} = \sqrt{4x^2}\sqrt{5x} = 2x\sqrt{5x}$

2. c must be equal to zero.

        If c were negative, then $x^2 = c$ would have no solutions.

        If c were positive, then $x^2 = c$ would have two solutions.

3. Since the square of every real number is greater than or equal to zero, this set is simply the set of all real numbers. $x^2 \geq 0$ is always true.

4. This is also the set of real numbers. If you take any number and square it and then square root it, the answer will always be non-negative.

   For example: $\sqrt{77^2} = 77$ which is ≥ 0

$$\sqrt{(-3.6)^2} = 3.6 \text{ which is} \geq 0$$
$$\sqrt{0^2} = 0 \text{ which is} \geq 0$$

5. $8^{5/3} = (8^{1/3})^5 = (\sqrt[3]{8})^5 = 2^5 = 32$

If you went the other direction the arithmetic would be harder . . .

$$8^{5/3} = (8^5)^{1/3} = \sqrt[3]{8^5} = \sqrt[3]{32768} = 32$$

6. If the black square has an area of 200 square miles, then each side will have a length of $\sqrt{200}$ miles. $\sqrt{200} = \sqrt{100}\sqrt{2} = 10\sqrt{2}$

     Similarly, the length of a side of the gray square is $\sqrt{50}$ miles. $\sqrt{50} = \sqrt{25}\sqrt{2} = 5\sqrt{2}$

     The length of a side of the white square $= \sqrt{72} = \sqrt{36}\sqrt{2} = 6\sqrt{2}$

Add to get the perimeter. $10\sqrt{2} + 5\sqrt{2} + 6\sqrt{2} = 21\sqrt{2}$ miles

7.

$$4x = \sqrt{2x+6} + 3$$

Isolate the square root on one side

$$4x - 3 = \sqrt{2x+6}$$

Square both sides

(This may introduce extraneous roots.)

$$16x^2 - 24x + 9 = 2x + 6$$

Solve by factoring

$$16x^2 - 26x + 3 = 0$$

A harder trinomial

Two numbers that multiply to $48x^2$ and

  add to $-26x$

$$16x^2 - 24x - 2x + 3 = 0$$
$$8x(2x-3) - (2x-3) = 0$$
$$(2x-3)(8x-1) = 0$$
$$2x - 3 = 0 \quad \text{OR} \quad 8x - 1 = 0$$
$$x = \frac{3}{2} \quad \text{OR} \quad x = \frac{1}{8}$$

Checking $x = \dfrac{3}{2}$ $4(\dfrac{3}{2}) \overset{?}{=} \sqrt{3+6} + 3$

$6 \overset{?}{=} 3 + 3$ yes

Checking $x = \dfrac{1}{8}$ $4(\dfrac{1}{8}) \overset{?}{=} \sqrt{\frac{1}{4}+6} + 3$

$1/2 \overset{?}{=} \sqrt{6.25} + 3$ no $x = \dfrac{1}{8}$ is extraneous

The final answer is $x = \dfrac{3}{2}$

8. $\sqrt{63} + \sqrt{7} + \sqrt{64} = \sqrt{9}\sqrt{7} + \sqrt{7} + 8 = 3\sqrt{7} + \sqrt{7} + 8 = 4\sqrt{7} + 8$

9. $\dfrac{6}{\sqrt{a}+\sqrt{b}} = \dfrac{6}{\sqrt{a}+\sqrt{b}} \dfrac{\sqrt{a}-\sqrt{b}}{\sqrt{a}-\sqrt{b}} = \dfrac{6\sqrt{a}-6\sqrt{b}}{a-b}$

10. $a^2 + b^2 = c^2$ becomes $12^2 + b^2 = 13^2$

$144 + b^2 = 169$

$b^2 = 25$

$b = \pm 5$

We omit the –5 because we are dealing with lengths.
The other leg is five feet long.

11. There are lots of real numbers that are perfect squares. For example, 0, 1, 4, 9, 16, 25, 36. . . .

12. $18\sqrt{3xy} - \sqrt{3xy} + 3\sqrt{xy} = 17\sqrt{3xy} + 3\sqrt{xy}$

You can't combine any further because the radicands (the part under the radical sign) are not alike.

13. $v = w^2$ Just square both sides of $\sqrt{v} = w$ and you get $v = w^2$.

14. $\sqrt{14xy}\,\sqrt{2} = \sqrt{28xy} = \sqrt{4}\,\sqrt{7xy} = 2\sqrt{7xy}$

15. one.

16. Problem 6 may have been a help in getting the answer to this problem. Taking a zillion square roots of 1382 will get you roughly to the same spot as taking a zillion square roots of 893679. These numbers all come falling out of the sky heading toward one spot.

Now, where might we find numbers that would head *upward* instead of downward? What about those numbers that are less than one? What's the square root of one-fourth? It's one-half. The square root of 0.01 is 0.1 (since $0.1^2 = 0.01$).

17. $\dfrac{x}{\sqrt{x}-\sqrt{y}} \dfrac{\sqrt{x}+\sqrt{y}}{\sqrt{x}+\sqrt{y}} = \dfrac{x(\sqrt{x}+\sqrt{y})}{x-y}$ or $\dfrac{x\sqrt{x}+x\sqrt{y}}{x-y}$

18. $4 = \sqrt{9-x} \Rightarrow 16 = 9 - x \Rightarrow x = -7$

Checking $x = -7$ $4 \overset{?}{=} \sqrt{9-(-7)} \Rightarrow 4 \overset{?}{=} \sqrt{16}$ yes

1. What is the name we give this set: {x | x is a real number and x *cannot* be expressed as y/z where y is an integer and z is a natural number}?

2. Solve $5x^2 = 500$

3. Suppose you've inherited this triangular piece of Texas. The lengths of the sides of your property are eight miles, five miles, and seven miles. How many square miles do you own? Give the exact answer and, if you own a calculator, give the approximate square miles (to the nearest tenth).

   Here's a formula that may help. It's called **Heron's formula**. It says that the area of *any triangle* (whose sides are a, b, and c) is equal to $\sqrt{s(s-a)(s-b)(s-c)}$ where s is the semi-perimeter. $s = \frac{1}{2}(a + b + c)$.

4. $\sqrt{6xy^2}\,\sqrt{2x^2y}$

5. $\sqrt{64} + \sqrt{8} + \sqrt{18}$

6. Simplify $\dfrac{9}{1 - \sqrt{y}}$

7. Solve $x + \sqrt{x + 10} = 2$

8. What is the conjugate of $70 + 3\sqrt{11}$?

9. The two legs of a right triangle have lengths of 6" and 8". What is the length of the hypotenuse?

10. Name a number that is irrational, whose square is rational.

11. $(3\sqrt{y} - 2)(2\sqrt{y} - 3)$

12. Simplify $\sqrt{12} + 7\sqrt{3}$

13. What is the area of a triangle whose sides are 5, 12, and 13 feet? (Hint: You may want to look at question number 3 above.)

14. $64^{1/3}$

15. Solve $y^2 - 16 = 1$

16. Solve $\sqrt{x + 3} = \dfrac{x - 1}{3}$

17. Simplify $\dfrac{3 + 5\sqrt{w}}{\sqrt{wz}}$

18. Name a number that is irrational whose square is also irrational.

answers

1. $\{x \mid x$ is a real number and x *cannot* be expressed as y/z where y is an integer and z is a natural number$\}$ are the irrational numbers

2.
$$5x^2 = 500$$
$$x^2 = 100$$
$$x = \pm 10$$

3. $a = 8$, $b = 5$, $c = 7$
$$s = \tfrac{1}{2}(a + b + c) = \tfrac{1}{2}(8 + 5 + 7) = \tfrac{1}{2}(20) = 10$$
$$\text{area} = \sqrt{s(s-a)(s-b)(s-c)} = \sqrt{10(2)(5)(3)} = \sqrt{300} = \sqrt{100}\sqrt{3}$$
$= 10\sqrt{3} \approx 17.3$ square miles.

4. $\sqrt{6xy^2}\,\sqrt{2x^2y} = \sqrt{12x^3y^3} = \sqrt{4x^2y^2}\,\sqrt{3xy} = 2xy\sqrt{3xy}$

5.
$$\sqrt{64} + \sqrt{8} + \sqrt{18} = 8 + \sqrt{4}\sqrt{2} + \sqrt{9}\sqrt{2}$$
$$= 8 + 2\sqrt{2} + 3\sqrt{2}$$
$$= 8 + 5\sqrt{2}$$

6. $\dfrac{9}{1-\sqrt{y}} = \dfrac{9}{1-\sqrt{y}}\,\dfrac{1+\sqrt{y}}{1+\sqrt{y}} = \dfrac{9+9\sqrt{y}}{1-y}$

7.
$$x + \sqrt{x+10} = 2$$
Isolate the square root $\quad\quad \sqrt{x+10} = 2 - x$
Square both sides $\quad\quad\quad x + 10 = 4 - 4x + x^2$
$$0 = x^2 - 5x - 6$$
$$0 = (x-6)(x+1)$$
$$x - 6 = 0 \quad \text{OR} \quad x + 1 = 0$$
$$x = 6, -1$$

Checking x = 6 in the original equation $\quad 6 + \sqrt{16} \overset{?}{=} 2 \quad$ no

Checking x = –1 in the original equation $\quad -1 + \sqrt{9} \overset{?}{=} 2 \quad$ yes

The final answer is x = –1.

8. $70 - 3\sqrt{11}$

9. $a^2 + b^2 = c^2$ becomes $6^2 + 8^2 = c^2$
$$36 + 64 = c^2$$
$$100 = c^2$$
$$\pm 10 = c$$

We omit the –10 since we are dealing with length.

The length of the hypotenuse is 10 inches.

10. Numbers like $\sqrt{2}$ or $\sqrt{3}$ or $\sqrt{5}$ are irrational and their squares are rational.

11. $(3\sqrt{y} - 2)(2\sqrt{y} - 3) = 6y - 9\sqrt{y} - 4\sqrt{y} + 6 = 6y - 13\sqrt{y} + 6$

12. $\sqrt{12} + 7\sqrt{3} = \sqrt{4}\sqrt{3} + 7\sqrt{3} = 9\sqrt{3}$

13. Using Heron's formula, $a = 5$, $b = 12$, and $c = 13$.
$$s = \tfrac{1}{2}(a + b + c) = \tfrac{1}{2}(5 + 12 + 13) = 15$$
$$\text{area} = \sqrt{s(s - a)(s - b)(s - c)} = \sqrt{15(10)(3)(2)} = \sqrt{900} = 30 \text{ sq ft}$$

14. $64^{1/3} = \sqrt[3]{64} = 4$ since $4^3 = 64$

15. $y^2 - 16 = 1$
$$y^2 = 17 \quad \text{This is called a pure quadratic.}$$
$$y = \pm\sqrt{17}$$

16.
$$\sqrt{x + 3} = \frac{x - 1}{3}$$

Square both sides (This may add extraneous roots.)
$$x + 3 = \frac{x^2 - 2x + 1}{9}$$

This is a fractional equation. Multiply both sides by 9
$$9(x + 3) = x^2 - 2x + 1$$
$$9x + 27 = x^2 - 2x + 1$$
$$0 = x^2 - 11x - 26$$
$$0 = (x - 13)(x + 2)$$
$$x - 13 = 0 \quad \text{OR} \quad x + 2 = 0$$
$$x = 13, -2$$

Checking $x = 13$ in the original equation
$$\sqrt{13 + 3} \overset{?}{=} \frac{13 - 1}{3}$$
$$4 \overset{?}{=} 4 \qquad \text{yes}$$

Checking $x = -2$ $\qquad \sqrt{-2 + 3} \overset{?}{=} \dfrac{-2 - 1}{3}$
$$\sqrt{1} \overset{?}{=} -1 \quad \text{no}$$

17. $\dfrac{3 + 5\sqrt{w}}{\sqrt{wz}} = \dfrac{3 + 5\sqrt{w}}{\sqrt{wz}}\dfrac{\sqrt{wz}}{\sqrt{wz}} = \dfrac{3\sqrt{wz} + 5w\sqrt{z}}{wz}$

18. There are many possible answers such as $\sqrt{\pi}$ or $2^{1/3}$ or $5^{1/8}$.

Fred and Jack hopped on the bus that took them west on Highway 190 right into Lampasas. The bus entered town on Del Norte Street and turned right on Walnut. They got off near Dee's Old Time Pit Bar-B-Q. It was starting to get dark.

Having spent most of his life on the KITTENS campus, Fred was a little apprehensive in this new situation. He had never had to figure out how to entertain himself for six hours on a Sunday night in an unfamiliar town.

They started walking east-northeast along W. 1st Street and it turned into E. 1st Street. After three blocks they hit the Lampasas Public Library (201 S. Main Street). Fred was delighted. He loved books and had heard of this library from the chaplain on the army base. It has over 22,000 books, magazines, and audio tapes. Fred's eyes and fingers were itching to examine them all.

The only thing that Fred was worried about was their library rule: "Parents should not leave young children unattended in the library." Last Friday he had turned six. Was he still a young child? He didn't think so, having taught at a university for five years. He finally resolved the issue in his mind when he realized that no parent was leaving him unattended. Hence the rule wasn't being violated.

Fred ran on ahead to the front door of the library. It was locked. In Lampasas their library is opened for four hours on Saturday (10 a.m. to 2 p.m.) and closed all day Sunday. "Rats!" Fred thought to himself. The books on *his* shelves in *his* office at KITTENS were always available to him.

Jack said to him, "I know what we'll do. There's a great bar that I know. Just down the street. It's called the CJ Bar.* You'll love it."

Fred was thunderstruck. Jack LaRoad had proclaimed·that health

---

* Citizens of Lampasas recognize Dee's and the library on Main Street, but probably have never heard of the CJ Bar. There are some things in this odyssey that might be thought of as fictional, but be assured, those items have been kept to a bare minimum.

was one of his top priorities, and now he's talking about spending Sunday evening at a bar! Fred shook his head. When they got to the front of the CJ Bar, Fred said that he wanted to stand outside for a moment and that it was okay for Jack to go inside.

"Shall I wait for you before I order?" Jack asked.

"No. Go ahead and do your drinking," Fred answered.

Jack couldn't figure out why Fred wanted to stay outside for a while, but he guessed that Fred just needed a little fresh air before going in. He headed inside and ordered the house specialty, Carrot Juice.

The CJ Bar was named for this specialty. (*Eponymous,* as they like to say in parts of Lampasas. If you want to say it aloud: eh-PON-eh-mes.)

Jack strode up to the bar, "Give me a quart of carrot juice." The man standing next to him, who had a hairier chest but smaller muscles, ordered one for himself also.

Given the competitiveness that testosterone induces, they both stuck their straws in their glasses at the same time and swallowed as hard as they could.

Jack finished two minutes before Mr. Hairy.

They ordered another round and this time, by accident, they both stuck their straws in the same glass. It took them five minutes to drink this quart. After they realized their faux pas (pronounced foh-PAH, which means an embarrassing social blunder) they both raced to the door explaining that they needed to talk with someone outside.

Fred was standing outside. Waiting.

Jack wondered how long it would take him to drink a quart. He told Fred that Mr. H had taken two minutes longer than he had. Jack also explained that if *somehow* they had *somehow* (Jack was embarrassed) *accidently* and, *not on purpose*, had worked on ("worked" was the most masculine word Jack could think of to describe a joint endeavor) the same quart, it *hypothetically* would take five minutes for them to complete the job.

Fred couldn't understand why Jack was fumbling with his words. Fred, Betty, and Alexander had once put three straws in a single milkshake and shared it, and no one was embarrassed.

Fred drew in the dust on the sidewalk:

Let x = the number of minutes it takes Jack to drink a quart.

Then x + 2 = the time for Mr. H to drink a quart.

5 = the time if they both worked on the same quart.

Jack knelt and drew a double underline under one of Fred's words.

It now read: 5 = the time <u>if</u> they both worked on the same quart.

Fred continued, "If it takes you, Jack, x minutes to drink a quart then you can drink $\frac{1}{x}$ of a quart in one minute.

"And you, Mr. H., can drink $\frac{1}{x+2}$ of a quart in one minute.

"And together, speaking hypothetically, you can drink $\frac{1}{5}$ of a quart in one minute."

So we have
$$\frac{1}{x} + \frac{1}{x+2} = \frac{1}{5}$$

It's a fractional equation. The first step is to eliminate the denominators. So we multiply through by $5x(x + 2)$

and obtain
$$\frac{1 \cdot 5x(x+2)}{x} + \frac{1 \cdot 5x(x+2)}{x+2} = \frac{1 \cdot 5x(x+2)}{5}$$

All the denominators disappear $5(x + 2) + 5x = x(x + 2)$

Distributive property $5x + 10 + 5x = x^2 + 2x$

Transpose everything to one side of the equation $0 = x^2 - 8x - 10$

Your Turn to Play

1. Can $0 = x^2 - 8x - 10$ be solved by factoring?

2. Jack could finish a bowl of air-popped popcorn (with no added butter) in x minutes. Mr. Hairy takes 5 minutes longer than Jack to finish a bowl. If they worked on a single bowl, they could finish it in 7 minutes. Find the quadratic equation. *You need not solve it.* We'll do that in the next lesson.

3. The CJ Bar has a pool table that is 40 inches longer than it is wide. The area of the table is 2736 square inches. Let x = the width of the table. Find the equation. *You need not solve it.*

4. When Jack looked at the balls on the pool table, he noticed nothing unusual.

Mr. Hairy said, "Let's make this a little more interesting." He took another #7 ball out of his pocket and rolled it onto the table.

Regular-sized ball with radius r

Jack couldn't believe his eyes. The new #7 ball was Texas-huge. Its radius was one inch larger than normal.

The combined surface area of a regular #7 and this extra-large #7 was 70 square inches. Find the equation. (The surface area of a sphere is $S = 4\pi r^2$.)

Extra-large variety with radius r + 1

5. Jack reached for a can of Diet Sluice. (No sugar or additives—just pure . . . water.) Mr. Hairy wasn't going to be outdone. His can had a radius that was one inch larger than Jack's. In addition, Jack's can was 8 inches tall and Hairy's was 9.

Hairy's held 125 cubic inches more than Jack's.

(The volume of a cylinder is $V = \pi r^2 h$ where r is the radius and h is the height.)

Find the equation for r. *You need not solve it.* Put it in the form $ar^2 + br + c = 0$ where a, b, and c are numbers.

Diet Sluice

. . . . COMPLETE SOLUTIONS . . . .

1. I can't think of two numbers that multiply to –10 and add to –8. It cannot be factored.

2.
$$\frac{1}{x} + \frac{1}{x+5} = \frac{1}{7}$$

$$\frac{7x(x+5)}{x} + \frac{7x(x+5)}{(x+5)} = \frac{7(x)(x+5)}{7}$$

$$7(x+5) + 7x = x(x+5)$$

$$0 = x^2 - 9x - 35$$

3. *Let x = the width of the table.*

   Then x + 40 = the length of the table.

   Since the width times the length is equal to the area . . .

$$x(x+40) = 2736$$

$$x^2 + 40x = 2736$$

$$x^2 + 40x - 2736 = 0$$

4. The surface area of the regular-sized ball is $4\pi r^2$.
The surface area of the larger ball is $4\pi(r+1)^2$.

$$4\pi r^2 + 4\pi(r+1)^2 = 70$$

$$4\pi r^2 + 4\pi(r+1)(r+1) = 70$$

$$4\pi r^2 + 4\pi(r^2 + 2r + 1) = 70$$

$$4\pi r^2 + 4\pi r^2 + 8\pi r + 4\pi = 70$$

$$8\pi r^2 + 8\pi r + 4\pi - 70 = 0$$

5. Volume of Jack's Diet Sluice = $8\pi r^2$.
Volume of Hairy's = $9\pi(r+1)^2$ which is $9\pi(r^2 + 2r + 1) = 9\pi r^2 + 18\pi r + 9\pi$.
The difference between these two volumes is 125 cubic inches.

$$(9\pi r^2 + 18\pi r + 9\pi) - 8\pi r^2 = 125$$

$$\pi r^2 + 18\pi r + 9\pi - 125 = 0$$

F red looked at $x^2 - 8x - 10$ and tried to think of two numbers that would multiply to -10 and add to -8. He couldn't think of any and announced, "It doesn't factor."

By this time Mr. H had grown impatient and had turned away without a word and wandered back into the CJ Bar.

Jack, however, who loved to learn new things, looked at Fred and said, "So we can't solve it?"

Fred smiled. "I have a trick that you've never seen. Just watch."

He took the equation that couldn't be solved as a pure quadratic and couldn't be solved by factoring $\qquad 0 = x^2 - 8x - 10$

and first switched the sides and used the symmetric property of equality* $\qquad x^2 - 8x - 10 = 0$

transposed the -10 $\qquad x^2 - 8x \quad = 10$

and added $+16$ to both sides $\qquad x^2 - 8x + 16 = 10 + 16$

Jack exclaimed, "Where'd you get that 16 from? Why?"

"I'll show you in a minute," was Fred's reply. "But first I'll finish the problem."

The left side of the equation, which was $x^2 - 8x + 16$, now factors nicely into $(x - 4)(x - 4)$ $\qquad (x - 4)^2 = 26$

_____

* *if a = b then b = a*

Taking the square root of
both sides

$$x - 4 = \pm\sqrt{26}$$

Transpose the -4 and we're done

$$x = 4 \pm \sqrt{26}$$

(Discard the $4 - \sqrt{26}$ since we want a positive answer. We were looking for the number of minutes it takes Jack to drink a quart of carrot juice.)

Jack stood there in silence. This was much more complicated than solving pure quadratics, where $z^2 = 7$ had a solution in one step: $z = \pm\sqrt{7}$.

He looked back over all the steps that Fred had written in the dust on the sidewalk, and they all seemed logical enough, except for where Fred got that $+16$ that he added to both sides of the equation.

"Let me try one!" Jack insisted.

Jack wrote

$$x^2 + 6x - 13 = 0$$

and said, "This is not a pure quadratic since it has the $6x$ in it and it doesn't factor, so we gotta use your new method."

Transpose the -13

$$x^2 + 6x = 13$$

Now he was stuck. "Hey, what's the magic number you stick in?"

Fred looked at the equation (and took half of $+6$ and squared it) and said, "$+9$ is the number you want."

So Jack added $+9$
to each side

$$x^2 + 6x + 9 = 13 + 9$$

And now the left side of
$x^2 + 6x + 9$ factored nicely
into $(x + 3)(x + 3)$

$$(x + 3)^2 = 22$$

Square root of both sides

$$x + 3 = \pm\sqrt{22}$$

Transposed the $+3$

$$x = -3 \pm \sqrt{22}$$

"But, tell me, how do you get that magic number that makes the left side a perfect square?" Jack asked.

Fred said, "Let's work backward for a moment and multiply out something like $(x + 7)^2$. We get $x^2 + 14x + 49$." He wrote in the dirt ⇨

$$x^2 + 14x + 49$$
$$(x + 7)^2$$

"How do we get from the 14x to the magic +49 ?" Fred asked Jack.

Jack studied the box and the previous two examples, which we write out here in boxes so you don't have to turn back to the previous pages:

$x^2 - 8x + 16$	$x^2 + 6x + 9$	$x^2 + 14x + 49$
$(x - 4)^2$	$(x + 3)^2$	$(x + 7)^2$

"I get it," Jack announced. "Give me one and see if I can get the magic number."

"Okay," said Fred, "How about $x^2 + 10x$?"

Jack grinned. "That's easy. The first step is to write $(x + 5)^2$ underneath, since 5 is half of the 10. The second step is to square the 5 to get that magic number. So I'd have $x^2 + 10x + 25$."

Jack shouldn't be having all the fun. Now it's . . .

Your Turn to Play

1. Solve $x^2 - 8 = -4x$

2. Solve $y^2 + 5y - 3 = 0$

3. Solve $5x^2 + 13x + 6 = 0$
   (Hint: First, divide through by 5 in order to get $1x^2$.)

. . . . COMPLETE SOLUTIONS . . . .

1. $x^2 - 8 = -4x$ is not a pure quadratic (since it has the $-4x$ in it). Let's see if we can solve it by factoring. We always try that first since it's much simpler than this new method.

Put everything on one side $x^2 + 4x - 8 = 0$

Does it factor? No. So we have to use this new method, which is called **completing the square**.

Transpose the -8 $x^2 + 4x \quad = 8$

| What's the magic number to add to both sides? We take half of $+4$ and then square it. |

Add the magic number to both sides $x^2 + 4x + 4 = 8 + 4$

$$(x + 2)^2 = 12$$
$$x + 2 = \pm\sqrt{12}$$
$$x = -2 \pm \sqrt{12}$$

And one final simplification since

$\sqrt{12} = \sqrt{4}\sqrt{3} = 2\sqrt{3}$ $x = -2 \pm 2\sqrt{3}$

2. $y^2 + 5y - 3 = 0$ is not a pure quadratic and it doesn't factor. So we need to solve it by completing the square.

$$y^2 + 5y \quad = 3$$

| Half of 5 is 5/2, which when squared is 25/4 |

$$y^2 + 5y + \frac{25}{4} = 3 + \frac{25}{4}$$

Doing the arithmetic on the right side of the equation:

$3 + \frac{25}{4}$

$= \frac{3 \cdot 4}{1 \cdot 4} + \frac{25}{4}$

$= \frac{12 + 25}{4} = \frac{37}{4}$

$$\left(y + \frac{5}{2}\right)^2 = \frac{37}{4}$$

$$y + \frac{5}{2} = \frac{\pm\sqrt{37}}{2}$$

$$y = \frac{-5}{2} + \frac{\pm\sqrt{37}}{2} \text{ or } \frac{-5 \pm \sqrt{37}}{2}$$

3. The new thing in solving $5x^2 + 13x + 6 = 0$ is that we don't have just x^2 (where the coefficient is 1). We have a $5x^2$.

To make it like the previous examples
we divide through by 5 $x^2 + \dfrac{13}{5}x + \dfrac{6}{5} = 0$

Now the rest of the steps are
just like we've done before.
Transpose the $\dfrac{6}{5}$ $x^2 + \dfrac{13}{5}x = -\dfrac{6}{5}$

Find the number to add to both sides so that
the left side becomes a perfect square.
Half of $\dfrac{13}{5}$ (which is $\dfrac{13}{10}$) square that,
and add it to both sides $x^2 + \dfrac{13}{5}x + \dfrac{169}{100} = \dfrac{-6}{5} + \dfrac{169}{100}$

The left side factors automatically. We just use the $\dfrac{13}{10}$
It's the right side, which is just arithmetic, that
is more work. Doing some interior decorating
we get $\dfrac{-6 \cdot 20}{5 \cdot 20} + \dfrac{169}{100}$ which is $\dfrac{-120 + 169}{100}$

$$(x + \dfrac{13}{10})^2 = \dfrac{49}{100}$$

Take the square root of both sides
and remember the ± $x + \dfrac{13}{10} = \pm\dfrac{7}{10}$

Transpose $x = -\dfrac{13}{10} \pm \dfrac{7}{10}$

Using the plus sign we have $x = \dfrac{-13+7}{10} = -\dfrac{6}{10} = -\dfrac{3}{5}$

and using the minus sign $x = \dfrac{-13-7}{10} = -\dfrac{20}{10} = -2$

Hey, that came out nicely: $x = -\dfrac{3}{5}$ or $x = -2$. There were no square roots in the final answer as we've had every other time that we have solved a quadratic by completing the square.

Isn't that nice? I, your reader, ask.

Yes and no. Yes, in the sense that answers without square roots in them are a little easier to look at. No, in the sense that we goofed. Answers without square roots in them indicate that the problem could have been solved by factoring.

So what?

Solving quadratics by factoring is usually a lot shorter and easier.

We'll do the same problem by factoring: $5x^2 + 13x + 6 = 0$

Factor $(x + 2)(5x + 3) = 0$

Set each factor equal to zero $x + 2 = 0$ OR $5x + 3 = 0$

Solve $x = -2$ OR $x = -3/5$

Look back and compare.

Completing the square took a whole page.

Factoring took three lines.

small essay
Making $50/hour

Back in Lesson 80, I mentioned that developing your mental strength might make your financial future brighter. If you exercised your brain by not just reading the questions in the cities and then reading the answers, you will probably be a much more desirable employee or a more successful business owner.

This is Lesson 83, and what I wrote in Lesson 80 is still true.

But readers who don't read carefully might think that I was saying that getting rich is the most important thing in life. That's silly.

Here are the facts of life . . .

Getting a lot of money is not the object of life . . . but it can help.

Breathing is not the object of life . . . but it helps.

Pizza is not the object of life . . . but for many, it is a joy.

Wait a minute! I, your reader, want to know what is the object of life!

It will mean a lot more to you if you discover it rather than my just telling you.   end of small essay

It was getting dark outside. Completing the square to solve quadratic equations was old hat for Fred. He had taught it at KITTENS for the last five years and the process was very mechanical. You transpose the number to the right side. Divide through by the coefficient of the x^2. Find the magic number by taking half of the coefficient of the x term and squaring it. Etc., etc., etc.

Work! Work! Work!

Factoring was more fun. When you had something like $x^2 + 22x + 40$, you had to figure out what two numbers multiply to 40 and add to 22.

Jack had gone back inside the bar. Fred knew that at the age of six, it was wrong for him to go into a bar. He had always tried to do the right thing simply because it was the right thing to do. His parents had never punished him because they never really paid much attention to him. Being drafted wasn't really punishment for something he had done wrong. His secretary Belinda had made the error in accidentally tossing out the army induction letters. The bad things that happened in life were just a part of life, and Fred just accepted them.

These moments outside the CJ Bar were to be the last pain-free seconds in Fred's life for the next several days.

Fred pulled his camouflage T-shirt around him more closely as the temperature began to drop. (If you've got a blue pencil and if this is your book, feel free to color the exposed parts of Fred. Many banal algebra books are printed with lots of color on the text pages, but this drives up the cost. By printing in black and white, you pay less for the book and you get the fun of coloring.)

Five kids—they must have been nine or ten years old—wandered down the street toward Fred. They were huge. Each weighed 80 or 90 pounds. Some of them were chewing gum with their mouths open.

"Hey! Army brat!" one of them called out to Fred. The rest of them repeated the taunt, "Army brat! Army brat!" One of them knocked Fred's coffee-cup helmet off his head.

They made fun of his square head, of his shortness, of his little dotty eyes. They ripped his shirt, and Fred fell down. (Any one of these bullies alone might have just walked by Fred, but the definition of a mob is a lot of bodies and no brain.)

This was all very new and scary. Fred had never read any martial arts books. The only thing that came to mind was a Bear Country book that said that if a bear is attacking you, you are to lie on your tummy and protect the back of your neck with your hands.

Fred did that.

One of the boys, who had never read *any* books but had seen a lot of violent movies, kicked Fred in the ribs. Fred almost passed out from the pain as the bullies continued their walk down the street.

Things wouldn't have been so bad if that boy had been wearing tennis shoes, but, instead, he was wearing his cowboy boots—the kind with the pointy toes. Fred's seventh rib was broken.

Fred tried calling out to Jack, but he couldn't raise his voice above a whisper, and the noise from the bar made it impossible for Jack to hear him anyway.

As many inhabitants of Lampasas know, the police department is located at 403 S. Main Street, which is two blocks from the library. An officer in his patrol car passed the CJ Bar. He saw the body lying on the sidewalk and stopped his car. He turned on his "Christmas Tree" (as some law enforcement officers call their overhead car lights) and hurried to Fred.

He had had first-aid training and knew that it was important not to move the victim. He put his coat over Fred and radioed for an ambulance. He asked, "Where's your mommy?" and Fred responded, "In heaven." He asked again, "Where's your daddy?" It hurt to talk, but Fred answered, "I don't know. I haven't seen him since I was six months old."

The ambulance arrived. Fred was starting to lose it, and he kept muttering, "My heart is beating. My heart is beating." What he was really worried about was that they would start doing CPR on him. That's why he kept saying, "My heart is beating." Someone pushing on his chest was the last thing he wanted.

There really was no danger that the ambulance guys would do that. They were well trained. As gently as possible they put Fred on a stretcher and into the ambulance.

With their lights and siren on, they drove to the nearest hospital. One medic held Fred's hand and said, "I know this hurts. You've got a big boo-boo, but we can't give you any medicine to take away the ouch. Try to think of something to take your mind off the hurty."

Fred couldn't speak because even breathing sent stabs of pain through his body. If he could have answered, he would have said something like:

I'm not a little baby. I'm six years old. I understand the medical complications that might accompany the administration of analgesics to one of my stature. Please be assured that I recognize the danger and frequency of iatrogenic injuries in modern medical treatment.

For readers who don't own dictionaries . . .
    analgesic (ann-el-GEE-zik) Something that decreases pain. Is pizza an analgesic?
    iatrogenic (i-a-treh-GEN-ik) Bad stuff accidentally caused by doctors.
    Fred knew a lot of words since he read a lot of good books.

Fred tried to think of something else to take his mind off his injury.
    The medic said, "It's bleeding," and pulled up Fred's shirt and said, "The assailant must have had metal-toed shoes." He washed the wound down with something that looked like brown iodine. It made Fred's eyes spin in his head.

Fred thought of something to divert his attention from his present circumstances. He thought of a way *to forever end the use of completing the square to solve quadratic equations.* (Are you interested?)
    One last time to complete the square . . . and then never again.

Here were Fred's thoughts enclosed by pain circles . . .

I'm going to do it one last time, but do it in such general terms that it'll be done once and for all.

I start with $ax^2 + bx + c = 0$.
You couldn't get more general than that.
The a, b, and c can be any real numbers.

Transpose the c $ax^2 + bx = -c$

Divide by a $x^2 + \dfrac{b}{a}x = \dfrac{-c}{a}$

Take half of the $\dfrac{b}{a}$ (which is $\dfrac{b}{2a}$)
and square it. Then add
$\dfrac{b^2}{4a^2}$ to both sides $x^2 + \dfrac{b}{a}x + \dfrac{b^2}{4a^2} = \dfrac{-c}{a} + \dfrac{b^2}{4a^2}$

Factor the left side
Add the right together $(x + \dfrac{b}{2a})^2 = \dfrac{b^2 - 4ac}{4a^2}$

Square root both sides $x + \dfrac{b}{2a} = \dfrac{\pm\sqrt{b^2 - 4ac}}{2a}$

Transpose $x = \dfrac{-b \pm \sqrt{b^2 - 4ac}}{2a}$

He had derived the **quadratic formula**. It is *the* formula in algebra. If you had to prove to someone that you know algebra, reciting the quadratic formula is probably the way to do it.

If this book were ever made into a movie, the opening credits might look like:

You can now solve any quadratic equation in one step. If you were given $5x^2 + 7x + 3 = 0$, you could just write the answer:

$$x = \frac{-7 \pm \sqrt{49 - 4(5)(3)}}{10}$$

Most students memorize the formula. You have it pretty well memorized if you can say $x = \dfrac{-b \pm \sqrt{b^2 - 4ac}}{2a}$ in a single breath. I had a student who could recite it three times in one breath.

The trick is that you recite the formula and stick in the numbers (like the 5, 7, and 3 in the above example) *as you recite it*. That way you could solve 50 equations in the time it takes your sister to floss and brush her teeth.

But! you exclaim, ***I don't wanna memorize no formula!***

You don't have to. Just do as Fred did in the "pain circles" on the previous page and derive the formula whenever you need it using the completion of the square technique. (Most people do the memorizing.)

Let's start with just five quadratic equations to solve. Tell your sister to go floss and brush her teeth and then ignore her response of "What! Are you crazy or something?!" Here's the formula, so you don't

have to turn back to a previous page: starting with $ax^2 + bx + c = 0$,

$$x = \frac{-b \pm \sqrt{b^2 - 4ac}}{2a}$$

Your Turn to Play

Solve these five:

1. $4x^2 + 9x + 2 = 0$
2. $3x^2 + 11x + 5 = 0$
3. $8x^2 - 2x - 12 = 0$
4. $x^2 + 10x - 7 = 0$
5. $5x^2 = -4x + 13$

. . . . COMPLETE SOLUTIONS . . . .

1. $4x^2 + 9x + 2 = 0$ $x = \dfrac{-9 \pm \sqrt{81 - 4(4)(2)}}{8}$

2. $3x^2 + 11x + 5 = 0$ $x = \dfrac{-11 \pm \sqrt{121 - 4(3)(5)}}{6}$

3. $8x^2 - 2x - 12 = 0$ First, I'm going to divide through by 2. It'll make the arithmetic easier: $4x^2 - x - 6 = 0$. $x = \dfrac{1 \pm \sqrt{1 - 4(4)(-6)}}{8}$

4. $x^2 + 10x - 7 = 0$ $x = \dfrac{-10 \pm \sqrt{100 - 4(1)(-7)}}{2}$

5. $5x^2 = -4x + 13$ This isn't in the general form $ax^2 + bx + c = 0$, so we'll transpose some terms, and we get $5x^2 + 4x - 13 = 0$.

$$x = \frac{-4 \pm \sqrt{16 - 4(5)(-13)}}{10}$$

One additional advantage to having her clean her teeth (rather than, say, brush her hair) is that she can't talk while you're trying to concentrate on these quadratics. After you've got the formula memorized by heart, you can take the ultimate test and solve quadratics while she's yelling at you because you used her toothbrush to clean. . . .

Maybe we shouldn't continue with this example. You get the idea. This is the formula to memorize.

A change of scene. Back at the army base, the chaplain and his wife were just sitting down to a late dinner. They had put their kids to bed and were in the process of celebrating their thirteenth wedding anniversary.

She had baked a pair of Cornish game hens, and that, together with some green beans and pecan pie for dessert, made the traditional dinner on their anniversary. It was the same meal as the first one she had made for him about fifteen years ago when they had started dating.

Normally they had dinner with their kids, but this one night each year, they had dinner alone—just the two of them. Earlier in the evening the kids, who never liked those "funny chickens," got the chance to make whatever dinner they wanted. They chose macaroni and cheese. Their mom never was able to figure out how one piece of cheesy macaroni ended up in the toaster. It really didn't matter that much to her. It was one time each year, and she knew that her kids were generally very well behaved. She wanted them to emerge, years from now, from their childhood with many happy memories.

The candles were lit and their wine glasses were filled. They raised their glasses and looked at each other and repeated the grace that had always begun their anniversary meals: "To Jesus, to life, to you." They knew that it was not necessary to close their eyes or bow their heads in order to say a prayer.

The phone rang.

It was the hospital.

The personnel at the hospital had learned over the years that, when they had a sick or injured soldier, it did no good to call Sergeant Snow or Col. C.C. Coalback. Their response had always been, "What do I care? Send the soldier back here, and we'll have our doctor look at him in the morning."

"We have a child here," the call began. "He says he's from base, but we can't figure out who his parents are. He says his name is Fred Gauze or something."

The chaplain knew who they were talking about: "That's Gauss—rhymes with house. I'll be right there."

The chaplain's wife had heard her husband say, "I'll be right there" many times over the last thirteen years. She headed to the fridge and popped a carton of chocolate milk and an apple into a paper bag and handed it to him as he headed out the door. The kids would find "funny chicken" sandwiches in their lunch boxes tomorrow.

The chaplain was pulling out of his driveway as another car zoomed past him with its lights flashing and its horn sounding. It was Pat, Chris, and Carol heading out. It was about 9:30, and they had two-and-a-half hours until their passes expired at midnight. A window was rolled down, and the entire camp heard their radio blasting out another stanza from the #1 country western song:

> ♪ ♬   She ran away and took the dog,
> And left me with the kid.
> I found a place, a foster home,
> And now of Sid I'm rid.   ♪ ♬

An empty beer can flew out of the window, and in several seconds it was quiet again. The chaplain finished backing out of his driveway and headed to the hospital.

When he got to the hospital, the doctor told him that Fred was sleeping comfortably now and that he could be transported back to the army base tomorrow around noon. They had put an IV in his arm and, along with saline and glucose, had introduced a tiny bit of sedative. With such a small body weight, it didn't take much. Fred was out like a light.

During the intake procedure, the hospital nurse had asked Fred what he had eaten during the last couple of days, and he answered: a part of a hot dog, part of a meatball, a sip of orange juice, and a whole lot of caffeinated cola, but the cola didn't count since he had thrown that up.

The chaplain walked into the ward where Fred was sleeping. They had put Fred in a little nightshirt, one decorated with blue and green frogs. There was a red stain on the side of the shirt. The chaplain held Fred's hand for a moment and thought, "How senseless," and then thought of the phrase that Fred had used when he had seen something that seemed to have no meaning. Fred would mutter to himself, "Division by zero." The chaplain reflected that there were a lot of "divisions by zero" in life that we may never know the meaning of on this side of the grave.

He headed back to his car and drove home.

As the chaplain pulled into his driveway, he noticed the lights were still on in the chow hall. The eight guys who were assigned to KP by Sergeant Snow were still working. They would have been done hours ago, except that they really weren't working that hard. In fact, at this point, they weren't working at all. They were playing.

It was a little after four in the afternoon when the eight of them had reported to the chow hall for duty. They were assigned to peel potatoes—the job you always see in the movies when you see some guys assigned to work in a military kitchen. The truth was that the two soldiers who normally ran the chow hall didn't need any help at all. Running all those beans and hot dogs through the microwave conveyer belt system took hardly any effort at all.

The truckload of potatoes that the base received each week had only one real use: peeling the potatoes was something that soldiers on penalty KP could do. The potatoes were never cooked. Sergeant Snow didn't like potatoes, so he ordered that they never be served. They were all just put in the garbage cans at the end of every week.

How those same potatoes were removed from the garbage cans each week, made into Giant French Fries at King of French Fries, and sold to those same soldiers is a story that we won't relate.

Instead we'll see how the eight guys in the kitchen were having fun. First they had stacked the potatoes up into tall pyramidal towers. Then, they played tag using potatoes, but stopped when they found out it hurt too much to be hit by a potato. Then they decided that they should peel some potatoes, since they had been told that they couldn't leave until the whole truckload was peeled.

They weren't very careful in their work. Using kitchen knives instead of potato peelers, they hacked their way through each potato. When they were done, each "peeled" potato was roughly in the shape of a cube, one inch on each side. Then they played poker using those potato hunks as poker chips.

An inexpertly
peeled potato

Bored with that, one guy started to lay out the hunks in a rectangular matrix. The others joined in this almost mindless activity. It

was something to do to pass the time. After an hour and a half, those potato hunks were a sight to behold. One guy said that it looked like a flag. Another said it reminded him of chairs in an auditorium. One said that there must be a zillion potatoes in that rectangle. Another, making fun of what he had overheard from Fred, said that there are exactly $6x^3 + 19x^2 + 22x + 30$ potatoes in that rectangle. Someone chimed in with, "Yeah, and there's $2x + 5$ rows."

Then it was quiet. They realized that they were facing the question: How many potatoes in each row?

One guy said, "We gotta subtract." Another suggested division of $6x^3 + 19x^2 + 22x + 30$ by $2x + 5$. Fred wasn't there to settle the question, so they had to resort to thinking.

> The easiest way to figure out whether to add, subtract, multiply, or divide is to take a simple example.

"If we had 35 potatoes and there were five rows," one thinker began, "then we'd have seven in each row. So it's division."

Another wrote on the wall with his felt tip pen: $(6x^3 + 19x^2 + 22x + 30) \div (2x + 5)$ and

then rewrote it as: $\dfrac{6x^3 + 19x^2 + 22x + 30}{2x + 5}$

"That's easy," said one of the guys who had had the first half of beginning algebra. "You factor the top, factor the bottom, and cancel like factors."

"Okay, smarty," said his pal, "How do you factor $6x^3 + 19x^2 + 22x + 30$. It's got four terms."

441

"That's still easy. I learned that in my beginning algebra course. You do it by grouping." He wrote $6x^3 + 19x^2 + 22x + 30$ on the wall and started to factor x^2 out of the first two terms and 2 out of the last two terms. He got $x^2(6x + 19) + 2(11x + 15)$ and then was totally stuck.

More silence ensued.

"Well, I dunno, guys," said Thud, whom no one acknowledged as the genius of the group. "If you say it's $6x^3 + 19x^2 + 22x + 30$ divided by $2x + 5$, why don't you just divide it? You know. Like regular division. Long division."

Thud wrote with his pencil on the wall:

$$2x + 5\overline{)\,6x^3 + 19x^2 + 22x + 30}$$

Let's see. 2x goes into
$6x^3$ how many times?
$3x^2$ times. So I put the $3x^2$
above the $6x^3$

$$\begin{array}{r} 3x^2 \\ 2x + 5\overline{)\,6x^3 + 19x^2 + 22x + 30} \end{array}$$

Then, as in regular long division,
you multiply the $3x^2$ times the
$2x + 5$ and
stick it down below

$$\begin{array}{r} 3x^2 \\ 2x + 5\overline{)\,6x^3 + 19x^2 + 22x + 30} \\ 6x^3 + 15x^2 \end{array}$$

and subtract

$$\begin{array}{r} 3x^2 \\ 2x + 5\overline{)\,6x^3 + 19x^2 + 22x + 30} \\ 6x^3 + 15x^2 \\ +\ 4x^2 \end{array}$$

and bring down the
next term

$$\begin{array}{r} 3x^2 \\ 2x + 5\overline{)\,6x^3 + 19x^2 + 22x + 30} \\ 6x^3 + 15x^2 \\ +\ 4x^2 + 22x \end{array}$$

Then repeat the whole process over and over again until you're done.

You divide the 2x into the 4x²
and get an answer
of 2x, which you put on top

$$
\begin{array}{r}
3x^2 + 2x \\
2x + 5{\overline{\smash{\big)}\,6x^3 + 19x^2 + 22x + 30}} \\
\underline{6x^3 + 15x^2 } \\
+\ 4x^2 + 22x
\end{array}
$$

Then you multiply the 2x
times the 2x + 5 and stick
the answer down below

$$
\begin{array}{r}
3x^2\ + 2x \\
2x + 5{\overline{\smash{\big)}\,6x^3 + 19x^2 + 22x + 30}} \\
\underline{6x^3 + 15x^2 } \\
+\ 4x^2 + 22x \\
\underline{+\ 4x^2 + 10x }
\end{array}
$$

Subtract and bring down the
next term

$$
\begin{array}{r}
3x^2\ + 2x \\
2x + 5{\overline{\smash{\big)}\,6x^3 + 19x^2 + 22x + 30}} \\
\underline{6x^3 + 15x^2 } \\
+\ 4x^2 + 22x \\
\underline{+\ 4x^2 + 10x } \\
+\ 12x + 30
\end{array}
$$

And then you repeat the whole process over and over again until you're done.

2x into 12x

$$
\begin{array}{r}
3x^2\ + 2x\ \ + 6 \\
2x + 5{\overline{\smash{\big)}\,6x^3 + 19x^2 + 22x + 30}} \\
\underline{6x^3 + 15x^2 } \\
+\ 4x^2 + 22x \\
\underline{+\ 4x^2 + 10x } \\
+\ 12x + 30
\end{array}
$$

Finally, you multiply the 6 times the 2x + 5 and stick your answer down below. Subtract and you're done.

$$
\begin{array}{r}
3x^2 + 2x + 6 \\
2x + 5\,\overline{)\,6x^3 + 19x^2 + 22x + 30} \\
\underline{6x^3 + 15x^2} \\
+\,4x^2 + 22x \\
\underline{+\,4x^2 + 10x} \\
+\,12x + 30 \\
\underline{+\,12x + 30} \\
0
\end{array}
$$

"Hey! It worked Thud!," they all shouted. They dumped all the potatoes into the garbage cans, turned out the lights, and headed to the bunkhouse to get some sleep.

Some notes on (Thud's) **long division of polynomials**:

♪#1: Long division is great when you can't factor the numerator. A fraction like

$$\frac{37x^{23} + 16x^{19} - 5x^7 + 99x^6 + 4x^4 + 11}{8x + 3}$$

just calls out for long division.

♪#2: Arrange the terms in descending powers. If you've got

$$\frac{6x^2 + 2x^5 + 9}{7x - 1} \text{ make it } \frac{2x^5 + 6x^2 + 9}{7x - 1}$$

♪#3: If some powers are left out, stick them back in. It'll make your division a lot easier.

Instead of $\quad 7x - 1\,\overline{)\,2x^5 + 6x^2 + 9}$

use $\quad 7x - 1\,\overline{)\,2x^5 + 0x^4 + 0x^3 + 6x^2 + 0x + 9}$

♪#4: What if there's a remainder? Do the same as you did in

arithmetic:
$$
\begin{array}{r}
14\,\tfrac{3}{5} \\
5\,\overline{)\,73} \\
\underline{5} \\
23 \\
\underline{20} \\
3
\end{array}
$$
and put the remainder up as a fraction.

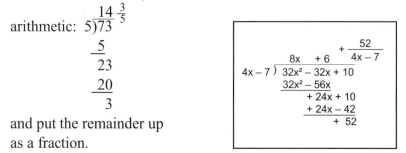

Your Turn to Play

1. $3x + 2 \overline{)\, 15x^3 + 19x^2 + 18x + 8}$

2. $7x - 2 \overline{)\, 21x^3 + 8x^2 + 3x + 40}$

3. $x + 4 \overline{)\, 3x^5 - 42x^3 - 2x + 376}$

4. You Are the Teacher.™

Here is the student's work. Find the (one) error.

I'm trying to solve $x^2 - 6 = -8x$

Transposing, I get $x^2 + 8x = 6$

Completing the square gives me $x^2 + 8x + 16 = 22$

Factoring $(x + 8)^2 = 22$

Square root $x + 8 = \pm\sqrt{22}$

Transposing $x = -8 \pm\sqrt{22}$ ☺

5. What do you call two ratios that are equal to each other?

E.g., $\dfrac{4}{5} = \dfrac{x}{19}$ [*E.g.* means *for example.*]

6. Thud could peel potatoes at the rate of 80/hour. Chris, at the rate of 400/hour. Each did half of the truckload of potatoes. Of course, Chris finished her half long before Thud. They worked a total of 9 hours. How long did Thud work?

7. The camp bells peal five minutes longer for lunch than for breakfast. They peal twice as long for dinner as for breakfast. They peal for a total of 85 minutes. How long do they peal for breakfast?

·······COMPLETE SOLUTIONS·······

1.

$$
\begin{array}{r}
5x^2 + 3x + 4 \\
3x+2 \,\overline{)\, 15x^3 + 19x^2 + 18x + 8} \\
\underline{15x^3 + 10x^2} \\
9x^2 + 18x \\
\underline{9x^2 + 6x} \\
12x + 8 \\
\underline{12x + 8} \\
0
\end{array}
$$

2.

$$
\begin{array}{r}
3x^2 + 2x + 1 + \frac{42}{7x-2} \\
7x-2 \,\overline{)\, 21x^3 + 8x^2 + 3x + 40} \\
\underline{21x^3 - 6x^2} \\
14x^2 + 3x \\
\underline{14x^2 - 4x} \\
7x + 40 \\
\underline{7x - 2} \\
42
\end{array}
$$

3.

$$
\begin{array}{r}
3x^4 - 12x^3 + 6x^2 - 24x + 94 \\
x+4 \,\overline{)\, 3x^5 + 0x^4 - 42x^3 + 0x^2 - 2x + 376} \\
\underline{3x^5 + 12x^4} \\
-12x^4 - 42x^3 \\
\underline{-12x^4 - 48x^3} \\
6x^3 + 0x^2 \\
\underline{6x^3 + 24x^2} \\
-24x^2 - 2x \\
\underline{-24x^2 - 96x} \\
94x + 376 \\
\underline{94x + 376}
\end{array}
$$

4. The fourth line (the factoring) should have been $(x + 4)^2 = 22$. This, of course, changes all the lines below the error.

5. The equality of two ratios is called a proportion.

6. *Let x = the number of hours that Thud worked.*

   Then 9 − x = the number of hours that Chris worked. If you wonder how we got the 9 − x, see the box on the first page of Lesson 26.

   Then 80x = the number of potatoes that Thud peeled.

   Then 400(9 - x) = the number of potatoes that Chris peeled.

   Since they peeled the same number of potatoes . . . $80x = 400(9 - x)$

   $x = 7.5$ hours $= 7$ hours and 30 minutes

7. *Let x = the number of minutes that the bells peal for breakfast.*

   Then x + 5 = the number of minutes they peal for lunch.

   Then 2x = the number of minutes they peal for dinner.

   $x + (x + 5) + 2x = 85$

   $x = 20$ minutes

peel potatoes
bells peal

Marshalltown, Iowa "In the center of the state and in the middle of all things good"

1. Solve $x^2 - 4x = 12$

2. Pat and Carol together can mow the camp lawn in seven hours. Working alone, Carol takes four hours longer than Pat to do the whole job. How long would it take Pat to do the whole job alone? Simplify your answer.

3. Suppose you have $\sqrt{w}$ where w is a whole number that is evenly divisible by 9. Can $\sqrt{w}$ be simplified?

4. Simplify and combine if possible: $\sqrt{32xy^2} + 4y\sqrt{x}$

5. Suppose you have to solve a quadratic equation $ax^2 + bx + c = 0$ where $b^2 - 4ac$ is equal to -16. What can you say about the number of solutions to this equation?

6. What is one million to the one-sixth power?

7. $x^2 + 10x + \xi = (x + \theta)^2$ What are the real numbers ξ and θ? This problem is a little different. If xi (ξ) and theta (θ) bother you, then replace them by *a* and *b*: $x^2 + 10x + a = (x + b)^2$. What are the real numbers *a* and *b*?

8. Using long division, find $(x^3 - 1) \div (x - 1)$.

9. Using the results of the previous problem, factor $x^3 - 1$. (This is something that we will do "officially" in advanced algebra.)

10. In a right triangle, one leg is three feet longer than the other. The hypotenuse is six feet long. How long is the shorter leg? Simplify your answer.

11. Solve $y^2 = 5$

12. Solve $z^2 = -25$

13. If you have $3x^3 - 5x^2 - 6x + 8$ chairs in an auditorium arranged in a rectangle, and there are $x - 2$ rows, how many chairs are in each row?

14. Solve $2x^2 - 5x - 3 = 0$ by completing the square. Show all the steps.

15. Solve $6x^2 - 7x = 2$

16. Suppose you have $\sqrt{y}$ where y is a natural number that is evenly divisible by 5. Can $\sqrt{y}$ be simplified?

447

answers

1.
$$x^2 - 4x = 12$$
$$x^2 - 4x - 12 = 0$$
$$(x - 6)(x + 2) = 0$$
$$x - 6 = 0 \quad \text{OR} \quad x + 2 = 0$$
$$x = -2, 6$$

2. This can be done by starting with *Let t = the number of hours it would take for Pat to do the job alone* or by using "six pretty boxes." Either way will work. The last time we used "six pretty boxes" for a job problem was back in Lesson 72.

	together	Carol	Pat
the whole job	7	$t + 4$	t
the part of the job that can be done in one hour	$\dfrac{1}{7}$	$\dfrac{1}{t + 4}$	$\dfrac{1}{t}$

The part of the job that can be done in an hour by Pat plus the part of the job that can be done in an hour by Carol is equal to the part of the job that can be done in an hour by both of them.

$$\frac{1}{t} + \frac{1}{t + 4} = \frac{1}{7}$$

$$\frac{7t(t + 4)}{t} + \frac{7t(t + 4)}{t + 4} = \frac{7t(t + 4)}{7}$$

$$7(t + 4) + 7t = t(t + 4)$$

$$7t + 28 + 7t = t^2 + 4t$$

$$0 = t^2 - 10t - 28 \quad \text{This doesn't factor.}$$

Using the quadratic formula
$$x = \frac{10 \pm \sqrt{100 - (4)(1)(-28)}}{2}$$

$$x = \frac{10 \pm \sqrt{212}}{2}$$

$$x = \frac{10 \pm \sqrt{4}\,\sqrt{53}}{2}$$

$$x = \frac{10 \pm 2\sqrt{53}}{2} = \frac{2(5 \pm \sqrt{53})}{2}$$

$$x = 5 \pm \sqrt{53} \quad \text{Discard } 5 - \sqrt{53} \text{ because we are}$$
looking for a positive answer. $x = 5 + \sqrt{53}$ hours

3. Yes. If w = 9c, where c is some whole number, then $\sqrt{w} = \sqrt{9c} = 3\sqrt{c}$

4. $\sqrt{32xy^2} + 4y\sqrt{x}$

$$= \sqrt{16y^2}\sqrt{2x} + 4y\sqrt{x}$$

$$= 4y\sqrt{2x} + 4y\sqrt{x} \quad \text{These cannot be added together.}$$

5. There aren't any. If you tried to solve $ax^2 + bx + c = 0$ using the quadratic formula, you'd end up with $x = \dfrac{-b \pm \sqrt{-16}}{2a}$ and this would have the square root of a negative number, which cannot be done.

6. That is the same as $\sqrt[6]{1,000,000}$ which equals 10 (since 10^6 = one million).

7. Here is my reasoning. $x^2 + 10x + \xi = (x + \theta)^2$

First, look at the right side

It is squared.

That means that the left side must also be a perfect square.

How do you make $x^2 + 10x + \xi$ into a perfect square?

Here are the three boxes from Lesson 83 . . .

$x^2 - 8x + 16$	$x^2 + 6x + 9$	$x^2 + 14x + 49$
$(x - 4)^2$	$(x + 3)^2$	$(x + 7)^2$

So to make $x^2 + 10x + \xi$ into a perfect square, $\xi = 25$.

And $x^2 + 10x + \xi$ will factor into $(x + 5)^2$.

Now look at the right side $(x + \theta)^2$. ⟵ $\theta = 5$

8.
$$
\require{enclose}
\begin{array}{r}
x^2 + x + 1 \\
x - 1 \enclose{longdiv}{x^3 + 0x^2 + 0x - 1} \\
\underline{x^3 - x^2} \\
x^2 + 0x \\
\underline{x^2 - x} \\
x - 1 \\
\underline{x - 1}
\end{array}
$$

9. $x^3 - 1 = (x - 1)(x^2 + x + 1)$ In arithmetic we know that if $3\overline{)21}$ (with 7 above) then we know that $3 \times 7 = 21$.

10.   *Let x = the length of the shorter leg.* ← This is what we are looking for.

Then x + 3 = the length of the other leg.

By the Pythagorean theorem, $x^2 + (x + 3)^2 = 36$

$$x^2 + x^2 + 6x + 9 = 36$$

$$2x^2 + 6x - 27 = 0$$

This doesn't factor.

Use the quadratic formula $x = \dfrac{-6 \pm \sqrt{36 - (4)(2)(-27)}}{4}$

$$x = \frac{-6 \pm \sqrt{252}}{4} = \frac{-6 \pm \sqrt{4}\sqrt{63}}{4} = \frac{-6 \pm \sqrt{4}\sqrt{9}\sqrt{7}}{4}$$

$$= \frac{-6 \pm 6\sqrt{7}}{4} = \frac{2(-3 \pm 3\sqrt{7})}{4} = \frac{-3 + 3\sqrt{7}}{2}$$

We discard $\dfrac{-3 - 3\sqrt{7}}{2}$ because it is negative and we are looking for length.

11. $y = \pm\sqrt{5}$

12. This has no real number solution since the square root of a negative number doesn't exist.

13. Looking at an example from arithmetic: If there were 18 chairs in 6 rows, we know that each row would contain $\dfrac{18}{6}$ chairs.

Similarly, if there are $3x^3 - 5x^2 - 6x + 8$ chairs in $x - 2$ rows, there would be $\dfrac{3x^3 - 5x^2 - 6x + 8}{x - 2}$ chairs in each row.

To simplify that fraction, factor top, factor bottom, cancel like factors.

To factor $3x^3 - 5x^2 - 6x + 8$

$x^2(3x - 5) - 2(x - 4)$ It doesn't factor by grouping.

Use long division instead.

$$
\begin{array}{r}
3x^2 + x\ \ -4 \\
x - 2{\overline{\smash{\big)}\,3x^3 - 5x^2 - 6x + 8}} \\
\underline{3x^3 - 6x^2} \\
x^2 - 6x \\
\underline{x^2 - 2x} \\
-4x + 8 \\
\underline{-4x + 8}
\end{array}
$$

14.

$$2x^2 - 5x - 3 = 0$$
$$2x^2 - 5x = 3$$
$$x^2 - (5/2)x = 3/2$$
$$x^2 - (5/2)x + 25/16 = 3/2 + 25/16$$
$$(x - 5/4)^2 = 24/16 + 25/16 = 49/16$$
$$x - 5/4 = \pm 7/4$$
$$x = 5/4 \pm 7/4 = 3, -1/2$$

15.

Try solving by factoring first

It doesn't factor. ☹

Use the quadratic formula

$$6x^2 - 7x = 2$$
$$6x^2 - 7x - 2 = 0$$

$$x = \frac{7 \pm \sqrt{49 - (4)(6)(-2)}}{12}$$

$$x = \frac{7 \pm \sqrt{97}}{12}$$

16. There's not enough information to say.

If y were equal to 25 or 75, then $\sqrt{y}$ could be simplified.

But if y were equal to 15 or 35, then $\sqrt{y}$ couldn't be simplified.

Copperopolis, California   Mark Twain slept and wrote here

1. Using long division find $(x^3 + 1) \div (x + 1)$.

2. Using the results of the previous problem factor $x^3 + 1$. (In advanced algebra, we'll "officially" learn how to factor the sum of cubes.)

3. If v were a whole number evenly divisible by 8, could $\sqrt{v}$ be simplified?

4. Give an example of a whole number that is not a natural number.

5. If $x^2 + 20x + \xi = (x + \theta)^2$ what are the real numbers ξ and θ?

6. If the two legs of a right triangle are equal and the hypotenuse is 6 yards long, how long is each of the legs?

7. You are given a radical with an index equal to 7 and a radicand equal to 23. Write this as a number raised to an exponent.

8. Solve $3x^2 - 11x - 4 = 0$

9. Solve $3x^2 - 11x - 5 = 0$

10. Find $(x^3 - 8) \div (x - 2)$

11. Using the results of the previous problem, factor $x^3 - 8$.

12. One leg of a right triangle is three inches shorter than the hypotenuse. The other leg is equal to 8 inches. How long is the first leg?

13. One leg of a right triangle is seven miles longer than the other. The hypotenuse is equal to 13 miles. How long is the shorter leg? (This can be solved by factoring.)

14. Solve $2x^2 - \pi x - 3 = 0$. (π is a real number just like 2 and 3.)

15. Together Carol and Jack can paint all the buildings at the camp in ten months. Carol, working alone, would take two more months than Jack working alone to paint all the buildings. How long would it take Jack working alone to paint all the buildings? Simplify your answer.

16. Solve $18x^2 + 324 = 0$

17. Solve $18x^2 = 324$ and simplify your answer.

18. If you were to write $33^{1/4}$ as a radical, what would the index on the radical be?

answers

1.

$$\begin{array}{r} x^2 - x + 1 \\ x + 1{\overline{\smash{\big)}\,x^3 + 0x^2 + 0x + 1}} \\ \underline{x^3 + x^2} \\ -x^2 + 0x \\ \underline{-x^2 - x} \\ x + 1 \\ \underline{x + 1} \end{array}$$

2. $x^3 + 1 = (x + 1)(x^2 - x + 1)$

3. Yes. Since v is divisible by 8, it could be written as $v = 4 \cdot 2w$ where w is a whole number. Then $\sqrt{v} = \sqrt{4 \cdot 2w} = \sqrt{4}\sqrt{2w} = 2\sqrt{2w}$.

4. Zero is the only possibility.

$$\text{The whole numbers} = \{0, 1, 2, 3, 4 \ldots\}.$$
$$\text{The natural numbers} = \{1, 2, 3, 4, 5 \ldots\}.$$

5. Here is my reasoning. $\qquad x^2 + 20x + \xi = (x + \theta)^2$

First, look at the right side: $(x + \theta)^2$.

It is squared.

That means that the left side must also be a perfect square.

So to make $x^2 + 20x + \xi$ into a perfect square, $\xi = 100$.

And $x^2 + 20x + \xi$ will factor into $(x + 10)^2$.

Now look at the right side $(x + \theta)^2$. $\qquad \theta = 10$

6. *Let x = the length of each of the two legs.*

Pythagorean theorem

$$x^2 + x^2 = 36$$
$$2x^2 = 36$$
$$x^2 = 18$$
$$x = \sqrt{18} = \sqrt{9}\sqrt{2} = 3\sqrt{2} \text{ yards}$$

$\pm 3\sqrt{2}$ is not correct since we are looking for a length.

7. $23^{1/7}$

8.

$$3x^2 - 11x - 4 = 0$$

Does it factor?

$$(3x + 1)(x - 4) = 0$$
$$x = -\frac{1}{3} \text{ or } x = 4$$

9. $3x^2 - 11x - 5 = 0$

Does it factor? We want two things that multiply to $-15x^2$ and add to $-11x$.

It does not factor. Use the quadratic formula. $x = \dfrac{11 \pm \sqrt{181}}{6}$

10.
$$
\begin{array}{r}
x^2 + 2x + 4 \\
x - 2\overline{)\,x^3 + 0x^2 + 0x - 8} \\
\underline{x^3 - 2x^2} \\
2x^2 + 0x \\
\underline{2x^2 - 4x} \\
4x - 8 \\
\underline{4x - 8}
\end{array}
$$

11. $x^3 - 8 = (x - 2)(x^2 + 2x + 4)$

12. Using the diagram as the Let statements . . .

Pythagorean theorem $8^2 + x^2 = (x + 3)^2$

$64 + x^2 = x^2 + 6x + 9$

$55 = 6x$

$\dfrac{55}{6}$ inches $= x$

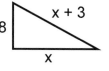

13. *Let x = the length of the shorter leg. Then x + 7 = the length of the longer leg.*

Pythagorean theorem $x^2 + (x + 7)^2 = 13^2$

$x^2 + x^2 + 14x + 49 = 169$

$2x^2 + 14x - 120 = 0$

Divide both sides by 2 to make things easier $x^2 + 7x - 60 = 0$

$(x - 5)(x + 12) = 0$

$x = 5, -12$

Discard the -12 since we are dealing with length. $x = 5$ miles.

14. By the quadratic formula $x = (\pi \pm \sqrt{\pi^2 + 24})/4$

15.

	together	Carol	Jack
the whole job	10	$t + 2$	t
one month's worth	1/10	$1/(t+2)$	$1/t$

$\dfrac{1}{t} + \dfrac{1}{t+2} = \dfrac{1}{10} \Rightarrow \dfrac{10t(t+2)}{t} + \dfrac{10t(t+2)}{t+2} = \dfrac{10t(t+2)}{10} \Rightarrow$

$10(t + 2) + 10t = t(t + 2) \Rightarrow 10t + 20 + 10t = t^2 + 2t \Rightarrow 0 = t^2 - 18t - 20$

$\Rightarrow t = \dfrac{18 \pm \sqrt{324 - 4(1)(-20)}}{2} = \dfrac{18 \pm \sqrt{404}}{2} = \dfrac{18 \pm \sqrt{4\sqrt{101}}}{2} = 9 + \sqrt{101}$

16. This has no solution in the real numbers since the square root of a negative number doesn't exist.

17. $\pm\sqrt{18} = \pm\sqrt{9}\sqrt{2} = \pm3\sqrt{2}$

18. It would be equal to 4. $33^{1/4} = \sqrt[4]{33}$

1. Using long division, find $(x^3 + 4x + 39) \div (x + 3)$

2. What natural number equals the radical whose radicand is 8 and whose index on the radical is three?

3. Solve $x^2 + x + 1 = 0$

4. Solve $x^2 + x = 1$

5. Together Pat and Carol can wash all the windows in camp in four hours. Carol working alone takes three more hours than Pat working alone to do that job. How long does it take Pat to do the whole job alone?

6. Find the values of ξ and θ that make the following true:
$$x^2 - 16x + \xi = (x + \theta)^2$$

7. Solve $-7x^2 + 49 = 0$

8. What is one-fourth raised to the one-half power?

9. Showing all your work, solve $6x^2 + 7x - 20 = 0$ by factoring.

10. Showing all your work, solve $6x^2 + 7x - 20 = 0$ by completing the square.

11. Showing all your work, solve $6x^2 + 7x - 20 = 0$ by the quadratic formula.

12. One leg of a right triangle is five feet longer than the other. The hypotenuse is six feet long. How long is the shorter leg? Simplify your answer.

13. Alexander can correct a set of homework papers in five hours. If he helps Betty correct a set, it takes two hours less than if Betty worked alone. How long does it take for Betty to correct a set of homework papers alone? Simplify your answer. Also, if you have a calculator, estimate to the nearest tenth of an hour how long she would take working alone.

14. We know that $x^{1/7}$ means $\sqrt[7]{x}$, which is a radical with an index of seven. There is one integer that can't be used as an index. What is it?

15. Suppose you have $\sqrt{u}$ where u is a whole number evenly divisible by 20. Can $\sqrt{u}$ be simplified?

16. Using long division, find $(2x^3 - 5x + 5) \div (2x - 1)$

answers

1.
$$\begin{array}{r} x^2 - 3x\ + 13 \\ x+3{\overline{)\ x^3 + 0x^2 + 4x + 39}} \\ \underline{x^3 + 3x^2} \\ -3x^2 + 4x \\ \underline{-3x^2 - 9x} \\ 13x + 39 \\ \underline{13x + 39} \end{array}$$

2. $\sqrt[3]{8} = 2$

3. Using the quadratic formula $\quad x = \dfrac{-1 \pm \sqrt{1 - 4(1)(1)}}{2} = \dfrac{-1 \pm \sqrt{-3}}{2}$

This has no solution since we get the square root of a negative number.

4. Using the quadratic formula on $x^2 + x - 1 = 0$ $\qquad x = (-1 \pm \sqrt{5})/2$

5.   *Let t = the number of hours it would take Pat to do the whole job alone.*

       Then t + 3 = the number of hours for Carol to do the whole job alone.

Pat can do $\dfrac{1}{t}$ of the job in one hour. Carol can do $\dfrac{1}{t+3}$ of the job in one hour.

Since they can do the whole job in 4 hours, they can do $\dfrac{1}{4}$ of the job in one hour.

$$\frac{1}{t} + \frac{1}{t+3} = \frac{1}{4}$$

$$\frac{4t(t+3)}{t} + \frac{4t(t+3)}{t+3} = \frac{4t(t+3)}{4}$$

$$4(t+3) + 4t = t(t+3)$$

$$4t + 12 + 4t = t^2 + 3t$$

$$0 = t^2 - 5t - 12 \qquad \text{This doesn't factor.}$$

$$t = \frac{5 \pm \sqrt{25 - 4(1)(-12)}}{2}$$

$$t = \frac{5 \pm \sqrt{73}}{2}$$

We are looking for a positive answer, and so we omit $\dfrac{5 - \sqrt{73}}{2}$

6. $\qquad\qquad\qquad\qquad x^2 - 16x + \xi = (x + \theta)^2$

First, look at the right side: $(x + \theta)^2$. It is squared.
That means that the left side must also be a perfect square.
So the left side will be $x^2 - 16x + 64$. $\qquad \xi = 64$
$x^2 - 16x + 64 = (x - 8)^2$. $\qquad \theta = -8$

7.
$$-7x^2 + 49 = 0$$
$$49 = 7x^2$$
$$7 = x^2$$
$$\pm\sqrt{7} = x$$

8. $(1/4)^{1/2} = \sqrt{1/4} = 1/2$

9. By factoring

$$6x^2 + 7x - 20 = 0$$
$$(2x + 5)(3x - 4) = 0$$
$$2x + 5 = 0 \quad \text{OR} \quad 3x - 4 = 0$$
$$x = -5/2 \quad \text{OR} \quad x = 4/3$$

10. By completing the square

$$6x^2 + 7x - 20 = 0$$
$$6x^2 + 7x = 20$$
$$x^2 + (\frac{7}{6})x = \frac{20}{6}$$
$$x^2 + (\frac{7}{6})x + \frac{49}{144} = \frac{20}{6} + \frac{49}{144}$$
$$(x + \frac{7}{12})^2 = \frac{20(24)}{6(24)} + \frac{49}{144}$$
$$(x + \frac{7}{12})^2 = \frac{529}{144}$$
$$x + \frac{7}{12} = \pm\frac{23}{12}$$
$$x = -\frac{7}{12} \pm \frac{23}{12}$$
$$x = 4/3 \text{ or } -5/2$$

11. By the quadratic formula

$$6x^2 + 7x - 20 = 0$$
$$x = \frac{-7 \pm \sqrt{49 - 4(6)(-20)}}{12}$$
$$x = \frac{-7 \pm \sqrt{529}}{12}$$
$$x = \frac{-7 \pm 23}{12}$$
$$x = 4/3 \text{ or } -5/2$$

Now that you have solved the same equation three different ways, take a second and look back over your work and see if you have a favorite method.

12. Using the diagram as the Let statements . . .

Pythagorean theorem

$$x^2 + (x + 5)^2 = 36$$
$$x^2 + x^2 + 10x + 25 = 36$$
$$2x^2 + 10x - 11 = 0$$

It doesn't factor.

Use the quadratic formula.

$$x = \frac{-10 \pm \sqrt{100 - 4(2)(-11)}}{4}$$

$$x = \frac{-10 \pm \sqrt{188}}{4} \qquad \sqrt{188} = \sqrt{4}\sqrt{47} = 2\sqrt{47}$$

$$x = \frac{-5 \pm \sqrt{47}}{2}$$

We omit $\dfrac{-5 - \sqrt{47}}{2}$ since we are looking for a positive length.

13. *Let t = number of hours to correct the papers if Betty works alone.* Alexander can do $\dfrac{1}{5}$ of the job in one hour. Together they can do the job in $t - 2$ hours. Together they can do $\dfrac{1}{t-2}$ of the job in one hour.

$$\frac{1}{t} + \frac{1}{5} = \frac{1}{t-2} \ \Rightarrow \ \frac{5t(t-2)}{t} + \frac{5t(t-2)}{5} = \frac{5t(t-2)}{t-2} \ \Rightarrow$$

$$5(t-2) + t(t-2) = 5t \ \Rightarrow \ 5t - 10 + t^2 - 2t = 5t \ \Rightarrow \ t^2 - 2t - 10 = 0$$

$$\Rightarrow t = \frac{2 \pm \sqrt{4 - 4(1)(-10)}}{2} = \frac{2 \pm \sqrt{44}}{2} = \frac{2 \pm \sqrt{4}\sqrt{11}}{2} = \frac{2 \pm 2\sqrt{11}}{2}$$

$= 1 + \sqrt{11}$ hours which is approximately 4.3 hours (to the nearest tenth).

14. Zero, since $\sqrt[0]{x}$ would mean $x^{1/0}$ which is division by zero which is not allowed.

15. Yes. If $u = 20v$ where v is a whole number, then $\sqrt{u} = \sqrt{20v} = \sqrt{4 \cdot 5v}$ $= 2\sqrt{5v}$

16.

$$
\begin{array}{r}
x^2 + x/2 - 9/4 + \dfrac{11/4}{2x - 1} \\
2x - 1 \overline{)\, 2x^3 + 0x^2 - 5x + 5\,} \\
\underline{2x^3 - x^2} \\
x^2 - 5x \\
\underline{x^2 - x/2} \\
-(9/2)x + 5 \\
\underline{-(9/2)x + 9/4} \\
11/4
\end{array}
$$

Arithmetic time!
$-5 - (-\tfrac{1}{2}) = -5 + \tfrac{1}{2}$
$= -10/2 + 1/2 = -9/2$

Arithmetic time!
$+5 - (+9/4) = 5 - 9/4$
$= 20/4 - 9/4 = 11/4$

Monday morning. The first official day of boot camp for the thirteen draftees. In the movies, the first day of boot camp would start around 4:30 with a drill instructor going through each bunkhouse, banging on a metal garbage can to wake up all the soldiers. They would then have to rush outside in the cold and dark and line up for inspection. This would be followed by a three-mile jog.

Having seen a lot of those boot camp movies, Thud got up at 3:30 a.m. and headed to the bathroom to wash the smell of potatoes off his hands. He brushed his teeth and his hair. He was determined to disprove what all the guys said about him: "Thud isn't the genius of our group."

Back at his bed, he took out his boots, shined them, and put them on. Then he tried to put on his pants but found they wouldn't go on over his boots. He noted that putting on pants and boots is not commutative in contrast to addition, which is commutative: $3 + 4 = 4 + 3$.

He took off his boots and put on his pants and then his boots.

"Oh good," he thought to himself. "I remembered to put on my socks first." It had always confused him as a child when his mom told him to "Put on your shoes and socks." She should have said it the other way around.

Then he put on his undershirt and his shirt and stood in front of the mirror. Something felt funny, but he couldn't figure out what it was until he got back to his bed and found a pair of underpants lying there.

By 4:30 a.m. he had all his clothes on. Next, Thud started to put stuff in his pack for the three-mile jog that he knew they always did on the first day of boot camp. He could hear the other men snoring as he got his things out of his locker. For once, he thought, he would not be the least prepared of the group.

What to take? His little bear was important, in case he got lonely. He gave it a squeeze and it went "Grrrr!"

His next item was the horn that his parents had given him for his You've-Been-Drafted party. That might come in handy on the jog if they needed to signal for help or something. He blew into it and it gave a little, "Toot!" One of the other guys in the room yelled, "Knock it off Thud and go back to bed!" That didn't bother Thud very much since people were always yelling at him. Besides, he thought, they'll be grateful to him if his horn saves their lives.

What else might he need on his three-mile jog? He felt he should take only essential items so his pack wouldn't get too heavy. One last item he found in his locker. It was his blender. You never know when that might come in handy. Someone on the jog might need to have a milkshake or something.

He plugged it in and pressed a button. It made a loud whirring sound. Several more guys yelled at Thud.

It was 4:45 a.m. and Thud was ready. He sat on the edge of his bunk and waited for reveille.

By 5:30 a.m. he was getting a little bored and opened his pack and looked at the three items in it. Thud was about to reinvent the concept of **function**.

He noticed that every time he blew on his horn he heard "Toot!" and never "Grrrr!" and never "Whirrr!" There was only one sound associated with his horn.

Each item in the pack was associated with a single sound.

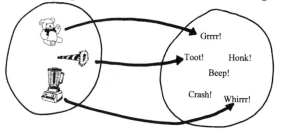

He thought of two sets: (1) the set of the three things that were in his pack and (2) the set of all sounds. He noticed that *each element in the first set was associated with exactly one element in the second set.* It's this association that we call a function.

Start with two sets. Make a rule (draw an arrow) that maps each member of the first set to exactly one member of the second set, and you have a function. A function isn't the first set. It isn't the second set. It's the rule (the arrows) that relates the first set to the second set.

Thud looked around the room, feeling very proud that he had invented the idea of function. He wanted to invent some more examples of functions.

> Thud's second example of a function

He thought of all the men who were sleeping in

the room and made that his first set. He let the second set be all beds in the room. One function could be to assign to each man the bed he was now in. Then each man was assigned *exactly one* element in the second set (since every man was in one bed and no man was in two beds).

This was getting to be fun. Let the first set be the set of all beds in the room, and let the second set be the set of all men in the room. There were many empty beds in the bunkhouse, so Thud couldn't make a function by assigning to each bed the man in it. Instead, Thud made the rule: Assign each bed to Thud. This was a great function. It worked. Each element of the first set (the beds) was assigned to *exactly one* element of the second set.

> Thud's third example of a function

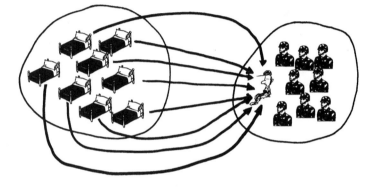

Thud was trying to get really creative. He knew that he could pick any first set (which is called the **domain**) and any second set (which is called the **codomain**) and any rule so long as each element of the domain had exactly one **image** in the codomain.

> Thud's fourth example of a function

For his first set, he chose the set of all magazines in the room. For the second set he picked the set of all music composers. The rule Thud used was that if the cover of the magazine had the letter M on it, then that magazine was assigned to the composer Johann Sebastian Bach. If it didn't have an M anywhere on the front cover, then it was assigned to Giuseppe Verdi. If the cover was missing, the magazine was assigned to Irving Berlin.

Why the letter M? A function is *any* rule that assigns to each element of the domain, exactly one element of the codomain. It turns out that M is Thud's favorite letter, so he used it. When you create functions you get to use your favorite stuff—just like my favorite Greek letter is ξ.

Using Thud's function, all the copies of *Modern Romance* would be assigned to Johann Sebastian Bach. The issue of *Creative Cooking for Cool Cats* that listed on its cover the article, "How I Made a Taffy Pulling Machine in My Basement," had plenty of M's in it, so it was also assigned to Johann Sebastian Bach.

It wouldn't have been a function if Thud had assigned all magazines with the letter M on their front covers to Johann Sebastian Bach and all magazines with the letter P on their front covers to Igor Fedorovich Stravinsky. Why not? Where do you assign an issue of *Modern Plumbing*? A function is a rule that assigns to each element of the first set, *exactly one* element in the second set. *Modern Plumbing* would be assigned to both Bach and Stravinsky.

When Fred was much younger, he played a game with his mother that he called Guess the Function. He would think of a function and give her lots of examples, and she was supposed to guess what rule he was thinking of.

One time he gave her:

dog $\rightarrow$ 4
cat $\rightarrow$ 4
snake $\rightarrow$ 0   (where "$\rightarrow$" means "maps to.")
bird $\rightarrow$ 2
Mom $\rightarrow$ 2
spider $\rightarrow$ 8
worm $\rightarrow$ 0
insect $\rightarrow$ 6
our kitchen table $\rightarrow$ 4

and his mom looked long and hard at those examples but couldn't figure it out. Fred tried to give some hints by dancing and throwing his legs up in the air like he was doing a cancan. He said, "Something attached to feet," and she responded, "Yards?"

He said, "Remember when we saw the *Moby Dick* movie. What did Captain Ahab have only one of?"

She guessed, "He had only one white whale."

Finally Fred told her that the rule he was thinking of was: Take the object and count the number of legs it has.

It's time you got a chance to play with functions.

Your Turn to Play

1. From now on, you will see examples of functions in every math course you take. Even in elementary school you saw lots of functions. Here is an example of a function with domain equal to pairs of numbers, like (5, 7), and codomain is numbers. Can you guess the common name of this function?

$(3, 4) \rightarrow 7$ $(8, 88) \rightarrow 96$ $(2, 4) \rightarrow 6$ $(10, 10) \rightarrow 20$ $(1, 1) \rightarrow 2$

2. Is this a function?

domain   codomain

3. Invent a function where the domain is the set of all 50 states and the codomain is the set consisting of just the number four (= {4}).

4. Let's play Guess the Function. Here are your examples. You are to try to figure out what rule I'm using. $5 \rightarrow 11$ $20 \rightarrow 41$ $3 \rightarrow 7$
$13 \rightarrow 27$ $-8 \rightarrow -15$

5. "Take the principal square root of" is a good example of a rule that could be used as a function. What might the domain for this rule be?

6. "Take a square root of" is *not* a good example of a rule that could be used as a function. Why not?

7. Is this a function?

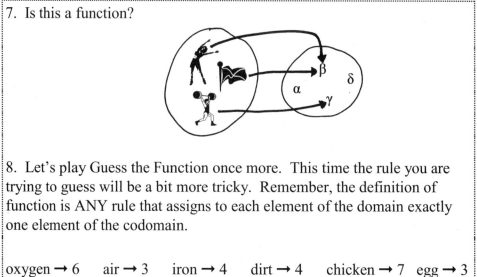

8. Let's play Guess the Function once more. This time the rule you are trying to guess will be a bit more tricky. Remember, the definition of function is ANY rule that assigns to each element of the domain exactly one element of the codomain.

oxygen → 6   air → 3   iron → 4   dirt → 4   chicken → 7   egg → 3
hair → 4   comb → 4   brush → 5   broom → 5   mop → 3
stocking → 8   Jeremiah → 8   something → 9   nothing → 7
pie → 3   pi → 2

Some families play Guess the Function at the dinner table. It's much better than watching television while you eat.

The best functions are those that are obvious once you hear the rule. For example, Fred's rule of "count the number of legs."

A bad choice is a rule that is so hard that nobody could ever guess it. For example: Smith → 21 Lee → 6 Gauss → 0   Anderson → 10   Wright → 3 Bliss → 2 Samson → 8

.... **COMPLETE SOLUTIONS** ....

1. Where I come from, we call that addition.

2. The definition of function is a rule that assigns to *each* element of the first set exactly one element of the second set. The horn didn't receive an assignment. It has no image in the second set. It is not a function.

3. There's only one possible function. It is the rule that assigns each state to 4.

4. My rule is that I double the number and then add one to it. In symbols: $x \rightarrow 2x + 1$.

5. The largest domain I can think of would be the non-negative real numbers since you can take the principal square root of any positive real number, and you can take the principal square root of zero. (You can't take principal square roots of negative numbers—at least not in this course.) You would have also been right if you had suggested sets like the natural numbers, the whole numbers, the positive real numbers, or the positive rational numbers. Each of these would be good domains for the rule, "Take the principal square root of."

6. "Take a square root of" can give us two answers. There are, for example, two square roots of 49. They are +7 and –7. Since a function is defined as a rule that associates to each element of the domain *exactly one* element of the codomain, this rule wouldn't work.

7. A function is defined as any rule that associates to each element of the first set exactly one image in the second set. This diagram fits that definition. The fact that the person stretching and the flag of Scotland are both mapped onto the Greek letter beta (β) doesn't violate the definition. (Recall Thud's third example of a function in which all the beds were mapped to him.)

8. The best functions are those that are obvious once you know the rule. Did you notice in this example that the longer words have bigger numbers attached to them? The rule is: To each word assign the number of letters in that word.

In case you are curious, the rule for Smith → 21 Lee → 6 Gauss → 0
Anderson → 10   Wright → 3 Bliss → 2 Samson → 8 is the number of
people with that last name in Rake, Iowa, in April, 1923.
   This is not a rule that people would call obvious.

❧❧❧❧❧❧❧❧❧❧❧❧

It was now 6 a.m., and Thud put on his packback. He knew that at any moment Sergeant Snow would burst through the doorway, wake everyone up, and announce the three-mile jog. Thud was going to be the first one out the door. He would feel so proud.

What Thud didn't understand was that Sergeant Snow didn't enjoy getting up before dawn. In fact it was nearly 8 in the morning before he rolled out of bed and headed to the fridge in his room for a little pre-breakfast snack. Then, after wiping the bean sauce off his face, he headed to the control panel that was installed in his room.

A.A.A.A.

Central Control Panel

When Snow pressed the first button on the panel, the lights went on in the chow hall. The second button turned on the microwave conveyer belt system. Button #3 turned the lights on in the bunkhouse. Button #4 played reveille in the bunkhouse. (Actually it was a tape recording of a Strauss waltz played on a saxophone.)

Button #5 turned on a TV program of a girl doing exercises. Watching this program for five minutes each morning was the extent of Sergeant Snow's attempts at physical fitness. He didn't do the exercises but just watched. He knew that you shouldn't exercise just after you've eaten.

Since pushing each button gave a single unique unvarying response, his control panel was a good example of a function. If pushing button #3 sometimes turned on the lights in the bunkhouse and sometimes turned on the lawn sprinklers, then it wouldn't be a function. If pushing button #8 did nothing at all, then it wouldn't be a function. Each button, one response—that's what we're looking for.

The fellows in the bunkhouse heard the waltz music and figured it was time to get up. They got on their bathrobes and headed to the chow hall. Thud didn't exactly know what was happening and followed them.

When Sergeant Snow wandered in to say good morning to the men and tell them what they'd be doing today, he noticed that some of the chairs were unoccupied. He decided to take roll and noticed that Fred, Pat, Chris, and Carol were not there. Jack said that he had been with Fred last night, but that they had been separated. No one knew the whereabouts of Pat, Chris, and Carol.

This was serious. Four unexcused absences. They had to be found. Sergeant Snow told the men that they would have to search the camp after breakfast. Each man received his assignment. Thud was assigned to go to the colonel's house and ask there. Another was to check the women's bunkhouse. Another to check the minefield. Jack was to ask at the chaplain's house.

Thud thought to himself, "Here's another function. If the first set was the set of these nine men and the second set was the set of all spots at the camp, we have a function whose rule associates to each member of the first set the place that each man was assigned to search."

If the first set had been the nine men plus Sergeant Snow, then we wouldn't have a function since Sergeant Snow wasn't doing any searching. He would not have an image in the codomain.

He gave the men plenty of time to do their searching and announced that their first class (on "How to Make a Grenade Work") would be held at 11 a.m. After breakfast Sergeant Snow was heading back to bed for his first regular nap of the day. By setting the first class at eleven o'clock he assured himself an uninterrupted sleep.

step 1: pull pin
step 2: run

Going to the colonel's house was the only search mission that was physically demanding. The mansion was set on a hill at the highest point in the camp. There were great views when you got to the top, but the walk up the hill was tough after ingesting a quart of beans and hot dogs.

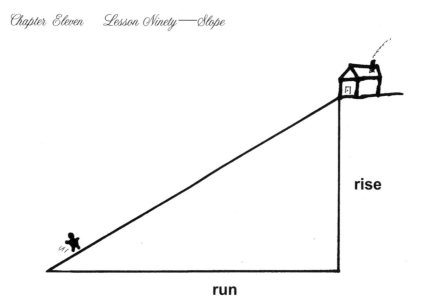

The steepness of a hill depends on two distances: the rise and the run. Neither number alone tells you much. For example, if you know that the rise is 100 feet, you really don't know how steep the hill is, but if you also know that the length of the run is 5280 feet, then you know that the hill isn't very steep at all.

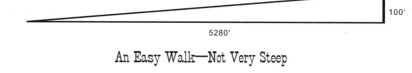

An Easy Walk—Not Very Steep

But if the run is only 200', then you'd better get out your mountaineering equipment.

100'
200'

A Tough Climb—Pretty Steep

The mathematical word for the steepness of a line is **slope**.

There are easy ways to define slope—and there are **hard** ways. Some banal algebra books offer difficult definitions.

468

Here are some of them that I have found in other books. I am going to put them in tiny type in a box. Please ignore these definitions.

Other Ways to Define **Slope**

Definition #1: Take the change in the y coordinate and divide it by the change in the x coordinate.

Definition #2: Slope = $\dfrac{y_2 - y_1}{x_2 - x_1}$ where (x_1, y_1) and (x_2, y_2) are two points on the line.

Definition #3: Slope is defined as the vertical displacement divided by the horizontal displacement.

Definition #4: m = $\dfrac{\Delta y}{\Delta x}$ where "Δ" stands for "change in" or "difference." The letter Δ is the Greek capital letter delta. Delta (Δ) was chosen because *delta* sounds like *difference*. "m" is the letter used for slope.

Wait! Stop! I, your reader, want to know why we use "m" to indicate slope.
We use m for slope, because we use s for arc length.
Okay. Why do we use s for arc length?
We use s for arc length, because we use a for acceleration. Logical isn't it?
No.

You can ignore these definitions.
They're not fit for human reading.

Why make it so complicated? We simply say that **slope** $= \dfrac{\text{rise}}{\text{run}}$

Quick test: What's the slope of this line?

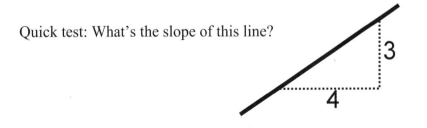

If you shouted out $\dfrac{3}{4}$ you were right. Both slope and function are easy concepts, but they are so important that they are mentioned in lots of the math courses that you'll take after beginning algebra.

After beginning algebra, comes:
> ➤ advanced algebra
>> ➤ geometry
>>> ➤ trig
>>>> ➤ calculus.

Advanced algebra will teach slope in Chapter 4 and spend all of Chapter 7 discussing functions. (There are ten chapters in the book.)

Geometry—dealing with lines and squares and stuff—certainly will include the concept of slope.

Trig—dealing with right triangles and the ratios of the sides—will introduce things like the tangent function. (If you look on your scientific calculator, you'll see a button labeled "tan." That stands for the trig function "tangent.") In the trig class, we tie it all together when we show that tan θ = slope, but you're not allowed to know that now.

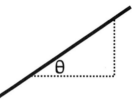

After doing all the slope and function business in advanced algebra, geometry, and trig, we get to calculus and there we find more uses for the idea of slope and tangent. In *Life of Fred: Calculus Expanded Edition* the first chapter is entitled, "Functions" and the fourth chapter, "Slope." The new trick we do in calculus is find the slopes on curves and not just on straight lines.

All this begins with the little formula:

slope = rise/run

If this is your book, then you may want to color this formula with your highlighter. The money you've saved because we've kept the price of this book down (by printing only in black) can be spent on the necessities of life like pizza.

470

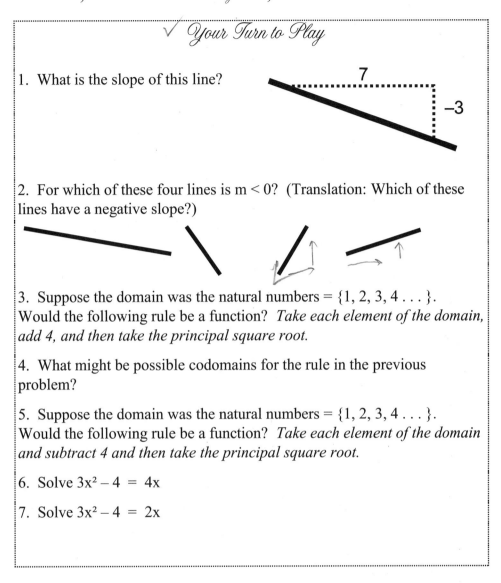

✓ *Your Turn to Play*

1. What is the slope of this line?

2. For which of these four lines is m < 0? (Translation: Which of these lines have a negative slope?)

3. Suppose the domain was the natural numbers = {1, 2, 3, 4 . . . }. Would the following rule be a function? *Take each element of the domain, add 4, and then take the principal square root.*

4. What might be possible codomains for the rule in the previous problem?

5. Suppose the domain was the natural numbers = {1, 2, 3, 4 . . . }. Would the following rule be a function? *Take each element of the domain and subtract 4 and then take the principal square root.*

6. Solve $3x^2 - 4 = 4x$

7. Solve $3x^2 - 4 = 2x$

. . . . COMPLETE SOLUTIONS . . . .

1. Slope = m = $\dfrac{\text{rise}}{\text{run}}$ = $\dfrac{-3}{7}$ [m is the letter which stands for slope.]

2. The first two have a negative slope—think *Going downhill.*

3. Under that rule the number 5, for example, would be mapped to the number 3. (Five plus four is nine and the principal square root of 9 is 3.) Under that rule the number 6 would be mapped to $\sqrt{10}$. Each element in the domain would be assigned to exactly one element in the codomain. Yes, it is a function.

4. Your answer might vary from mine. The largest possible set that might work as a codomain would be the set of every possible number and everything in the physical universe. This set would contain the real numbers; it would contain all the postage stamps issued by Argentina in 1963; it would contain all the films of Jeanette MacDonald.

   The smallest possible set which could be the codomain would be the set that just contained the images of {1, 2, 3, 4 . . . } under the rule of *add 4 and take the principal square root*. This would be the set { $\sqrt{5}$, $\sqrt{6}$, $\sqrt{7}$, $\sqrt{8}$, 3, $\sqrt{10}$, $\sqrt{11}$, $\sqrt{12}$ . . . }.

5. What would be the image of 1 under this rule? I would first subtract 4 and then take the principal square root. You would get $\sqrt{-3}$ which doesn't make any sense. Since 1 does not have an image, this rule is not a function. The definition of function says that *every* element in the domain must be assigned.

6. $3x^2 - 4 = 4x$ transposing $3x^2 - 4x - 4 = 0$
   factoring $(3x + 2)(x - 2) = 0$ $x = -2/3, 2$

7. $3x^2 - 4 = 2x$ transposing $3x^2 - 2x - 4 = 0$
   using the quadratic formula $x = \dfrac{2 \pm \sqrt{4 - 4(3)(-4)}}{2(3)} = \dfrac{2 \pm \sqrt{52}}{6}$

   (If you noticed that $\sqrt{52} = \sqrt{4}\sqrt{13} = 2\sqrt{13}$, then after several steps you could have x $= \dfrac{1 \pm \sqrt{13}}{3}$)

Jack was the only one to find out any news about where the four missing soldiers were. When Jack knocked on the chaplain's door, he was greeted by the chaplain's wife. She said that her husband was on the roof repairing some storm damage.

Jack headed around to the side of the house and climbed the ladder to the roof. He noticed that the slope of the ladder was equal to $\frac{15}{5}$ or 3. He liked that slope for a ladder.

15'

5'

Not a good slope for a ladder

If the slope had been something like 1/2, he would have been worried since there would have been the danger of the ladder sliding out from under him.

"Hello Jack!" called out the chaplain. "How have things been?"

"We've got trouble at the camp," Jack began. "Four of our soldiers are missing: Fred, Pat, Chris, and Carol. We're searching the whole camp for them."

The chaplain said that he didn't know where Pat, Chris, or Carol were, but he did know about Fred. He told Jack all about the injury that Fred had received last night. "It's a real shame that no one was with that little fellow when he went into town last night."

Jack bit his tongue for a moment but then said, "I was with him, but we got separated."

"It happens," the chaplain said as he set down the roofing shingle and hammer that he had in his hands. They got off the roof and headed off to see the sergeant.

When they got to the sergeant's door, they found a sign posted on it:

Jack and the chaplain looked at each other. Without saying a word, they each knew that this was important and headed outside to go to the colonel's mansion. They passed the big tree. Thud was sitting under it playing with his blender.

The chaplain was no slouch when it came to being physically fit, and when he and Jack climbed the hill up to the mansion, neither one had to slow down for the other.

"I sure wish I knew how steep this hill is!" puffed Jack. "I don't know the rise or the run. All I know is that Fred once graphed it when he was drawing a map of the camp. He said that the line was $y = 2x + 3$."

"Well," the chaplain puffed back, "That's enough to know the slope."

Jack knew he had a math challenge. He loved thinking about math. It gave him real pleasure to figure things out. He didn't want the chaplain to give him any more hints—this was now *his* problem to play with.

Okay, you also don't want to be left out. It's . . .

Your Turn to Play

1. Let's look at the line $y = 2x + 3$. If we were graphing it, we'd first name possible values for x and then find the corresponding values of y. Fill in the values: (2, ?) and (5, ?).

2. Continuing the previous problem, plot the two points and draw the line.

3. Continuing the previous problem, we know two points on the line. In

going from the point (2, 7) to the point (5, 13), how much does the value of x change?

4. Continuing the previous problem, how much does the y value change in going from (2, 7) to (5, 13)?

5. Continuing the previous problem, what is the name of the change in y in the solution to problem 4?

6. Concluding the previous problem, what is the slope of the line whose equation is $y = 2x + 3$?

7. What is the slope of $y = 3x + 1$?

(Hint: Follow the steps of problems 1–6.)

8. You may have noticed that the slope of $y = \mathbf{2}x + 3$ is 2.

　　　　　　 that the slope of $y = \mathbf{3}x + 1$ is 3.

Now, without doing those six yucky steps, make a wild guess:

What's the slope of $y = 4x + 9$?

9. Now, if m and b are numbers, what's the slope of $y = mx + b$?

10. What's the slope of the line $-50x + 10y = 17$?

11. The slope of $y = (-2/5)x + 3$ is $-2/5$. So far that's easy. But that's a negative slope. Draw a line with a slope of $-2/5$. The rise will be -2 and the run will be 5.

12. Fill in the blank: If we think of lines with a positive slope as going uphill, we would consider lines with negative slopes as going __?__.

13. How would you describe the graph of a line with zero slope?

14. What is the slope of $y = 0x + 5$?

. . . . COMPLETE SOLUTIONS . . . .

1. If x equals 2 then the equation $y = 2x + 3$ becomes $y = 2(2) + 3 = 7$. So (2, 7) is one point on the line. If x equals 5, then $y = 2x + 3$ becomes $y = 2(5) + 3 = 13$, and we have the point (5, 13).

2.

3. Going from (2, 7) to (5, 13), the value of x changes from 2 to 5, so it increases by 3. The dashed line on the graph shows the change in x.
The dashed line is the run.

4. To go from (2, 7) to (5, 13), the value of y changes from 7 to 13, so it increases by 6.

On the graph the dashed line shows the change in y.

5. That vertical dashed line is the rise.

6. The rise is 6 and the run is 3. Therefore, the slope (which is $\frac{\text{rise}}{\text{run}}$) is $\frac{6}{3}$ which is equal to 2.

7. $y = 3x + 1$. First I name a couple of x values and find the corresponding y values.

   If x = 1, then $y = 3x + 1$ becomes $y = 3(1) + 1 = 4$. The point (1, 4).

   If x = 4, then $y + 3x + 1$ becomes $y = 3(4) + 1 = 13$. The point (4, 13).

So we have the points (1, 4) and (4, 13).

   The change in x is the change going from 1 to 4, which is +3.

   The change in y is the change going from 4 to 13, which is +9.

The slope is $\dfrac{\text{change in y}}{\text{change in x}} = \dfrac{9}{3} = 3$.

The slope of $y = 3x + 1$ is equal to 3.

8. The slope of y = **4**x + 9 is **4**.

    Continuing in that spirit, note that . . .

    The slope of y = **29**x + 3 is **29**.
    The slope of y = **5.83**x + 98 is **5.83**.
    The slope of y = **π**x + √2̄ is **π**.

9. The slope of y = mx + b is m. This is great! No need to go through the six steps to find the slope of a line. Just put the equation in the form y = mx + b, and you can instantly say what the slope is.

10. You must first put –50x + 10y = 17 into the form y = mx + b.
Transpose the –50x   10y = 50x + 17
Divide by 10   y = 5x + 1.7
The slope is equal to 5.

11. The rise is –2 and the run is 5.

12. Looking at the graph above, we'd say that lines with negative slopes go downhill.

13. If the slope = m = $\frac{rise}{run}$ and m is zero, then we have $\frac{0}{run}$ (since fractions are equal to zero only when their numerators are equal to zero). It's difficult to draw a triangle where one side has zero length. I'll draw a line with a slope very, very close to zero. How about m = 0.00001, which is $\frac{1}{100000}$?

When the slope is zero, the line is horizontal.

14. y = 0x + 5 has a slope of zero. So y = 5 is a horizontal line. It's easy to name points on that line, for example, (2, 5), (73, 5), (0, 5), (–9, 5), (π, 5), (√7̄ , 5).

J ack rang the bell at the colonel's mansion, and they were greeted by the robot maid with her breathy voice, "Hi, boys!" This use of "boys" seemed strange to Jack since he was 22 years old and she seemed about 20 years younger than he.

"May we see the colonel? It's important business concerning four of our soldiers who are absent without leave," the chaplain explained.

Let x = age of maid.
Then age of Jack = x + 20
Given: age of Jack = 22
Therefore $x + 20 = 22$
Add −20 to each side: $x + 20 + (-20) = 22 + (-20)$
Simplify $x = 2$

Do you remember when this looked hard?

She escorted Jack and the chaplain through the room with the hundreds of trophy animal heads, through the room with an Olympic-size swimming pool surrounded by a jungle of green tropical plants, and then asked them to wait in the library while she went to inform the colonel of his visitors.

"What a library!" exclaimed Jack. It was bigger than the Lampasas Public Library. Jack headed over to the brand new computer and sat in the leather chair in front of the 48-inch screen. There were stacks of computer games next to the machine. "There must be a thousand dollars' worth of games right here!" Jack uttered in amazement.

The chaplain wandered around the library looking at the books. It was what you might call an unbalanced collection. There were no books that you would call literature—no poems of Christina Rossetti, no books mentioned in Fadiman's third edition of *Lifetime Reading Plan*.

There were books with titles like *How to Get Rich Quick*; *Secrets of Making Dough*; and *Lie, Cheat, and Steal Your Way to the Top*.

One leather-bound book was entitled *The Thousand Verses of She Ran Away*. Curious, he opened it and read:

♪ ♬   From foster home a letter came,
    "Oh send my stuff" the kid
     Did plead, but I wrote back "I've sold
     Your junk. So sorry Sid."   ♪ ♬

"A good example of enjambment," the chaplain thought to himself.* He closed the book and put it back on the shelf.

The maid came back and announced, "The colonel will see you now. I'm afraid he only sees one person at a time. He doesn't like crowds."

Jack looked at the chaplain and said, "I think you ought to go. I'll stay here and find something to do."

As the chaplain left, Jack settled down in front of the computer and fired up the Giant Pollywog Monsters game.

The robot maid and the chaplain headed down another long hallway. They passed a garage. The door was open, and the chaplain could see seven cars inside—seven identical black limos. The first one had *Sunday* painted on the door. The second had *Monday.*

The maid pointed to a door, and she headed back to the library. The doorway was guarded by two helmeted and armed soldiers. He entered into the colonel's inner sanctum. Col. C.C. Coalback sat at his desk with his head supported by both hands. He uttered, "What!" which was halfway between "What do you want?" and "Hurry up and tell me why you are here."

The chaplain looked into the colonel's eyes. They weren't exactly focused since the left one was drifting ever so slightly toward the ceiling. He figured that this wasn't the time to pass on to the Colonel what he and Jack had been discussing on the way up to the mansion about using $y = mx + b$ to graph lines.

No use telling the colonel that m is the slope of the line and *b is the place where the line hits the y-axis.* (If you didn't know that b was the **y-intercept**, here's an example: $y = 4x + 5$. When does it hit the y-axis? When the x coordinate is 0. Points on the y-axis are points like (0, 8) or (0, –32). When x is zero, $y = 4x + 5$ becomes $y = 5$.)

✶ Enjambment (en-JAM-ment) has nothing to do with fruit preserves. In poetry, enjambment is the running on of a thought from one line to the next without a break. In what the chaplain was reading, it occurred between the second, third, and fourth lines.

Instead, the chaplain put aside all the small talk about the **slope-intercept** form of the line ($y = mx + b$) and got right to the point. He explained how Pvt. Fred Gauss had been assaulted while on leave and. . . ."

"Jeez, gimme a break," the colonel interrupted. (That really wasn't an interjection that should have been inflicted on any Christian. The word *jeez* entered American English in about 1920 as a shortened form of *Jesus*.)

The colonel continued, "What in blazes do I care about some Pvt. Goose? Can't you see I'm bishy!" He meant "busy" but his speech was a little slurred. "Go talk to my maid. Get outta here." The colonel picked up his phone and waved the chaplain out of the room.

The chaplain headed back to the library. Just before he entered the library he heard, "Stop it! Can't you see that you made me miss my shot at the Giant Polywog monster! I could have gotten to level four." The maid had been telling Jack how much faster her robotic hands were than Jack's and that she had made it up to level 87 yesterday.

"The colonel said I should talk to you," the chaplain told the maid.

She told the chaplain that she had full power-of-attorney signed by the colonel. "Hey," she went on, "the old man is so busy, someone's gotta run the camp. The only thing he told me was, 'Don't pay more than ten grand for any bribe.' For the rest of the show, I'm just supposed to use my best judgment."

The chaplain explained to her that Fred was in the Lampasas hospital after he had been attacked last night by some nine-year-olds.

"I remember that shrimp," the maid said. "Yesterday I took that midget up to see the colonel. How come he couldn't stand up to a bunch of kids?"

"He's not an undersized person," the chaplain responded. "He's only six years old and was drafted by mistake."

He went on to explain that even after Fred's injury healed, he wouldn't be of much use to the army. It would probably take at least eleven or twelve years before he became the "lean, mean, fighting machine" that the army could use.

"Well," she thought out loud, "I could extend his draft time to twelve years and then the army would have what they wanted. I have the power, you know, since I've got the colonel's power of attorney."

"Wouldn't it make more sense," the chaplain pleaded, "to just let the kid go and let him enjoy his childhood? The army wouldn't waste twelve years of expense raising the child. They can always draft lots of 18-year-olds who have already had their childhood years paid for."

It was the money angle that got to the maid. She knew that if she kept expenses down, the colonel would be happy. "Okay, I'll sign the H.D."

The chaplain received the discharge paper and started to head out the door. Jack indicated that he wanted to continue playing at the computer since his grenade class wasn't until 11.

Honorable Discharge

Be it hereby known and proclaimed that the herein named individual,

__Fred Gauss__

is hereby honorably discharged from the Army.   __the Maid__
                                    power of attorney from the Colonel

---

Your Turn to Play

1. Place these equations in slope-intercept form:
The slope-intercept form is described on the top of the previous page.

$$y - 80 = 2x$$
$$7y = 3x + 9$$
$$y = 4$$
$$2x + y = -8.3$$

2. What we did in the previous problem was a function. The domain was the set of all linear equations in the variables x and y. What was the codomain?

3. The y-intercept of a line is the point where the line hits the y axis. If I'm told that (α, β) is a y-intercept, what could I say about either α or β? [α is alpha, the first letter of the Greek alphabet. β is beta, the second letter of the Greek alphabet. The most common letter of the Greek alphabet that you encounter is probably π.]

4. Where do the graphs of each of these lines intercept the y axis:

$$y - 80 = 2x$$

$$7y = 3x + 9$$

$$y = 4$$

$$2x + y = -8.3$$

5. What is the equation of the line that has a slope equal to 38.9 and which hits the y axis at –9?

6. If a circle has a radius of 5 inches, what is its exact circumference?

. . . . COMPLETE SOLUTIONS . . . .

1. $y - 80 = 2x$ → $y = 2x + 80$

   $7y = 3x + 9$ → $y = \dfrac{3}{7}x + \dfrac{9}{7}$

   $y = 4$ → $y = 0x + 4$

   $2x + y = -8.3$ → $y = -2x + (-8.3)$

2. The codomain was the set of all linear equations in the form $y = mx + b$ where m and b are any real numbers.

3. If (α, β) is on the y axis, then α must be zero. β could be any number.

4. Looking back to the answers to question one where we placed these equations in the slope-intercept form, we can just read off the y-intercepts: 80, 9/7, 4 and –8.3.

5. $y = 38.9x - 9$

6. $C = \pi d$. If the radius is equal to 5" then the diameter is 10" and the circumference is 10π inches.

The chaplain headed down the steep hill, got into his car, and headed to the Lampasas hospital to tell Fred the good news. The passage, "He executes justice for the orphan and the widow, and shows His love for the alien . . ." went through his head. Fred was the closest thing to an orphan that the chaplain had ever encountered on the army base.

As he drove toward the hospital he saw a billboard advertising the board game FunctionLand. From Fred's invention of the game Guess the Function that he played with his mother when he was young, a clever entrepreneur had morphed it into a board game with dice, cards, and play money. It was a big hit across the nation.

You'd roll the dice to find out who would start first and then draw a card that looked like:

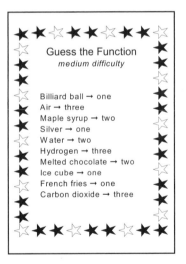

Guess the Function
medium difficulty

Billiard ball → one
Air → three
Maple syrup → two
Silver → one
Water → two
Hydrogen → three
Melted chocolate → two
Ice cube → one
French fries → one
Carbon dioxide → three

The chaplain wondered if he could invent a game like FunctionLand. "Invent a game . . . invent a game," he thought to himself. "I know! We'll call it Invent a Function." He visualized the rules for his new game:

Rules for Invent a Function
1. I name the domain and codomain.
2. You invent a function.

That was simple. Was the chaplain on his way to fame and fortune? Let's see. It's . . .

Your Turn to Play

1. Suppose the domain is the set of all giraffes at the San Francisco Zoo. Let the codomain be the set of all people who live at A.A.A.A. army camp. Invent a function with this domain and codomain.

.......**COMPLETE SOLUTIONS**.......

1. There are many possibilities. One function would be to map the heaviest giraffe to the Colonel. The second heaviest to Sergeant Snow and the rest to Thud.

A second possibility would be to associate the male giraffes to Jack and the female giraffes to Chris.

A third possibility would be to define the function by mapping all of the giraffes to Carol Alfredo Thumbs.

Wait! There's something wrong with that game. It's toooo easy. Do you see why?

Your Turn to Play . . . Again!

1. Why?

.......**COMPLETE SOLUTIONS**.......

1. Anyone could win with no thought at all. Suppose the domain was the letters of the Arabic alphabet and the codomain was {apple, pencil, Golden Gate Bridge}. Even if you can't name any of the letters of the Arabic alphabet, that's not a problem. Just map all the letters to apple and you're done.

Here's one way to make the game tougher: introduce the concept of the **range** of a function. The range of a function is the set of images in the codomain. The range is the elements of the codomain that are "hit" by elements of the domain.

484

If you look at the first function that Thud created earlier in this chapter, the range would be the set {Grrrr!, Toot!, Whirrr!}.

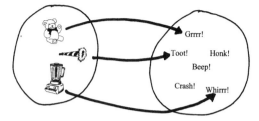

The range of a function
is always part of (or all of)
the codomain. In mathematical language, this is expressed as *the range is always a* **subset** *of the codomain.* Every member of the range is a member of the codomain.

Now, using the idea of range, the game Invent a Function might be worth playing. Let's give it a try.

Your Turn to Play

1. Invent a function where the domain is the set of all people now in Lampasas and the range is {Mercury, Venus}.

2. What would be wrong with answering question 1 with "Map all people who live in one-story houses to Mercury and all people who live in two-story houses to Venus"?

3. Invent a function whose domain is {4, 5, ✳} and whose range is {Pat, Chris, Carol, Thud}.

4. Fill in the blank: The number of elements in the ___?___ can't be more than the number of elements in the domain.

5. What about the number of elements in the codomain? Suppose you have six elements in the domain. How many elements must be in the codomain?

. . . . . . .COMPLETE SOLUTIONS . . . .

1. There are many possibilities. One might be to map all males to Mercury and all females to Venus. A second possibility is to map all the people who are currently standing up to Mercury and the rest to Venus. A third possibility would be to map all people having driver's licenses that show that their last name begins with B to Mercury and everyone else to Venus.

2. A function is defined as a rule that associates to *each* element of the domain, one element of the codomain. What about the people who live in three-story houses? Or the people who don't live in houses? Those people would not have an image in the codomain.

3. That's a toughie. If the problem had read, "Invent a function whose domain is {4, 5, ✳} and whose *codomain* is {Pat, Chris, Carol, Thud}" it would have been easy. You could, for example, map 4 and 5 to Chris and ✳ to Thud.

   But this problem asked that the *range* be {Pat, Chris, Carol, Thud}. If we mapped 4 to Pat and 5 to Chris and ✳ to Carol, then Thud wouldn't be the image of any element in the domain. I can't see a way of answering this question.

4. The number of elements in the *range* can't be more than the number of elements in the domain.

5. Let's draw a picture ➝

The domain on the left has six elements. How many elements must be in the circle on the right (the codomain)?

domain   codomain

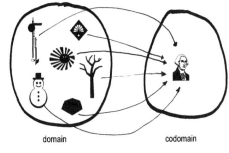

domain   codomain

We know there has to be *at least one element* in the codomain since each element in the domain must be assigned to something.

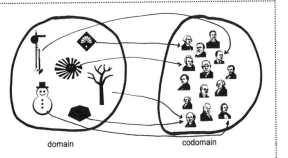

But there's no limit as to how many elements we could put in the codomain. Instead of just one president, we could have 14 of them and make a function:

So the final answer is that we must have at least one element in the codomain if we've got six elements in the domain.

By the way, the answer to the Guess the Function card, four pages ago, is that all solids are mapped to one, liquids to two, and gases to three.

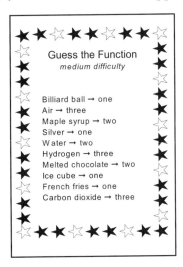

Guess the Function
medium difficulty

Billiard ball → one
Air → three
Maple syrup → two
Silver → one
Water → two
Hydrogen → three
Melted chocolate → two
Ice cube → one
French fries → one
Carbon dioxide → three

As the chaplain entered the hospital, a little kid in a wheelchair looked at him and asked, "Are you Zorro?" That really surprised the chaplain. Then he realized that he was dressed mostly in black and was carrying a black umbrella. As he turned the corner and was out of the sight of the little kid, the chaplain did a little pretend fencing using his umbrella. He made an imaginary Z in the air.

How wonderful it would be if the graph of some line like y = (2/3)x + 4 could be drawn in an instant—just like Zorro slashing his famous Z in the air.

The good news is that this can be done. It takes about two seconds to slash the line y = (2/3)x + 4 onto a pair of axes. Here's how: y = (2/3)x + 4 has a y-intercept of 4, so we know the line will hit the y-axis at 4.

Next, we know the slope is 2/3. Starting at the intercept point, draw the rise/run triangle.

Slash in the line. You're done!

Pleasantville, New York home to *The Reader's Digest*

1. Graph y = (1/2)x + 3

2. The graph of a line whose slope is zero is horizontal. Here's the graph of a line whose slope is equal to one.

With a slope of 4, the line is even steeper. What would the graph of a line look like if its slope were equal to a million?

3. Suppose I had a domain with three elements in it, say {✿, ✈, ☺} and a codomain with two elements in it, say {A, B}. How many possible functions could I have? (This question is not one of those questions you can instantly answer. This should take several minutes of thought to figure out.)

4. Is this a function?

5. What is the equation of the line whose graph is

6. Invent a function, if possible, whose domain is the set of all animals now at the Lampasas animal shelter and whose codomain is the rational numbers.

7. Why would the following *not* be a good answer to question 6 above: "The rule is associate all the animals to the number π"?

8. Graph $y = -2$

9. Graph $-2x + y = 3$

10. (Not easy) Guess the function:

    $(10, 2) \rightarrow 6$
    $(3, 1) \rightarrow 1$
    $(-8, 2) \rightarrow -12$
    $(0, 5) \rightarrow -10$
    $(5, 0) \rightarrow 5$
    $(40, 3) \rightarrow 34$
    $(30, 3) \rightarrow 24$

11. When you're playing Guess the Function, it's possible to create really hard rules that no one will be able to guess. For example:

    $32 \rightarrow$ Art
    $832 \rightarrow$ Lloyd
    $84 \rightarrow$ B.
    $100 \rightarrow$ Mike
    $29 \rightarrow$ Zeb

Probably no one would ever be able to guess the rule that I used, even if I gave a hundred examples instead of just five. Here's my rule: Take the 1954 San Francisco telephone book and turn to the page number that's the same as the element in the domain. (So in my first example, 32 → Art, I turned to page 32.) Take the 18th name on that page and the image will be the first name of that person. What is the range of this function?

12. Invent a function where the domain and range are the letters of the Russian alphabet.

13. What is the common name for this function:

    $(2, 4) \rightarrow 8$

    $(0, 3) \rightarrow 0$

    $(-2, -2) \rightarrow 4$

    $(-3, 4) \rightarrow -12$

14. How many functions could have a domain = {@} and a codomain equal to the natural numbers = {1, 2, 3, . . .}?

answers

1.

2. It would be almost vertical.

3. Here's one possible function:

    ✿ → A

    ✈ → A

    ☺ → B

Here's a second:

    ✿ → A

    ✈ → B

    ☺ → B

       Altogether I count eight possible functions. There are two possibilities for ✿. There are two possible images for ✈. There are two possible places that ☺ could be mapped. $2 \times 2 \times 2 = 8$.

       In general, if there are x ways you can do one thing and y ways you can do another, there will be xy ways to do them both. If I own 4 shirts and 3 pair of pants, there are 12 possible outfits I could wear.

4. It satisfies the definition of a function since each element of the domain has exactly one image in the codomain. It's a function.

5. $y = (3/7)x + 2$ From the graph, the y-intercept is 2. b = 2. From the graph, the slope is 3/7. m = 3/7.

6. There are many possible functions. For example, one function would

be to map all four-legged animals to 8, all two-legged animals to 3/23 and all the rest, if any, to –2/9. A second example would be to map all the animals to 90729.

7. π is not a rational number and, hence, is not in the codomain. (A rational number is any number that can be expressed as x/y where x is an integer and y is a natural number.)

8.

9. First it should be put in the slope-intercept form: $y = 2x + 3$. Then the graph will be

10. Take the first number in the ordered pair and subtract twice the second number. In symbols $(x, y) \rightarrow x - 2y$. About 4% of all readers could solve this problem.

11. The range, which is the set of all images, is the set of all first names of people appearing as the 18th name on any page in the 1954 San Francisco phonebook. It would have been incorrect to say that the range is the set of all first names or to say that the range was the set of all first names in the 1954 San Francisco phonebook. These sets would have been correct answers if the question were to name the *codomain* of this function.

12. Some of you who are unfamiliar with the letters of the Russian alphabet may be thinking, "How can I do this when I can't even name any of the letters?" The good news is that you don't need to know those letters. Just map each letter to itself. This mapping, which maps each element to itself, is called the **identity function**.

13. That function is called multiplication.

14. There is an infinite number of possible functions. One example would be @ → 1. A second example would be @ → 2. A third possible function would be @ → 3.

  If the question had asked how many possible functions with the domain equal to {@} and the *range* equal to the natural numbers, the answer would have been None.

Upper Sandusky, Ohio featured in the Jimmy Dorsey song *Dusk in Upper Sandusky*

1. Graph y = (–3/2)x + 5

2. Name four points that lie on the graph of the line x = $\sqrt{3}$.

3. Guess the Function:

> house → 0
> mouse → 0
> rat → 0
> flea → 0
> fly → 1
> moose → 0
> cat → 0
> bat → 1
> airplane → 1

4. Suppose there are two elements in the domain and two elements in the range. How many possible functions could there be? Let's say that the domain is {#, %} and the range is {M, N}.

5. Is this a function?

6. y = x^2 is a rule that maps each possible value of x to a value of y. If the domain were {3, 5, 98}, this would be a function. If the domain were all real numbers this would be a function. Would that rule be a function if the

domain were {5, }?

7. Invent a function, if possible, where the domain is equal to all the books in the Lampasas library and where the codomain is all the people now inside the Lampasas city limits.

8. When asked to graph y + 4x = 2, a student drew this:
But that's not the correct graph.
What did the student do wrong?

9. Graph y = 2.

10. Give an example of a function where the domain is the natural numbers, {1, 2, 3, . . . } and where the range is {horseshoe, cloud}.

11. Graph y = (–2/3)x – 1.

12. Thud and Carol were playing Invent a Function. Thud said that the domain should be {5, 6, 7} and that the codomain should be all rational numbers.

  Carol offered the following possible function: For each element in the domain, pretend it's the radius of a pizza and associate the corresponding area of the pizza. The formula is A = πr², where A is the area of any circle, and r is the radius. So, as Carol explained, 5 would be mapped to 25π, etc.

  Did Carol win the game?

13. Continuing the previous question: Give an example of a function that would have won the game.

14. Guess the Function:

> (2, 5)•(1, 6) → 32
> (3, 30)•(2, 20) → 606
> (5, 10)•(5, 20) → 225
> (0, 3)•(34,5) → 15
> (4, 0)•(7, 987) → 28
> (–8, 100)•(–7, 100) → 10056
> (1, 2)•(3, 4) → 11

answers

1.

2. Answers will vary. They'll all be of the form ($\sqrt{3}$, ?). For example: ($\sqrt{3}$, 0), ($\sqrt{3}$, 1), ($\sqrt{3}$, 6.07), ($\sqrt{3}$, π), or ($\sqrt{3}$, –389).

3. The rule is: if the element in the domain contains a "y" or a "b" or a "p", then its image is 1. Otherwise its image is 0.

   (Of course, there's the much more obvious rule: if it can fly, then its image is 1; otherwise, its image is zero.) When you play Guess the Function, it's possible to win with several different rules that fit all the examples.

4. One possible function is # $\rightarrow$ M and % $\rightarrow$ N. The only other possibility is # $\rightarrow$ N and % $\rightarrow$ M.

   # $\rightarrow$ M and % $\rightarrow$ M will not work because the range would not be {M, N} in that case.

5. We notice that has two images in the codomain, so this is not a function. In a function, each element of the domain must have *exactly one* image in the codomain.

6. Each element of the domain must be assigned an image in the codomain. Under the rule y = x^2, I have no idea what ()^2 would mean. If you can figure out what a carton of milk squared means, then it would be a function. But until then, we will have to say that it is not a function.

7. There are many possible answers, and some of them are wrong.

   The answer that is frequently given is, "Map each book to the person who has most recently touched it." But that won't work since the person who most recently touched the copy of *Gone With the Wind* is not currently inside the city limits.

   If you answered, "Associate to each book the person currently inside the city limits who has most recently touched that book," you would get half credit. It turns out that no one currently inside the city limits has ever touched the copy of *Making Your Bowling Ball a Table Centerpiece.* (The book gives you all kinds of suggestions for things you can paste to the ball and flowers to stick in the holes in the ball.) The only person who ever touched *Making Your Bowling Ball a Table Centerpiece* was the former librarian who put the book on the shelf four years ago. Unfortunately, she died about a year ago and is buried in the cemetery outside the city limits.

   If we knew the name of anyone who is currently inside the city limits, this would be easy. We could then map all the books to that person. Who? The mayor might be out of town right now. Here are some possibilities: (A) the oldest person currently inside the city limits; (B) the tallest person currently at Dee's Old Time Pit Bar-B-Q (assuming you're reading this during normal business hours); or (C) we know the actual name of an individual currently within the city limits: Fred! He's at the hospital. So one possible function would be to map all the books to Fred.

8. The first step is to put the equation in the slope-intercept form (y = mx + b). This would make the equation y = –4x + 2. The graph should have been:

9. y = 2

10. There are many possible answers. One possibility is to map the odd numbers to horseshoe and the even numbers to cloud. A second possibility is to map the number 7 to horseshoe and all the rest to cloud. A third possibility is to map all the numbers less than 100 to horseshoe and the rest to cloud.

11. y = (–2/3)x – 1

12. Carol Alfredo Thumbs did not win, because numbers like 25π are not rational. (We talked about irrational numbers at the end of Lesson 75.)

13. There are many possibilities. For example, map all the elements of the domain to the number 89.389765. Or, for example, map each number to itself (called the identity mapping). Or map each number to its square.

14. This function is called the inner product. You'll encounter this function when you study linear algebra (which is a course you take after calculus). The rule is (a, b)•(c, d) → ac + bd.

There are (at least) four different *algebra* courses—all with entirely different subject matter.

You're in *beginning algebra* now. Then comes *advanced algebra*. Then after geometry, trig, and calculus come two other algebra courses: *linear algebra* and *abstract algebra*. Linear algebra deals with vector spaces, and abstract algebra deals with groups, rings, and fields. Linear algebra and abstract algebra can be taken in your junior year of college after taking calculus.

Elizabethtown, Kentucky where Abraham Lincoln's father lived

1. Graph $y = \frac{1}{4}x + 1$

2. Name four points that lie on the graph of x = 7.

3. You take a letter to the post office where a clerk weighs it and tells you how much money you owe. There's an obvious function here. What are its domain and codomain?

4. Give an example, if possible, of a function where the domain is the set of all pizza toppings featured at Stanthony's Pizza, {olives, onions, pepperoni, anchovy, salami, pineapple, . . .} and where the range is the whole numbers {0, 1, 2, 3, 4, . . . }.

5. If the domain is {A, B, C} and the codomain is {L}, how many possible functions are there?

6. What is the equation of the line where the graph passes through (0, 4) and (5, 20)?

 (Hint: the point (0, 4) tells you what the y intercept is. All you have left to do is figure out the slope of the line that passes through (0, 4) and (5, 20). We did that in the answers to questions 3 through 6 in Lesson 91.)

7. You are given a function that maps each value of x to its corresponding y value. The function is given by the rule $y = x^2$. The x values are in the domain and the y values are in the codomain. What is the image of 7 under this function? What is the image of $\sqrt{13}$ under this function?

8. What is the slope of 2y = 6x + 17?

9. You are given an equation y = mx + b where m and b are numbers. You are told that the graph of this line passes through the origin. What, if anything, can you say about m or about b? The origin is the point (0, 0).

10. Guess the Function:

4 → 11	100 → 299	−10 → −31
0 → −1	15 → 44	7 → 20
2 → 5	9 → 26	5 → 14
−3 → −10	11 → 32	
30 → 89	10 → 29	

11. Invent a Function, if possible, where the domain is {7, 8, 9} and where the codomain is the real numbers.

12. Invent a Function, if possible, where the domain and where the range are both the set of all people now in Wyoming.

13. Is this a function?

14. Pat and Thud are playing Invent a Function. Pat said that the domain and codomain are both the real numbers. Thud suggested the following rule: $y^2 = x$, where the values of x are in the domain and the values of y are in the codomain. Did Thud win?

answers

1. $y = \frac{1}{4}x + 1$

2. Answers will vary. They'll all be of the form (7, ?).
    For example: (7, 0), (7, 53), (7, –3), (7, π).

3. The domain would be all possible weights of letters. The codomain would be any possible amount of money. (The range would be harder to specify. It would depend on what stamps are costing nowadays. If, for example, the first ounce cost 10¢ and additional ounces cost 3¢, then numbers like 12¢ would never be in the range.)

4. This can't be done. The range has an infinite number of elements in it, and the finite number of pizza toppings in the domain can't be mapped to an infinite number of images. The problem would have been doable if we had said that the *codomain* were the whole numbers. Then you could have, for example, mapped all the toppings to the number 837298.)

5. Just one. That function is the rule that maps A, B, and C all to L.

6. $y = (16/5)x + 4$
    The slope of the line that passes through (0, 4) and (5, 20) has a rise of 16 (computed as 20 – 4) and a run of 5 (computed as 5 – 0).

7. 49 and 13

If x = 7, then y = x² would mean that y = 49.

If x = √13, then y = x² would mean that y = √13 √13 = 13.

For any non-negative number p, √p √p = p.

8. m = 3

$2y = 6x + 17$ ⇨ $y = 3x + \dfrac{17}{2}$ which is now in the form y = mx + b

9. We know nothing about the slope m from the given information. Since the line passes through the origin (0, 0) we know that the y-intercept is equal to zero. Therefore b = 0.

10. Triple the number and subtract one. In symbols: x → 3x – 1.

11. There are many possibilities. For example, you could map 7, 8, and 9 all to π. Or you could map 7 →19, 8 → 32, and 9 →√7.

12. Map each person to himself or herself. That's about the only easy-to-name function that will work. It would not be correct to map them all to one particular person since we wanted the *range* to be all of the people in Wyoming.

13. No. Each element in the domain must have an image and ∇ doesn't have an image.

14. No Thud didn't win. y² = x where the values of x are in the domain and the values of y are in the codomain doesn't give us a function. Suppose, for example, that x was equal to 16. What would y equal? It might be 4 or it might be –4. One element in the domain, 16, is not supposed to give two images (4 and –4) in the codomain.

W hen the chaplain arrived at Fred's bed, he could hardly wait to tell Fred the good news. It's not everyone who gets out of the army after only four days. "Fred, you won't believe it," the chaplain began. "You've got a discharge!"

A nurse was at the side of Fred's bed. She examined his nightshirt. The red stain was still there amid the little blue and green frog decorations, but it was all dry by now. The nurse turned and said, "I wouldn't worry about it, chaplain. He stopped bleeding last night. We just haven't changed his nightshirt yet, since that'd be awfully painful for him—what with his broken rib."

They had propped Fred up in bed with a big pillow behind him. This, of course, meant that part of him was no longer under the covers. This was fine with him. His bare feet sticking out underneath his nightshirt weren't cold at all.

Earlier that morning he had asked the nurse what the temperature of the room was. She had told him, "We always keep it at least at 25°." Fred mentally translated that into: x ≥ 25°, where x is the temperature of the room. ("≥" means *greater than or equal to*.)

He realized that she was talking "medical talk" and what she meant was 25°C, which is the Celsius temperature scale (the scale that most of the world uses). Using the Celsius scale, water becomes ice at 0°C and boils at 100°C.

To convert from Fahrenheit the formula is:
$$C = \frac{5}{9}(F - 32)$$

So keeping the room at least at 25° could be written as C ≥ 25 where C is the temperature in Celsius. Or you could write it as 25 ≤ C. Either way is fine. But how warm is 25 ≤ C?

Start with   25 ≤ C

Replace the C by its equivalent, $\frac{5}{9}$ (F − 32)

$$25 \le \frac{5}{9}(F - 32)$$

Multiply both sides by 9 $\qquad 225 \le 5(F - 32)$

Divide both sides by 5 $\qquad 45 \le F - 32$

Add + 32 to each side $\qquad 77 \le F$

They kept the temperature of the room at a minimum of 77° Fahrenheit! No wonder Fred's toes weren't cold.

In fact, Fred was feeling kind of warm. His breakfast that morning hadn't helped him cool off at all. They had served him three tacos and two burritos—all loaded with hot sauce. "We gotta put some meat on them bones of yours," the nurse had announced. "Six years old and you hardly weigh anything at all. I got postage stamps that weigh more than you."

Your Turn to Play

1. Convert 68°F to Celsius.
2. Convert 30°C to Fahrenheit.
3. Solve C = $\frac{5}{9}$ (F − 32) for F.
4. Kids who aren't very hungry tend to play with their food. Fred was no exception. First he moved the burritos so that they were parallel with each other and then put the three tacos on top, making a little bridge.

Next, he took two of the tacos and pretended that they were race cars as he ran them around the plate.

Finally, he cracked open a taco and noticed that it was 30% meat.

He sliced open a burrito and noticed it was 40% meat.

The meat in one taco and one burrito weighed a total of 11.4 oz.

One taco and one burrito weighed a total of 33 oz.

How much does a taco weigh?

5. Solve $4x - 60 \leq 82$.

6. Solve $\dfrac{1}{x + 6} + \dfrac{2}{x - 2} = \dfrac{7}{12}$

·······COMPLETE SOLUTIONS·······

1. $C = \dfrac{5}{9}(F - 32)$

Letting F equal 68, we have $C = \dfrac{5}{9}(68 - 32) = \dfrac{5}{9}(36) = 20°$

2. $C = \dfrac{5}{9}(F - 32)$

Letting C equal 30, we have $30 = \dfrac{5}{9}(F - 32)$

Multiply by sides by 9	$270 = 5(F - 32)$
Distributive property	$270 = 5F - 160$
Add 160 to each side	$430 = 5F$
Divide both sides by 5	$86° = F$

3.
$$C = \dfrac{5}{9}(F - 32)$$
$$9C = 5(F - 32)$$
$$9C = 5F - 160$$
$$9C + 160 = 5F$$
$$\dfrac{9C}{5} + 32 = F$$

> If you wanted to be really tricky, you could do the whole thing in two steps.
>
> First, multiply both sides by 9/5 and get
> $$\dfrac{9}{5}C = F - 32$$
> Second, add 32 to both sides.

Using the symmetric law of equality $\quad F = \dfrac{9C}{5} + 32$

4. Let x = weight of a taco. ← We always start with a "Let x =" line.

    Then 33 − x = weight of a burrito. Since one taco and one burrito weigh 33 oz.

    Then 0.3x = weight of meat in a taco.

    Then 0.4(33 − x) = weight of meat in a burrito.

    $$0.3x + 0.4(33 − x) = 11.4$$ Since the meat in one taco and one burrito weigh 11.4 oz.

    x = 18 oz.   This is a fairly large taco.

Using "six pretty boxes" . . .

	taco + burrito	taco	burrito
total weights	33	x	33 − x
weight of meat	11.4	0.3x	0.4(33 − x)

5. $4x − 60 \leq 82$

Add 60 to both sides $4x \leq 142$

Divide both sides by 4 $x \leq 35.5$

6. $\dfrac{1}{x + 6} + \dfrac{2}{x − 2} = \dfrac{7}{12}$

$$\dfrac{1(x + 6)(x − 2)12}{x + 6} + \dfrac{2(x + 6)(x − 2)12}{x − 2} = \dfrac{7(x + 6)(x − 2)12}{12}$$

$$12(x − 2) \quad + \quad 24(x + 6) \quad = \quad 7(x + 6)(x − 2)$$

$$36x + 120 = 7x^2 + 28x − 84$$

$$0 = 7x^2 − 8x − 204$$

$$0 = (7x + 34)(x − 6)$$

$$7x + 34 = 0 \quad \text{OR} \quad x − 6 = 0$$

$$x = −34/7 \quad \text{OR} \quad x = 6$$

Checking x = −34/7 $\dfrac{1}{−34/7 + 6} + \dfrac{2}{−34/7 − 2} \overset{?}{=} \dfrac{7}{12}$

Skipping some of the arithmetic $\dfrac{7}{8} − \dfrac{14}{48} \overset{?}{=} \dfrac{7}{12}$ yes

Checking x = 6 $\dfrac{1}{12} + \dfrac{2}{4} \overset{?}{=} \dfrac{7}{12}$ yes

F red wasn't exactly pleased with all the attention he was getting. It reminded him of all the marriage proposals that he had received when his new teaching style had gotten national attention.

But three tacos and two burritos for breakfast! The whole thing must have weighed at least two pounds. At that rate, Fred figured he'd be as fat as Santa Claus by the end of June.

still life: *Three Tacos & Two Burritos*
by Pierre Stanoir (1841–1919)

If the three tacos and two burritos had weighed exactly two pounds, he could have graphed that in an instant.

$$3T + 2B = 2$$

and let's let T be x
and B be y, so that it'll
look more familiar

$$3x + 2y = 2$$

First put it into
the slope-intercept
$y = mx + b$ form

$$2y = -3x + 2$$
$$y = (-3/2)x + 1$$

503

and then graph it
"instantly" using the
Zorro method

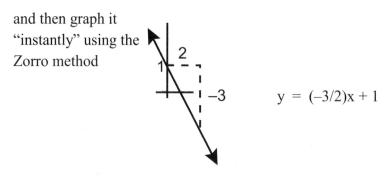

$$y = (-3/2)x + 1$$

So we can easily graph $3x + 2y = 2$, but what we'd like to graph is the fact that the three tacos and two burritos weighed *at least* two pounds. We would like to graph $3x + 2y \geq 2$.

How do you do it? How do you graph an *in*equality?

The first step is to do what we've already done. Namely, graph the equality. We did that and got the graph of $3x + 2y = 2$

There's only one more step in graphing $3x + 2y \geq 2$. We have to figure out which side of the line to shade in. The final answer will either look like: or it will look like:

We either shade to the left of the line or shade to the right of the line. The graph of an inequality such as $3x + 2y \geq 2$ is a half plane. (A **plane** is a term from geometry. A plane is like a giant flat sheet of paper.)
All the points that satisfy the *equality* $3x + 2y = 2$ are on the line that we graphed. All the points that satisfy the *in*equality $3x + 2y \geq 2$ lie in a half plane whose boundary is $3x + 2y = 2$.

Either all the points to the left of $3x + 2y = 2$ satisfy $3x + 2y \geq 2$ or all the points to the right of $3x + 2y = 2$ satisfy $3x + 2y \geq 2$.

So all we have to do is test *a single point* on each side of the
$3x + 2y = 2$ boundary and that will tell us everything. If that point satisfies
$3x + 2y \geq 2$, then all points on that side of the boundary "work."

What's a good point that's to the left of 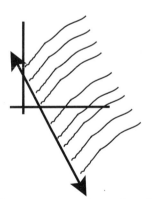the line? The easiest
one looks like the origin, $(0, 0)$.

Does $(0, 0)$ make $3x + 2y \geq 2$ true or does it make it false?

Put $(0, 0)$ into $3x + 2y \geq 2$ $\qquad\qquad 3(0) + 2(0) \overset{?}{\geq} 2$

$$0 \overset{?}{\geq} 2 \qquad \text{false}$$

Since zero is not greater than or equal to 2, that test point doesn't work,
and, therefore, we don't shade the left side of the boundary $3x + 2y = 2$.

Let's try a test point on the right of the boundary. How about $(100, 200)$?
We can be certain that point is to the right of the line.

Put $(100, 200)$ into $3x + 2y \geq 2$ $\qquad 3(100) + 2(200) \overset{?}{\geq} 2$

$$700 \overset{?}{\geq} 2 \qquad \text{true}$$

Since one test point $(100, 200)$ that is to
the right of the boundary line works,
every point to the right of the boundary
works, and we shade in the right side of
the line to obtain the graph of the inequality.

This is the graphical translation of: "Three tacos and two burritos
must have weighed at least two pounds."

> **So to graph an inequality, there are two steps:**
> ☞ **Graph the line and**
> ☞ **Test a point on each side of the line**
> **to find out which side to shade in.**

As Fred stared at each of those tacos he noticed that they were dripping with hot sauce. He told the nurse that he didn't like a lot of chili pepper in his food since he was only six years old. She assured him there wasn't any chili pepper in the hot sauce.

He also noted that since this was real life—and not some artificial situation dreamed up by some textbook writer—that the tacos, each of which weighed x pounds, had a positive weight. No taco had a negative weight. No taco was floating off the plate up to the sky.

That whole paragraph of English can be summarized as x > 0.

To graph x > 0, we follow the same two steps: we first graph the line x = 0, and then we test points on each side of x = 0 to determine which side to shade in.

To graph x = 0, we'll use the approach we learned for graphing any equation. We'll find some points that satisfy the equation and then connect the points. (Some books call that method **point-plotting**.) Here are some points that satisfy x = 0: (0, 3), (0, 19), (0, π), (0, –2), (0, $-\sqrt{3}$). They all lie on the y-axis.

The graph of x = 0 is   (The graph of x = 0 is the y-axis.)

The second step in graphing x > 0 is to find a test point on each side of x = 0 and test it in x > 0.

First, we have to pick some test point that is clearly to the left of the y- axis.

We try (–3, –4) in x > 0 $-3 \overset{?}{>} 0$ That's false. We don't shade in the left side of the y-axis.

We'll try (5, 7), which is clearly to the right of the y-axis.

We try (5, 7) in x > 0 $5 \overset{?}{>} 0$ That's true. So we shade in the right side of the y-axis.

The graph of x > 0 is

But that's not right.

506

That's the graph of x ≥ 0, not x > 0. We
need to show that x = 0 is *not* part of the graph.
We want only the stuff that's to the right of x = 0.
Here's what we should have drawn for x > 0:

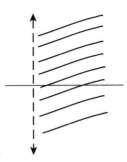

We use a dashed line for the boundary when
we're graphing strict inequalities (< or >) and use
a solid line when we're graphing either ≤ or ≥.

The dashed line tells us that we're drawing
the half plane up to *but not including* the boundary line.

Your Turn to Play

1. What about the burritos? They must also have a positive weight. That
means that we need to graph y > 0. What does that graph look like?

2. Two pages ago three tacos and two burritos must have weighed at least
two pounds was graphed as:

The fact that tacos must have a
positive weight was graphed as:

The fact that burritos must have a positive weight
was graphed in the answer to question 1. Its graph was:

All three of the above facts are true, so the points which correspond to
weights that the tacos and burritos could actually have is a combination of
these three graphs. If you could cut these graphs out and lay them one on
top of each other, the place where all three shadings occur is the final
solution.

Draw the region where all three shadings overlap.

3. As you noticed in the picture of Fred on the first page of this chapter,

he was sitting in his hospital bed holding a strawberry milkshake. This was actually an emergency strawberry milkshake. Fred had been assured by the nurse that "there wasn't any chili pepper in the hot sauce" on his tacos. And there wasn't. But a couple of tablespoons of habanero pepper sauce—how could that hurt anyone? (Habanero sauce is the stuff that can almost melt glass.)

Thinking that the sauce was like catsup, Fred had taken a mighty (for him) bite of the first taco. He almost went into respiratory arrest. It took a Code Blue, two gallons of ice water, and that emergency strawberry milkshake to quell the swelling in his throat. (Fred's the only person I know who calls a peppermint milkshake *spicy*.)

Wanting to make sure that the strawberry milkshake would cool Fred's throat, the doctor specified that it should be colder than 5°. Of course, he was talking about Celsius. So C < 5°, where C is the temperature of the drink in Celsius. What would that translate into on the Fahrenheit scale?

. . . . COMPLETE SOLUTIONS . . . .

1. To graph y > 0, we must first graph y = 0 (and use a dashed line since we have a strict inequality in the original problem). Graphing y = 0 is done almost instantly if we put y = 0 in the y = mx + b form. Then y = 0 becomes y = 0x + 0. So it's a line with a slope of zero (that means a horizontal line) with a y-intercept of 0. That's the x-axis. So we have as our first step

Since it's a strict inequality, y > 0, we'll make it a dashed line

The test point (5, –7), which is below the dashed boundary line, when substituted into the original inequality y > 0, yields –7 $\overset{?}{>}$ 0, which is false. Therefore, we don't shade below the boundary line.

The test point (3, 6), which is above the dashed boundary line, when substituted into the original inequality y > 0, yields 6 $\overset{?}{>}$ 0, which is true. Therefore, we shade the region above the dashed boundary line.

The graph of y > 0 is

2. Imagine making a sandwich with some lettuce, a slice of cheese, and a slice of bologna. It's those places in the sandwich where if you bite down and get all three ingredients that are the areas that we're looking for.

Imagine one of the three graphs were shaded with clear yellow plastic and the second graph were shaded with clear blue plastic. When you lay one of the graphs on top of the other, it's the area that appears green (since yellow + blue = green) that we would be looking for. If the third graph were shaded red, then the area of overlap of all three graphs would be almost black.

Imagine a bed with three blankets thrown haphazardly over the bed at weird angles so that parts of the bed are covered by all three blankets, some by only two blankets, some by one, and some by none. It's the spots in the bed that are warmest (under all three blankets) that we're looking for.

So when we combine   and   and

we get   Any point you pick in this shaded region will satisfy $3x + 2y \geq 2$ and $x > 0$ and $y > 0$.

3. We were told that $\qquad\qquad C < 5$

Replace C by $\frac{5}{9}$(F − 32) $\qquad \frac{5}{9}$ (F − 32) $< 5 \qquad$ The conversion formula is $C = \frac{5}{9}$ (F − 32)

Multiply both sides by 9 $\qquad 5(F − 32) < 45$

Divide both sides by 5 $\qquad\quad F − 32 < 9$

Add +32 to both sides $\qquad\quad\; F < 41$

So Fred's emergency strawberry milkshake was colder than 41° Fahrenheit.

F red was always willing to share. He asked the chaplain, "Would you care for a milkshake?" He pointed to the 15 frosty glasses on the table beside his bed. There was almost every flavor of milkshake you could dream of: chocolate, vanilla, peach, raspberry, fudge, ...

"The nurse," Fred continued, "felt really bad about the hot sauce thing. She said that no one at her house ever had such a reaction to a little habanero. She said that one of her daughters even puts habanero sauce on her cereal in the morning. The nurse went to the local ice cream store, got these for me and paid for them out of her own money. The only one that you have to be careful about is the shake on the end. It's peppermint."

The chaplain thanked Fred and chose the pineapple shake. A few moments passed in silence as they worked on their shakes. Then the chaplain told Fred the good news about his honorable discharge from the army.

"That means I get to go back to teaching at KITTENS!" Fred laughed. "Ouch!" he exclaimed. His side hurt when he laughed. He resorted to smiling.

The chaplain thought of Fred's expression, division by zero, that he had uttered when life had started to seem so senseless when his rib was broken. Now, that he was getting out of the army and going back to his favorite occupation of teaching, Fred's wound in his side took on a whole new meaning. It meant freedom.

The chaplain asked Fred why they called division by zero senseless in mathematics: "Shouldn't $\frac{5}{0}$ have some meaning?"

This made Fred very happy. He loved to talk about math with someone who was interested. And the chaplain was interested—both in math and in Fred.

"Do you know what $\dfrac{14}{2}$ means?" Fred asked rhetorically.* There are many ways to look at fourteen divided by two. One of them is to write down a whole bunch of 2s in addition and see how many it takes to add up to 14.

$$\begin{array}{r} 2 \\ 2 \\ +\,2 \\ \hline 6 \end{array}$$

This doesn't work. We need more 2s.

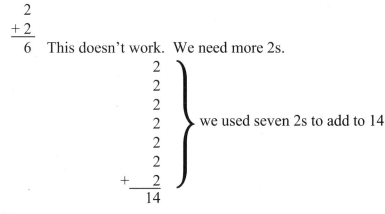

$$\left.\begin{array}{r} 2 \\ 2 \\ 2 \\ 2 \\ 2 \\ 2 \\ +\,2 \\ \hline 14 \end{array}\right\} \text{we used seven 2s to add to 14}$$

"So $\dfrac{14}{2}$ can be thought of as the question: How many twos does it take to add up to fourteen? Okay, back to looking at $\dfrac{5}{0}$ and seeing what that means. The expression $\dfrac{5}{0}$ asks: How many zeros does it take to add up to five? I'll start adding and you tell me when to stop."

Fred asked the chaplain to give him the clipboard at the foot of his bed, and he began to write in a blank space on the page:

Fred Gauss, age 6, Broken rib. Parents can't be located. Probably some kind of mascot at A.A.A.A. Pulse okay.

0 + 0 + 0 + 0 + 0 + 0 + 0 + 0 + 0 + 0
+ 0 + 0 + 0 + 0 + 0 + 0 + 0 + 0 + 0 + 0
+ 0 + 0 + 0 + 0 + 0 + 0 + 0 + 0 + 0 + 0
+ 0 + 0 + 0 + 0 + 0 + 0 + 0 + 0 + 0 + 0
+ 0 + 0 + 0 + 0 + 0 + 0 + 0 + 0 + 0 + 0
+ 0 + 0 + 0 + 0 + 0 + 0 + 0 + 0 + 0 + 0
+ 0 + 0 + 0

"Stop! I give up," the chaplain said. "No matter how many millions of zeros you write, they'll never add up to five. The expression $\dfrac{5}{0}$ couldn't be any number."

"My point exactly," responded Fred.

---

* Rhetorical questions are questions asked to make a point, not because you expect a reply.

Your Turn to Play

1. Find the value of:

$$\frac{27}{1}$$

$$\frac{27}{0.1}$$

$$\frac{27}{0.01}$$

$$\frac{27}{0.001}$$

2. Complete this sentence: *There are two ways to make the value of a fraction really large: (1) make the numerator really large or (2) . . . .*

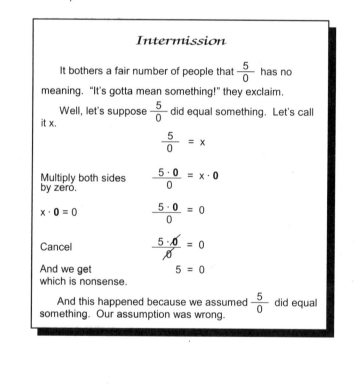

Intermission

It bothers a fair number of people that $\frac{5}{0}$ has no meaning. "It's gotta mean something!" they exclaim.

Well, let's suppose $\frac{5}{0}$ did equal something. Let's call it x.

$$\frac{5}{0} = x$$

Multiply both sides by zero.

$$\frac{5 \cdot \mathbf{0}}{0} = x \cdot \mathbf{0}$$

$x \cdot \mathbf{0} = 0$

$$\frac{5 \cdot \mathbf{0}}{0} = 0$$

Cancel

$$\frac{5 \cdot \cancel{\mathbf{0}}}{\cancel{0}} = 0$$

And we get which is nonsense.

$$5 = 0$$

And this happened because we assumed $\frac{5}{0}$ did equal something. Our assumption was wrong.

·······COMPLETE SOLUTIONS·······

1. $\dfrac{27}{1} = 27$

   $\dfrac{27}{0.1} = 270$

   $\dfrac{27}{0.01} = 2700$

   $\dfrac{27}{0.001} = 27000$

2. There are two ways to make the value of a fraction really large: (1) make the numerator really large or (2) make the denominator a non-negative number very close to zero.

   [It wouldn't have been quite correct to complete the sentence with . . . *or (2) make the denominator really small.* In algebra, we introduced negative numbers. No longer is zero the smallest number we know. Minus five is smaller than zero. And –3978293923934288 is even smaller.]

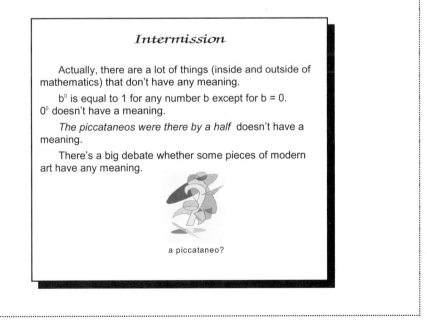

Intermission

Actually, there are a lot of things (inside and outside of mathematics) that don't have any meaning.

b^0 is equal to 1 for any number b except for b = 0.
0^0 doesn't have a meaning.

The piccataneos were there by a half doesn't have a meaning.

There's a big debate whether some pieces of modern art have any meaning.

a piccataneo?

❧❧❧❧❧❧❧❧❧❧❧❧❧❧❧❧❧❧❧❧❧

The chaplain had finished his twelve-ounce pineapple shake, and Fred had finished all but eleven ounces of his shake. The chaplain asked Fred where his stuff was, so that he could gather it up and leave the hospital.

Fred said he didn't have any things. The coffee cup that he was using as a helmet had been broken last night. When he had arrived at the hospital, they cut off his camouflage T-shirt rather than try to undress him since they didn't know the extent of his injury and didn't want to accidentally make it worse. Right now, Fred had fewer things with him than Gandhi did when he died.

The chaplain said that he was going to sign Fred out of the hospital and that he'd be right back.

Very slowly Fred got out of bed. It hurt to move and it hurt to laugh. A little kid in a wheelchair rolled up to Fred and asked, "Is Zorro really your daddy? It's neat that you got a dad who goes around with a sword saving people and stuff."

Fred thought to himself how wonderful it would have been if the chaplain had really been his dad. His life would have been so different. Fred smiled a bit and decided to change the subject, "Have you ever played Guess the Function? It's really fun."

The little kid in the wheelchair shook his head, but he was eager to learn. Anything that had the words "play" and "fun" had to be good.

"Here goes!" said Fred. He got a piece of chalk (teachers at KITTENS always seemed to have a piece of chalk handy), and he wrote on the floor:

$$5 \rightarrow 5$$
$$7 \rightarrow 7$$
$$1000 \rightarrow 1000$$

"I got it!" cried the kid.

514

"Just a second. It's not what you think," Fred said as he continued with more examples:

$$3.5 \rightarrow 3.5$$
$$0 \rightarrow 0$$
$$7887 \rightarrow 7887$$

A couple of nurses and a janitor gathered around. Some of them owned the FunctionLand board game and knew all about Guess the Function.

One of them said, "It's the identity function. That's way too easy."

Fred said, "Hold your horses, you guys" and continued writing:

$$\sqrt{7} \rightarrow \sqrt{7}$$
$$-2 \rightarrow 2$$

"Hey! You made a mistake. That should be –2," one of the nurses shouted.

Thinking that Fred had made a math error was kind of silly. He continued writing:

$$-73 \rightarrow 73$$
$$-\tfrac{1}{2} \rightarrow \tfrac{1}{2}$$
$$18 \rightarrow 18$$
$$-100 \rightarrow 100$$

Then suddenly, everyone claimed that they had it figured out. A nurse said that it was the **absolute value** function. The janitor said the rule was that you square the number and then take its principal square root. The kid said that if the number was negative, you took away the negative sign, and otherwise, if it was zero or positive, you left it alone.

All three guesses were correct.

One nurse said, "We had that absolute value stuff when I studied algebra for my R.N. training. They made it really mysterious by introducing all sorts of symbols. Instead of saying 'the absolute value of –9 is 9' my math teacher would write $|-9|$ and say that stood for 'the absolute value of –9.' He would write on the blackboard stuff like $|-4| = 4$. It took me the longest time to figure out why they used those absolute value symbols instead of regular English."

Fred didn't want to argue with her. He knew that in calculus he would be lost without those little absolute value symbols. The *central new idea in calculus* upon which both the differentiation half and the integration half of calculus are based is the idea of **limit**. The definition of the limit of a function (which takes pages and pages to explain) is that:

> The limit of a function f(x) as the value of x approaches some fixed number a is equal to some number L, if for every number e > 0, there exists another number d > 0 such that $0 < |x - a| < d$ implies that $|f(x) - L| < e$.

(Note to the reader: The above were Fred's thoughts. He never imagined that they might appear in a beginning algebra book. No other algebra book that we know of has ever contained this central definition of calculus. Beginning algebra students would have heart attacks if they saw that printed in their algebra books. So pretend that you never saw that definition. Stop! Before you do that pretending, let's take a tiny, tiny look at that definition at the heart of calculus: *The limit of a function f(x) as the value of x approaches some fixed number a is equal to some number L, if for every number e > 0, there exists another number d > 0 such that $0 < |x - a| < d$ implies that $|f(x) - L| < e$.*

Please notice that we have defined function already. You know what that is. We have showed you the less-than symbol (<) and you know what absolute value symbols are. The only thing you haven't

seen are the symbols "f(x)." That's new to you. Everything else you've had already. And all "f(x)" means is the image of x under a function whose name happens to be f. Sometimes we have functions that we call g, so we write g(x) when we want the image of x under some function named g. Sometimes we call a function h. In trig we have a function called the tangent function and we write tan(x). The f and g and h and tan are just names. They are not variables like x and y and z. Now, please pretend you never saw *The limit of a function f(x) as the value of x approaches some fixed number a is equal to some number L, if for every number e > 0, there exists another number d > 0 such that $0 < |x - a| < d$ implies that $|f(x) - L| < e$* printed in this book. It usually takes calculus students a week or so to understand this definition.)

Those were Fred's thoughts about the use of absolute value symbols. He knew that they're not used a whole lot until calculus, but they are so simple that everyone introduces them in beginning algebra.

What could be hard about: $|12| = 12$ or $|-1| = 1$?

   While we're waiting for the chaplain to get back, let's take a break and make it . . .

Your Turn to Play

1. $|-893| = ?$

2. $|\sqrt{34}| = ?$

3. $|\pi| = ?$

4. If there were an absolute value button on your calculator it might

look like $\boxed{|x|}$

   (There isn't such a button, so you don't have to look for it.) Now suppose your calculator display showed in little red letters: -283.4 and you hit the absolute value button. What would your display read?

5. If you had to plot $y = \dfrac{1}{x}$ which values of x could *not* be used?

6. Plot (by point-plotting) the curve $y = \dfrac{1}{x}$

7. For what value(s) of x is the fraction $\dfrac{x^2 - 36}{x^2 + 5x + 6}$ undefined?

8. Remember when the janitor defined the absolute value function by saying that you square the number and then take its principal square root. He was thinking $\sqrt{x^2}$. What a weird definition! Try it out for $x = 7$ and $x = -9$.

. . . . COMPLETE SOLUTIONS . . . .

1. 893

2. $\sqrt{34}$

3. π since π is a positive number, which is approximately equal to 3.1415 926535897932384626433832795028841971693993751058209749445923 07816406286208998628034825342117067982148086 5. . . .

---

4. ⧸ 283·4 They don't have such a button on calculators because who would want to pay the extra money to have a button that just knocked off minus signs if they appeared?

On some calculators there's a button that looks like

You can use that button to change the sign of whatever's in the display. If it's negative it will become positive, and if it's positive, it will be turned negative. You can use that button to eliminate minus signs that you don't want.

5. $y = \frac{1}{x}$ is undefined when the denominator is equal to zero since you can't divide by zero. So if you were to plot $y = \frac{1}{x}$ you would never get points where the first coordinate (known as the x coordinate or the abscissa) is equal to zero.

6. To plot $y = \frac{1}{x}$ by point plotting, we first find a bunch of points on the curve. We name a value for x and then find out what the corresponding value for y is. If x were equal to 4, for example, then y would equal 1/4. So the point (4, 1/4) is on the curve. If x were equal to 10, then y would equal 1/10. So the point (10, 1/10) would be on the curve. Let's not forget negative values for x. If x were equal to –1, then y would also be equal to –1, and we'd have the point (–1, –1). You'll probably have to name about a dozen values of x to figure out what the shape of this curve is since it's not a straight line. In calculus this curve is called a **hyperbola**.

7. A fraction can have any numerator it wants. It's division by zero that gets us into trouble. So in looking at $\frac{x^2 - 36}{x^2 + 5x + 6}$ we want to make sure that $x^2 + 5x + 6$ is never equal to zero. Let's find out when it does equal zero. We set $x^2 + 5x + 6$ equal to zero and we have the quadratic equation

$$x^2 + 5x + 6 = 0$$

This can be solved by factoring. $\qquad (x + 2)(x + 3) = 0$

$ab = 0$ implies $a = 0$ OR $b = 0$. $\qquad x + 2 = 0$ OR $x + 3 = 0$

$$x = -2 \text{ OR } x = -3.$$

$\frac{x^2 - 36}{x^2 + 5x + 6}$ is undefined when x equals either –2 or –3.

8. Putting x = 7 into $\sqrt{x^2}$ we get $\sqrt{(7)^2}$ which is $\sqrt{49}$ which is 7. That works.

Putting x = –9 into $\sqrt{x^2}$ we get $\sqrt{(-9)^2}$ which is $\sqrt{81}$ which is 9. That works. It's a great definition: $|x| = \sqrt{x^2}$, but it's really strange and not very often useful.

Want to hear a bizarre story? Many banal beginning algebra books have a really peculiar definition of absolute value. Nothing like "if the number's negative, make it positive." That would be too easy.

On page 52 in one real book (we're not naming names) and on page 26 in another book, they hit their readers who are just starting out in algebra with this definition:

$$|x| = \begin{cases} x \text{ if } x \geq 0 \\ -x \text{ if } x < 0 \end{cases}$$

In both books, this definition comes long before the students have solved their first little equation like 2x = 8.

It's a great way to kill off a lot of algebra students right at the start. Sure it's true that –x is a positive number if x itself is negative, but to do that on page 26! What's the hurry? The first *real use* for $|x|$ will come in calculus with maybe a passing mention of absolute value in studying logs in advanced algebra.

We have waited until the last chapter of this book to introduce the absolute value function so that you'll have a better chance of remembering it when you get to calculus.

For fun, the next time you're in the library, go to the math section (which is at "QA" or about "510" depending on which numbering system is used) and pull a beginning algebra book off the shelf. Look in the index under "absolute value" and see what definition they use.

If you were paranoid, you might think these books were out to "get you," but as one of my students remarked once, "It's not paranoia if it's reality."

F red was concerned about all those unconsumed milkshakes. The nurse had packed those 13 shakes in a large box so that he could take them with him. He knew that not only would the shakes be melted before he got a chance to drink them, but they would probably be evaporated by then. Fred had been working on his strawberry shake for over an hour before the chaplain had arrived, and you know how much of that he had finished.

The big wooden box with the 13 milkshakes in it weighed 48 pounds. Fred took three of the shakes out of the box and gave them away. The box with the milkshakes still weighed more than he did. It still weighed more than 36 pounds.
If x is the weight
of a shake, then

$$48 - 3x > 36$$

Now there's one little bitty thing I should mention to you about solving inequalities. At the beginning of this chapter, we just treated them the same as equalities. With equalities (like $5x = 12 - 2x$) we were free to do anything we wanted as long as we did the same thing to both sides. With inequalities that is *almost* true.

Let's solve the above inequality as if it were an equality and see what trouble we get into:

$$48 - 3x > 36$$

Subtract 48 from both sides $-3x > -12$

Divide both sides by -3 $x > 4$ ◄——— error on this line

Wait! That's impossible. If each milkshake weighed more than four pounds then all 13 shakes would weigh (4×13) 52 pounds, which is more than the milkshakes plus the box weighed. (See 8ᵗʰ line on this page.)

Where was our error? The little bitty thing that's different when you solve *in*equalities is that whenever you multiply (or divide) by a negative number you have to change the "sense" of the inequality. You take $>$ and change it to $<$. And $<$ will become $>$.

For example, say you started out with something true like $-7 < -5$ and you multiplied both sides by -2. If you forgot to change the sense of the inequality you'd get $+14 < +10$, which isn't true.

Let's do the problem correctly this time:

$$48 - 3x > 36$$

Subtract 48 from both sides $\qquad\qquad -3x > -12$

Divide both sides by -3 $\qquad\qquad\quad x < 4$ ◄——— no error this time.

Each of the milkshakes weighs less than four pounds.

The chaplain came back and reported, "We're all checked out now and we can go." Fred gave a kiss on the cheek to the nurse who had brought him all those wonderful milkshakes. Then Fred, in his little blue-and-green frog hospital shirt, and the chaplain headed out into the Texas sunshine.

The chaplain wasn't permitted by army regulations to drive Fred all the way to Kansas, so they headed to the bus station so Fred could "ride the dog" (as it's sometimes expressed) back to the university. The chaplain paid for the ticket and watched as Fred boarded the Greyhound.

Fred had a little tear in his eye as he saw the chaplain through the back window of the bus. They waved goodbye to each other as the bus pulled out.

Your Turn to Play

1. Solve $5 + 6x < 18$

2. Solve $26 < 6 - 4x$

3. Solve $5x \leq 3x$

4. Try various numbers in place of x to find out which values of x make $|x - 5| < 2$ true. (Experiment!)

5. Solve $-7x \geq 21$

6. For what values of x is $\dfrac{x^2 - 5x + 6}{15x^2 - 2x - 24}$ undefined?

.... **COMPLETE SOLUTIONS** ....

1.
$$5 + 6x < 18$$
$$6x < 13$$
$$x < \frac{13}{6}$$

2.
$$26 < 6 - 4x$$
$$20 < -4x$$
$$-5 > x \qquad \text{which can also be written } x < -5$$

3.
$$5x \leq 3x$$

Subtract 3x from both sides $\qquad 2x \leq 0$

Divide by 2 $\qquad x \leq 0$

4. Experimenting, we try x = 99 in $|x - 5| < 2$. That gives $|99 - 5| \overset{?}{<} 2$. This isn't true.

   We try x = 0 and get $|0 - 5| \overset{?}{<} 2$. This also isn't true.

   x = 6 works. $|6 - 5| \overset{?}{<} 2$ is true.

   x = 7 doesn't work. $|7 - 5| \overset{?}{<} 2$ isn't true.

   But x = 6.99 works. $|6.99 - 5| \overset{?}{<} 2$ is true.

   x = 3 doesn't work, but x = 3.00001 does.

   Our final answer for $|x - 5| < 2$ is $3 < x < 7$.

5.
$$-7x \geq 21$$

Divide both sides by –7 $\qquad x \leq -3$

6. $\dfrac{x^2 - 5x + 6}{15x^2 - 2x - 24}$ is undefined when $15x^2 - 2x - 24$ is equal to zero.

$$15x^2 - 2x - 24 = 0$$
$$(3x - 4)(5x + 6) = 0$$
$$3x - 4 = 0 \quad \text{OR} \quad 5x + 6 = 0$$

The fraction is undefined when x = 4/3, –6/5

Mechanicsville, Virginia site of several Civil War battles

1. My short-wave radio picked up a radio program from Germany. It announced that the low temperature in Berlin will be 15° tonight. Since all of continental Europe uses the Celsius temperature scale, they meant 15°C. What would that mean on the Fahrenheit scale?

2. Solve $-7x \leq x + 7$

3. If a and r are any numbers, then in advanced algebra we are told that $a + ar + ar^2 + ar^3 + \ldots$ is an **infinite geometric progression** that adds up to a nice finite sum of $\dfrac{a}{1-r}$ only when $|r| < 1$.

    In advanced algebra you might be asked to add up $1/2 + 1/6 + 1/18 + 1/54 + 1/162 + 1/486 + \ldots$. You would be adding up an infinite number of fractions. That is something you never had to do in regular arithmetic.

    Those fractions look like $a + ar + ar^2 + ar^3 + \ldots$ when $a = 1/2$ and $r = 1/3$. It's an infinite geometric progression. You then check to make sure that $|r| < 1$. Yes, it is true that $|\frac{1}{3}| < 1$.

    Now you are allowed to use the formula that says that the sum of those infinite number of fractions is equal to $\dfrac{a}{1-r}$

$$\frac{a}{1-r} = \frac{\frac{1}{2}}{1 - \frac{1}{3}} = \frac{3}{4} \quad \text{(after doing the arithmetic).}$$

Knowing when $|r| < 1$ is useful.

Name six other numbers that satisfy $|r| < 1$.

4. What values of x will make the following expression true:
$$|6x^4 - 7x^3 - \sqrt{173}\,x + 221| < -5$$

5. Graph $10x + y \leq 4$.

6. Solve $4 - 2x < 16$

7. Name five integers that satisfy $|x - 26| < 3$.

8. For what values of x will the fraction $\dfrac{6x - 12}{x^2 - 49}$ be undefined?

9. Thud sent the following letter to the editor of the A.A.A.A. camp newspaper. Where did he goof up?

Dear Editor,

    I have just proved something really cool. Here's my steps:

First start with a regular equation to solve: $\dfrac{x^2 - 5x + 6}{x - 3} = 1$

Next, since it's a fractional equation we eliminate the fractions by multiplying both sides by x – 3 and we get

$$x^2 - 5x + 6 = x - 3$$

Transpose: $x^2 - 6x + 9 = 0$

Factor: $(x - 3)(x - 3) = 0$

So: $x = 3$ *OR* $x = 3$

> Love,
> Thud

(Ignore the fact that Thud should have written: *Here <u>are</u> my steps.*)

answers

1. $C = \dfrac{5}{9}(F - 32)$ becomes $15 = \dfrac{5}{9}(F - 32)$

$$15 \cdot 9 = 5(F - 32)$$
$$135 = 5F - 160$$
$$295 = 5F$$
$$59^\circ = F$$

The temperature will be 59°F.

2. $-7x \le x + 7$
$$0 \le 8x + 7$$
$$-7 \le 8x$$
$$-\dfrac{7}{8} \le x \qquad \text{or } x \ge -\dfrac{7}{8}$$

3. $|r| < 1$ is satisfied by, for example, 0.3, $\dfrac{3}{7}$, -0.5, and $\pi/10$.
   $|r| < 1$ turns out to be equivalent to $-1 < r < 1$.

4. It's easy to get tied up in all the fancy coefficients and exponents. If you look at the big picture, you will notice that I'm asking when will the absolute value of some expression be less than -5. The absolute value of any expression is never less than zero, so there are no values of x that will make that true.

5. $10x + y \leq 4$

First graph $10x + y = 4$.

Place it in $y = mx + b$ form $y = -10x + 4$

The y-intercept is 4. The slope is -10.

Graph with a solid line since the original problem is $\leq$.

Test $(0, 0)$, which is the left of the line:

$$10 \cdot 0 + 0 \overset{?}{\leq} 4 \quad \text{yes}$$

Test $(25, 25)$, which is to the right of the line:

$$10 \cdot 25 + 25 \overset{?}{\leq} 4 \quad \text{no}$$

6. $4 - 2x < 16$

Subtract 4 from both sides $- 2x < 12$

Divide both sides by -2 and

change the sense of the inequality $x > -6$ or $-6 < x$

7. 24, 25, 26, 27, 28

8. The rule is that you can't divide by zero. When is $x^2 - 49$ equal to zero?

$$x^2 - 49 = 0$$
$$x^2 = 49$$
$$x = \pm 7$$

x = 7, –7 are the two values of x that will make the fraction undefined.

9. Thud forgot to check his answers in the original equation. The answer of x = 3 does not check.

Saint Augustine, Florida "the nation's oldest city"

1. If we know that the temperature of the hot dogs and beans, as served from the chow hall microwave, is somewhere between 40ºC and 55ºC, what would they be on the Fahrenheit scale?

(Hint: Start with $40 < C < 55$, which is a double inequality, and do the same thing to all three "sides." Start, as usual, by replacing C with $(5/9)(F - 32)$ and then multiply all three quantities by 9, etc., until you have just F in the middle.)

2. Graph $-2x + 3y > 2$

3. For what values of x will the following fraction be undefined:
$$\frac{x^2 - 64}{6x^2 + 7x + 2}$$

4. Thud liked sending in letters to the editor of the A.A.A.A. camp newspaper. What, if anything, is wrong with his letter?

Dear Editor,

Here's a neat way of showing that $\frac{5}{0}$ couldn't be a number.

Suppose it were. Say it were equal to ξ.
$$\frac{5}{0} = \xi$$

Multiply both sides by zero:
$$0\,\frac{5}{0} = 0\xi$$

Canceling the zeros on the left side:
$$5 = 0\xi$$
$$5 = 0$$

But zero times any number is equal to zero and not equal to 5.

Since that's not true, our assumption that $\frac{5}{0}$ was a number couldn't be true.

Neat, huh?

Love,
Thud

5. In Chapter 10, at 9:30 p.m. Pat, Chris, and Carol were heading out of

camp with passes that expired at midnight. They never showed up the next morning since the three of them decided on a whim to head to Mexico. Three weeks later they were captured and brought back for trial.

Pat and Chris were each sentenced to serve the same number of years. Carol, because he resisted arrest, was sentenced to one more year than Pat. Their three sentences added up to at least 19 years. Showing all your work, and letting x = the number of years that Pat was to serve in prison, find out how long Pat was to serve.

6. Showing your work, establish whether the following is true or false:
$$|-988| < |-40|$$

7. On Pat's trip back to Lampasas in the military prisoner transport, they stopped for lunch in San Antonio. A guard said she could buy whatever she wanted at the fast-food place up to $5. She got two of the Double-Bacon-and-Cheese-Whizzoreeno sandwiches (each costing $x) and one of the quart-sized containers of root beer (which cost $y). Remembering that x > 0 and y > 0, since food isn't free in San Antonio restaurants, graph the three inequalities.

8. $|-20| - |7| = ?$

9. For what values of y is the expression –y a positive number?

answers

1. $40 < C < 55$

Replace C by $\frac{5}{9}$ (F – 32) $40 < \frac{5}{9}$ (F – 32) < 55

Multiply all sides by 9 $40(9) < 5(F - 32) < 55(9)$
    $360 < 5F - 160 < 495$

Add 160 to all sides $520 < 5F < 655$

Divide all sides by 5 $104° < F < 131°$

2. First graph $-2x + 3y = 2$ as a dashed line. In y = mx + b form, it is $y = \frac{2}{3}x + \frac{2}{3}$
Testing (0, 0): $-2(0) + 3(0) \overset{?}{>} 2$ no
Testing (0, 8): $-2(0) + 3(8) \overset{?}{>} 2$ yes

3. The fraction is undefined when the denominator is equal to zero.

$$6x^2 + 7x + 2 = 0$$
$$6x^2 + 3x + 4x + 2 = 0 \qquad \text{Find two things that add to 7x and multiply to } 12x^2$$
$$3x(2x + 1) + 2(2x + 1) = 0$$
$$(2x + 1)(3x + 2) = 0$$
$$2x + 1 = 0 \quad \text{OR} \quad 3x + 2 = 0$$
$$x = -\frac{1}{2} \text{ or } -\frac{2}{3}$$

The fraction is undefined at $-\frac{1}{2}$ and at $-\frac{2}{3}$

4. Thud, despite the handicap of a name that has decidedly negative connotations, may have a half-way decent argument there. Thud's argument isn't exactly perfect, but, on the other hand, it isn't too shabby.

Thud, by the way, is his nickname. His given name is Thudeous Maximus. Perhaps he was named after a great Roman wrestler. No one is quite sure.

5.   *Let x = the number of years Pat was to serve in prison.*

Then x = the number of years Chris was to serve in prison.

Then x + 1 = the number of years Carol was to serve in prison.

Since the three sentences were to add up to at least 19 years . . .

$$x + x + x + 1 \geq 19$$
$$3x + 1 \geq 19$$
$$3x \geq 18$$
$$x \geq 6$$

Pat was to serve at least 6 years in prison.

6. $|-988| < |-40|$ means $+988 < +40$ which is false.

7. To graph $2x + y \leq 5$, we first graph $2x + y = 5$.
Putting that in the $y = mx + b$ form: $y = -2x + 5$

It is drawn with a solid line because the original problem has a $\leq$ and not a $<$ sign.

Testing $(0, 0)$: $2(0) + 0 \overset{?}{\leq} 5$ yes

Testing $(389, 500)$: $2(389) + 500 \overset{?}{\leq} 5$ no

The shaded area shows the possible prices for the two food items.

8. $|-20| - |7| = 20 - 7 = 13$

9. $-y$ is positive whenever y is negative.

If y were -77, then $-y$ would equal 77.

Fort Lauderdale, Florida "the Venice of America"

1. When Fred was a baby, he once caught the flu and had a fever. His mother consulted the only health book she owned. It was written in French with a title something like *Docteur Spoeque*. It said that it was serious only if the kid's temperature rose above 40°C.

> She got really worried since Fred's temperature was 99° (using her American-bought thermometer) and put Fred in the refrigerator. Should she have been *that* worried?

2. Solve $30 < -5x$

3. For what values of x is $\dfrac{66 - x}{3 + x}$ undefined?

4. $|-6| + |+6| = ?$

5. Solve $-3x \geq x + 16$

6. For what values of x is the fraction $\dfrac{x - 4}{2x - 3}$ undefined?

7. Graph $4x - y > 3$.

8. Name a number that will make $|x - 2| < -400$ true.

9. Mencius, who lived about three hundred years before Christ, wrote:

> It is the nature of things to be unequal. . . .   To think of them as equal is to upset the whole scheme of things. Who would make shoes if big ones were of the same price as small ones?

> Fred's shoes are what you might call smallish. Let's let x = the price of a pair of his shoes. If you bought six pairs of his shoes and $2 worth of doughnuts, you still would have money left over if you had started with a twenty-dollar bill. What can you say about the price of a pair of his shoes? (Please show all your work.)

10. $|-7.7| = ?$

answers

1. Converting 40°C to Fahrenheit . . . $40 = \frac{5}{9}(F - 32)$

$$40 \cdot 9 = 5(F - 32)$$
$$360 = 5F - 160$$
$$520 = 5F$$
$$104 = F$$

Fred's temperature was only 99°F, not 104°F. You should never put your child in the refrigerator under any circumstances.

2. $30 < -5x$

Divide both sides by −5 and
change the sense of the inequality $-6 > x$ or $x < -6$

3. x = −3 makes the denominator equal to zero and the fraction undefined.

4. $|-6| + |+6| = 6 + 6 = 12$

5. $-3x \geq x + 16$
        $-4x \geq 16$
        $x \leq -4$

6. The fraction is undefined when the denominator, 2x −3, is equal to zero. $2x - 3 = 0$ ⇨ $2x = 3$ ⇨ $x = 3/2$

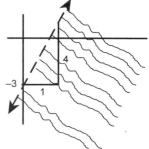

7. First graph 4x − y = 3 (as a dashed line since the original is > and not ≥.
4x − y = 3, in y = mx + b form, is y = 4x + −3.

Testing (0, 0) in 4x − y > 3: $4 \cdot 0 - 0 \overset{?}{>} 3$ no

Testing (77, 0) in 4x − y > 3: $4 \cdot 77 - 0 \overset{?}{>} 3$ yes

8. The question asks for the absolute value of an expression to be less than −400. That's silly. Absolute values are never negative.

9. $6x + 2 < 20$
                            $6x < 18$
                            $x < 3$

A pair of Fred's shoes must cost less than three dollars.

10. 7.7

A.R.T.

All **R**eorganized **T**ogether

A Super-Condensed and Reorganized-by-Topic Overview of Beginning Algebra
(Highly abbreviated)

Topics:
 Absolute value
 Adding like terms
 Arithmetic of the Integers
 Exponents
 Fractional Equations
 Fractions
 Functions
 Geometry
 Graphing
 Inequalities
 Laws
 Multiplying and Factoring Polynomials
 Numbers
 Order of Operations
 Quadratic Equations
 Radicals
 Sets
 Statistics
 Two Equations and Two Unknowns
 Word Problems
 Words/Expressions

Absolute value
 The absolute value of a number means take away the negative sign
 if there is one. $|-5| = 5$ $|0| = 0$ $|4| = 4$ (p. 515)

Adding like terms $6wz^3 + 2wz^3 = 8wz^3$ (p. 87)
 6 apples plus 2 apples is 8 apples.

A.R.T.

Arithmetic of the Integers

Going from -7 to $+8$ means $8 - (-7) = 15$ (p. 26)

$8 - (-7)$ becomes $8 + (+7)$ (p. 30)

To subtract a negative is the same as adding the positive.

For multiplication:

Signs alike ⇨ *Answer positive*

Signs different ⇨ *Answer negative* (p. 51)

Exponents

$x^2 x^3 = (xx)(xxx) = xxxxx = x^5$ (p. 251)

When the bases are the same, you *add* the exponents.

$(x^2)^3 = x^6$ When you have an exponent-on-an-exponent,

you multiply. (problem 9 on p. 265)

$\dfrac{x^m}{x^n} = x^{m-n}$ (p. 268)

x^{-3} equals $1/x^3$ (p. 268)

x^0 always equals 1 (except that 0^0 is undefined.)

Fractional Equations

$$\frac{1}{12} + \frac{1}{16} + \frac{1}{24} = \frac{1}{x}$$ (p. 328)

becomes $\dfrac{1 \cdot \mathbf{48x}}{12} + \dfrac{1 \cdot \mathbf{48x}}{16} + \dfrac{1 \cdot \mathbf{48x}}{24} = \dfrac{1 \cdot \mathbf{48x}}{x}$

which simplifies to $4x + 3x + 2x = 48$

Memory aid: Santa Claus delivering packages (p. 331)

Fractions

Simplifying: (p. 337)

$$\frac{x^2 + 5x + 6}{x^2 + 6x + 8} = \frac{(x+3)(x+2)}{(x+4)(x+2)} = \frac{(x+3)\cancel{(x+2)}}{(x+4)\cancel{(x+2)}} = \frac{x+3}{x+4}$$

Memory aid: factor top; factor bottom; cancel like factors.

One tricky simplification:

$$\frac{(x-3)(x-4)}{x(4-x)} = \frac{(x-3)(x-4)}{-x(x-4)} = \frac{x-3}{-x}$$ (p. 356)

Adding (p. 346), Subtracting (p. 347)

Multiplying and Dividing (p. 352)

Long Division by a binomial (p. 442)

For example: $2x + 5 \overline{)\, 6x^3 + 19x^2 + 22x + 30}$

A.R.T.

Functions

A function is any rule that associates
to each element of the first set (called the domain)
exactly one element of the second set (called the codomain).

Each element in the domain has an image in the codomain.
The set of images is called the range.
Examples of functions start on page 460.
The identity function maps each element to itself. (p. 491)

Geometry

diameter of a circle 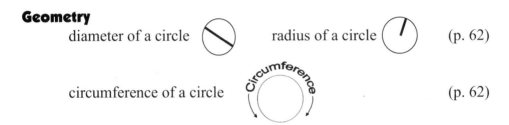 radius of a circle (p. 62)

circumference of a circle (p. 62)

π ≈ (p. 64)

3.14159265358979323846264338327950288419716939937510582097494459230781640628620899862803482534211706798
21480865132823066470938446095505822317253594081284811174502841027019385211055596446229489549303819644428
81097566593344612847564823378678316527120190914564856692346034861045432664821339360726024914127372458700
06606315588174881520920962829254091715364367892590360011330530548820466521384146951941511609433057270360
57595919530921861173819326117931051185480744623799627495673518857527248912279381830119491298336733624400
65664308602139494639522473719070217986094370277053921717629317675238467481846766940513200056812714526350
60827785771342757789609173637178721468440901224953430146549585371050792279689258923542019956112129021960
08640344181598136297747713099605187072113499999983729780499510597317328160963185950244594553346908302642
52230825334468503526193118817101000313783875288658753320838142061717766914730359825349042875546873115950
62863882353787593751957781857780532171226806613001927876611195909216420198938095257201065485863278865930
61533818279682303019520353018529689957736225994138912497217752834791315155748572424541506959508295331160
86172785588907509838175463746493931925506040092770167113900984882401285836160356370766010471018192955500
96198946767837449448255379774726847104047534646208046684259069491293313677028989152104752162056966024050
80381501935112533824300355876402474964732639141992726042699227967823547816360093417216412199245863150300
28618297455570674983850549458586926995690927210797509309295532116534498720275596023648066549911988183470
97753566369807426542527862655181841757467289097777279380008164706001614524919217321721477235014144197356
85481613611573525521334757418494684385233239073941433345477624168625189835694855620992192221842725502540
25688767179049460165346680498862723279178608578438382796797668145410095388376860950680006422512520511730
92984896084128488626945604241965285022210661186306744278622039194945047123713786960956364371917287467760
46575739624138908658326459958133904780275901 . . .

C = πd   C = circumference   d = diameter (p. 66)
perimeter = distance around the outside. (p. 94)
     rectangle of length ℓ and width ω has perimeter = $2\ell + 2\omega$

     triangle with sides a, b, and c has perimeter = a + b + c

Trapezoid (p. 94)
     area, A = ½h(a + b) (p. 95)

A.R.T.

Sector   (p. 94)

Area of a rectangle of width a and length b is A = ab   (p. 94)

Area of a triangle whose sides are a, b, and c   (p. 418)
is $\sqrt{s(s-a)(s-b)(s-c)}$ (Heron's Formula)
where s is the semi-perimeter. $s = \frac{1}{2}(a + b + c)$

Volume of a cylinder
$V = \pi r^2 h$ (p. 185)

Volume of a sphere
$V = (4/3)\pi r^3$ where r is the radius   (p. 260)

Volume of a cone
$V = (\frac{1}{3})\pi r^2 h$ (p. 276)

Pythagorean Theorem:
In any right triangle, $a^2 + b^2 = c^2$ (p. 385)

Graphing
Ordered pair: (3, 7)   (p. 201)
The four quadrants: (p. 202)

$$\begin{array}{c|c} \text{II} & \text{I} \\ \hline \text{III} & \text{IV} \end{array}$$

Origin = where the x- and y-axes meet = (0, 0)   (p. 207)

Abscissa = first coordinate = x-coordinate   (p. 203, problem 3)

Ordinate = second coordinate = y-coordinate   (p. 203, problem 3)

Linear equations are equations whose graphs are lines.
They can be put in the form: ax + by = c   (p. 217)

Graphing any equation (called point-plotting)   (p. 221)
Three steps: 1. name x values
2. find the corresponding y values
3. plot those points until you have enough
of them to "connect the dots."

534

A.R.T.

Rectangular coordinate system = graphing (x, y) on the   (p. 257)
    x- and y-axes.

Slope = rise/run and is denoted by the letter m.   (p. 469)

rise

run

Slope-intercept form of the line   (p. 480)
    $y = mx + b$ Has a slope of m and hits the y-axis at b
    Memory aid: the Zorro method for quickly
        drawing a line   (p. 488)

To graph an inequality: (p. 504)
    First, graph the equality.
    Second, test a point on each side of the line in the
        original inequality.
    Make the line solid if the original problem had $\leq$ or $\geq$.
        Make the boundary line dashed if the
        original inequality was $<$ or $>$.   (p. 507)

Inequalities
    $<$ less than $4 < 10$ $-8 < 3$ $-500 < -9$
    $>$ greater than $-2 > -18$ $100 > 7$

Solving $5 < 4x - 7$ is done exactly like solving $5 = 4x - 7$,
    except that if you multiply or divide by a negative number,
    you must change the "sense" of the inequality:
    $<$ becomes $>$.   (p. 520)

Laws
Reflexive property of equality: $a = a$ (p. 115)

Symmetric law of equality: if $a = b$ then $b = a$ (p. 95)

Commutative law of addition: $a + b = b + a$ (p. 356)

Commutative law of multiplication: $ab = ba$ (p. 264)

Distributive property: $a(b + c) = ab + ac$ (p. 23)

A.R.T.

Multiplying and Factoring Polynomials

$(a + b)(c + d) = ac + ad + bc + bd$ (p. 285)

memory aid: boys and girls meeting

Common factor: $4x^2y - 6xy = 2xy(2x - 3)$ (p. 296)

Easy trinomials: $x^2 + 5x + 6 = (x + 2)(x + 3)$ (p. 298)

Memory aid: 2 and 3 add to 5 and multiply to 6

Difference of squares: $49x^2 - 25y^2 = (7x + 5y)(7x - 5y)$ (p. 301)

Grouping: $6x^3 - 15x^2 + 4x - 10$ (p. 303)

$= 3x^2(2x - 5) + 2(2x - 5)$

$= (2x - 5)(3x^2 + 2)$

Harder trinomials: $6x^2 + 29x + 35$ (p. 306)

Split the 29x into two numbers that add to 29x and multiply to $(6x^2)(35)$

$6x^2 + 15x + 14x + 35$

and finish the factoring by grouping

$3x(2x + 5) + 7(2x + 5)$

$(2x + 5)(3x + 7)$

Numbers

Natural numbers $= \{1, 2, 3, 4, 5, \ldots\}$ (p. 21)

Whole numbers $= \{0, 1, 2, 3, 4, 5, \ldots\}$ (p. 21)

Integers $= \{\ldots -3, -2, -1, 0, +1, +2, +3, +4, \ldots\}$ (p. 22)

Rational numbers are all numbers that *can be* written as an integer divided by a (non-zero) integer.
Examples: 3/4, 19, –5/12   (p. 103)

Real numbers $=$ all the numbers on the number line   (p. 391)

Irrational numbers $=$ real numbers that are not rational   (p. 392)

Examples: $\pi, 6\pi, \sqrt{2}, \sqrt{17}, \pi - 3\sqrt{5}$

Perfect squares $= \{1, 4, 9, 16, 25, 36, 49, 64, \ldots\}$ (p. 392)

Order of Operations

❀ parentheses

❀ exponents

❀ multiplication and division (done left-to-right)

❀ addition and subtraction

A.R.T.

Quadratic Equations

Solving by factoring   (p. 292)
$$x^2 + 5x + 6 = 0$$
$$(x + 2)(x + 3) = 0$$
$$x + 2 = 0 \quad \text{OR} \quad x + 3 = 0$$
$$x = -2 \quad \text{OR} \quad x = -3$$

Solving pure quadratics: $x^2 = 81$ (p. 378)
$$x = \pm 9$$

Solving by completing the square   (pp. 427–428)

Solving by quadratic formula   (p. 435)

starting with $ax^2 + bx + c = 0$,

the quadratic formula is $x = \dfrac{-b \pm \sqrt{b^2 - 4ac}}{2a}$

Radicals

radical sign $\sqrt{}$

This radical sign $\sqrt[4]{}$ has an index of 4.

The radicand is the stuff under the radical sign.   (p. 396)

Simplifying radicals:

$$\sqrt{20} = \sqrt{4}\sqrt{5} = 2\sqrt{5}$$ (p. 397, problems 8–9)

$a^{1/2} = \sqrt{a}$ (p. 400)

$a^{1/3} = \sqrt[3]{a}$

$\sqrt[3]{\sqrt[7]{xyz}} = \sqrt[21]{xyz}$

A radical equation is an equation with the unknown under
a radical sign. Example: $5 = \sqrt{3 + x}$ (p. 404)

Solving radical equations: (p. 408)
Isolate the square root on one side of the equation
and square both sides

Extraneous (extra) roots may occur when you solve   (p. 408)
a radical equation. They are eliminated by
checking all the answers in the original equation.

Rationalizing the denominator   (p. 406)

◇Monomial case: $\dfrac{7}{\sqrt{x}}$

multiply top and bottom by $\sqrt{x}$

$$\frac{7\sqrt{x}}{\sqrt{x}\,\sqrt{x}} = \frac{7\sqrt{x}}{x}$$

A.R.T.

Rationalizing the denominator

✧Binomial case: $\dfrac{5}{3 + \sqrt{y}}$

multiply top and bottom
by the conjugate   (p. 409, problem 8)

$$\frac{5}{3 + \sqrt{y}} \cdot \frac{3 - \sqrt{y}}{3 - \sqrt{y}} = \frac{15 - 5\sqrt{y}}{9 - y}$$

Sets

Parentheses: ( )   Braces: { }   Brackets: [ ]

Order of listing the elements doesn't matter. {#, *} = {*, #}

(p. 37, problem 18)

Empty set = null set = { } = the set with no elements in it.   (p. 37)

Two sets are equal if they have the same elements in them.

Set builder notation
        "{ x |" is read, "The set of all x such that . . ."   (p. 104)

Union of two sets: {7, 8} ∪ {7, 11} = {7, 8, 11}   (p. 191)

A is a subset of B if every element in A is also in B.   (p. 485)

Statistics

Mean average = add them up and divide by the number of
        things you've added up. The average of 3, 8, 10 is
        $(3 + 8 + 10)/3 = 21/3 = 7$ (p. 211)

Median average = the number in the middle of the list.
        The average of 3, 4, 5, 225, 103899 is 5   (p. 211)

Mode average = the most popular number.
        The average of 3, 3, 3, 5, 5, 5, 5, 7, 10, 88 is 5   (p. 211)

Factorial
        $6! = 6 \times 5 \times 4 \times 3 \times 2 \times 1$ (p. 254)
        0! equals 1 (by a definition that we'll introduce in
                advanced algebra)
There are n! ways to line up n objects in a row.   (p. 254)

A.R.T.

Two Equations and Two Unknowns

✧Solution by elimination using addition: (p. 189)

to solve $\begin{cases} 12x + 5y = 3 \\ 7x + 2y = 6 \end{cases}$ multiply the first equation by 2 and the second equation by -5 and add them.

The y's will disappear.

✧Solution by graphing:

to solve $\begin{cases} x + y = 10 \\ x = y \end{cases}$ (p. 240)

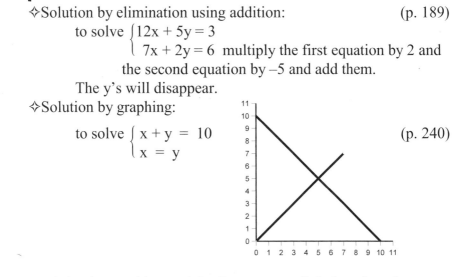

If you're solving by graphing and the lines are parallel, then there is no solution. (There are no values of x and y that make both equations true.) The equations are inconsistent. (p. 252, problem 9 and 10)

If you're solving by graphing and the two equations graph the same line, then the equations are dependent. Any values of x and y that satisfy one equation will satisfy the other. (p. 252, problem 11)

✧Solution by substitution:

to solve $\begin{cases} 23x + 39y = 4 \\ 15x + y = 7 \end{cases}$

solve the second equation for y and substitute that into the first equation. (p. 243)

Word Problems

With ratios or continued ratios (p. 84)
With consecutive numbers (p. 98)
With consecutive even (or odd) numbers (p. 99)
With distance-rate-time (d = rt) (p. 109)
Mixture problem (p. 154, problems 7 and 8)
Stamp problem (p. 153, problem 1)
Age problem (pp. 159–161)
Job problem (A can do job in 2 hours and B in 3) (pp. 323–325)

A.R.T.

Words/Expressions

Ratio means division. The ratio of 3:5 means $3 \div 5$ (p. 31)

Proportion = equality of two ratios $\dfrac{3}{5} = \dfrac{12}{20}$

Coefficient is the number in front of the letters
The coefficient of $4x^3yz$ is 4. (p. 63)

Consecutive numbers—for example: 44, 45, 46, 47. (p. 98)

Consecutive even numbers—for example: 22, 24, 26. (p. 99)

Like terms: ($7x^5y^3z$ and $4x^5y^3z$ are like terms) (p. 87 and 262)

Polynomials: formed by adding, subtracting, or multiplying
numbers and/or letters (p. 287)

monomials = one term $6x^2y$
binomials = two terms $3x + x/19$
trinomials = three terms $389z + \pi wy^3 + \sqrt{3389}$
These *aren't* polynomials: neither $\sqrt{x}$ nor $\dfrac{3}{y + 1}$

Index

abscissa 203
absolute value. 515-517, 519
adding fractions. . . . 340-343, 346
adding integers. 35, 36
Aesop's fable of the grasshopper
        and the ant. 207
age problems. 159-161, 163,
        164, 170, 176, 177, 191
alliteration. 159
area of a square. 258
area of a trapezoid. 95
Article 114 of the Weimar
      Constitution of Germany
    104
associative law of addition. . . . . .
    . . . . 24, 29
atomic weight. 248
Avogadro's number. 248
bases 262
binomial 287
braces. 21, 23, 24, 39
brackets. 21-23
cancel crazy. 373, 374
centimeter. 342
Christina Rossetti 336
circumference 24, 62, 241
codomain. 461, 481
coefficient 63
combining like terms. 87
commutative law of addition
    356
commutative law of multiplication
    264
completing the square. . . 428, 429
complex fractions. 352, 353
conjugate. 409
consecutive even integers. . . . 99
consecutive numbers. 98
consecutive odd integers. 99

continued ratio. 83-85, 120,
    . . . . 131, 139, 153
conversion factor. . . 342, 375, 394
coordinates 203
cube root. 400, 401
dependent equations. 252
developing your mental strength
    414
diameter. 24, 62, 241
distance-rate-time problems. . . . . .
      . . . . . 109, 111, 120, 121,
        123, 124, 130, 131,
        141-143, 163, 164,
        169, 312, 315
distributive law. 23, 29, 109
distributive law—the proof. . . . . .
    . . . . 137, 138
dividing by zero. 510-512
dividing fractions. 351-354
division of signed numbers. . . 51
domain. 461, 463, 481
eliding a word. 284
empty set. 23, 29, 37
enjambment. 479
epinephrine 108
eponymous. 422
Erasmus. 82
exponents. . 18-20, 29, 33, 47, 247
exponents—all the laws in one
    chart. 269
extraneous roots. 332, 408
factorial. 73, 254
factoring
    common factor. 296, 297
    difference of squares. . . . .
    . . . . 301, 302
    easy trinomials. 298-300
    grouping. 303-305
    harder trinomials. . . . 306-311

factors.................... 339
Fadiman's *Lifetime Reading Plan*
.... 266
Fahrenheit (the man)......... 25
Fahrenheit and Celsius.. 499-501, 526, 529
fenestration 137
finite 18, 29, 111
fractional equations. 328-332
fractional exponents........ 400
function—definition..... 460, 462
function—examples...........
.... 460-463, 466, 467
gram...................... 57
graph y = log x by point-plotting
.... 228
graphing any equation. . . 221, 222
graphing inequalities in two
unknowns... 503-509, 527
greater than >........... 53, 59
Greek alphabet 219
Guess the Function game.......
...... 463, 464, 489, 492, 493, 496, 514, 515
hebdomadal............... 214
heptathlon 74
Heron's formula........... 418
hyperbola 224, 518
hyperbole................. 102
hypotenuse............... 396
i before e, except after c..... 380, 381
iatrogenic injuries. 434
identity function........... 491
the *Iliad*. 263
image.................... 461
inconsistent equations. . . 252, 255
index.................... 400
inequalities in one unknown. ....
.... 520-522
infinite. 18, 29, 111

infinite geometric progression
.... 523
infinite numbers........... 398
infinite sets................ 19
integers................ 22, 29
interior decorating—adding
fractions........... 340
Invent a Function game..... 463, 483, 485, 490, 492, 497
irony 183, 403
irrational numbers. 392
job problems. . 324-326, 386, 423
less than <.............. 51, 59
limit—as defined in calculus
.... 516
linear equations. 217, 218
long division of polynomials
.... 442-446
Marx Brothers movies........ 57
mean average......... 211, 213
median average. 211, 213
Mencius. 529
mixture problems.. 154, 169, 170, 176, 177
mnemonics 345, 356
mode average......... 211, 213
monomials. 287, 288
Mt. Everest................ 26
multiplying binomials. . . 285, 286
multiplying fractions. ... 351, 352
multiplying signed numbers......
.... 48, 50, 51, 60
natural numbers.......... 21, 29
negative exponents...........
.... 267-269, 399
negative numbers........... 22
negative times a negative equal a
positive—the proof. .....
.... 115, 116, 119
null set. 23, 24, 29

number line. 22, 23

oral literature. 263

order of operations. 96, 273

ordered pair. 201

ordinate. 203

origin. 207, 226

parabola 224

parentheses. 21

passing a law today that made
       what you did yesterday
       illegal. 46

perfect square numbers. 392

perimeter. 94, 127, 415

Phillips screws. 25

pi π. 64-66, 295

plotting a point. . . . 201, 205, 210

point plotting. 221, 222, 506

polynomial. 287, 288

principal square root. . . . 380, 472

proportion 59

pure quadratic. 378, 379, 386

Pythagorean theorem. . . 385, 386,
                      450

quadrants. 202

quadratic 291

quadratic formula. 435-437

radical equation. . . 404, 406, 408

radicand. 396

range of a function. . . . .
          . . . . 484, 485, 491

ratio. 31, 58

rational expressions. 337

    adding. 340-342, 346

    dividing. 351-354

    multiplying. 351, 352

    simplifying. 337-339

rational numbers. . . 103, 104, 112

rationalizing the denominator
       . . . . 406, 407

real number line. 392

real numbers. 391

reciprocal 373

rectangle. 94

rectangular coordinate system
       257

rectangular parallelepiped. . . . 378

reflexive property of equality
       115

right triangle. 385

sector 94

set 21, 39

set builder notation. 104

sets are equal—definition. 39

seven famous words for
       simplifying fractions
       339

simplify a square root. 397

six pretty boxes. . . 123-126, 130,
         133, 134, 142, 157, 164,
         167, 171, 174, 209

slide rule. 63

slope. 468-472

slope-intercept form of the line
       . . . . 480, 481

solving quadratic equations
    by completing the square
       . . . . 426-430, 435
    by factoring. 291-293
    quadratic formula. . . . 435-437

solving systems of equations
    by graphing. 239-241
    elimination method. . . 189-193,
         196, 197, 228, 236
    substitution method. . 243, 244

stamps problem. 153

Stanthony 57

subset 485

subtracting fractions. . . . .
       . . . . 346-348, 351

subtracting negative numbers. . . .
. . . . 30, 31
surface area of a sphere. 424
symmetric law of equality. . . . . .
. . . . 95, 113, 386
terms 339
theorem. 385, 387
three signs in a fraction. . . . 334,
335
transposing. 183
trapezoid. 94
trigonometric
equations—graphing by
point-plotting. 224
trinomial 288
Trojan War. 263
union of two sets. 191, 236
Venn diagram. 106, 107
volume of a cone. 276
volume of a cube. 259
volume of a cylinder. . . 185, 246,
252, 294, 424
volume of a sphere. 260
whole numbers. 21, 29
word problems into
equations—the four steps
. . . . 85
x – 4 and 4 – x. 356, 376
x-coordinate. 203
y = mx + b. 480
y-intercept. 479, 481
zero exponent. 268
zero-sum game. 56

You have mastered all of beginning algebra. Next comes:
➤ *advanced algebra*
➤ *geometry*
➤ *trigonometry.*

After two years of college calculus, you will be a junior and ready to declare that you are a math major.

Being a math major! Yes. That's a lot more fun than being:
✿ an English major—and writing long term papers
✿ a chemistry major—and getting acid burns in the lab
✿ a psychology major—and dealing with all the abnormals
✿ a history major—and learning all those dates.

It's your choice.

To learn about other
books in this series visit

LifeofFred.com